AF344285

PROGRAMMING THE MATRIX ANALYSIS OF SKELETAL STRUCTURES

ELLIS HORWOOD SERIES IN CIVIL ENGINEERING

Series Editor: Professor P. HOLMES, Imperial College of Science and Technology

Bhatt, P	**PROGRAMMING THE MATRIX ANALYSIS OF SKELETAL STRUCTURES**
Blockley, D.I.,	**THE NATURE OF STRUCTURAL DESIGN AND SAFETY**
Britto, A. & Gunn, M.J.	**CRITICAL STATE SOIL MECHANICS VIA FINITE ELEMENTS***
Calladine, R.R.	**PLASTICITY FOR ENGINEERS**
Carmichael, D.G.	**STRUCTURAL MODELLING AND OPTIMIZATION**
Cyras, A.A.	**MATHEMATICAL MODELS FOR THE ANALYSIS AND OPTIMISATION OF ELASTOPLASTIC SYSTEMS**
Dowling, A.P. & Ffowcs-Williams, J.E.	**SOUND AND SOURCES OF SOUND**
Edwards, A.D. & Baker, G.	**PRESTRESSED CONCRETE***
Farkas, J.	**OPTIMUM DESIGN OF METAL STRUCTURES**
Graves-Smith, T.R.	**LINEAR ANALYSIS OF FRAMEWORKS**
Hendry, A.W., Sinha, B.A. & Davies, S.R.	**INTRODUCTION TO LOAD BEARING BRICKWORK DESIGN**
Heyman, J.	**THE MASONARY ARCH**
Holmes, M. & Martin, L. H.	**ANALYSIS AND DESIGN OF STRUCTURAL CONNECTIONS: Reinforced Concrete and Steel**
Irons, B. & Ahmad, S.	**TECHNIQUES OF FINITE ELEMENTS**
Irons, B. & Shrive, N.G.	**FINITE ELEMENT PRIMER**
Jordaan, J.J.	**PROBABILITY FOR ENGINEERING DECISIONS: A Bayesian Approach**
Kwiecinski, M.,	**PLASTIC DESIGN OF REINFORCED SLAB-BEAM STRUCTURES**
Lencastre, A.	**HANDBOOK OF GENERAL HYDRAULICS***
Megaw, T.M. & Bartlett, J.	**TUNNELS: Planning, Design, Construction**
Melchers, R.E.	**STRUCTURAL RELIABILITY; Analysis and Prediction***
Mrazik, A., Skaloud, M. & Tochacek, M.	**PLASTIC DESIGN OF STEEL STRUCTURES**
Pavlovic, M.	**THIN PLATES AND SHELLS: Theory and Applications**
Shaw, E.	**ENGINEERING HYDROLOGY***
Spillers, W.R.	**INTRODUCTION TO STRUCTURES**
Szabo, K. & Kollar, L.	**STRUCTURAL DESIGN OF CABLE-SUSPENDED ROOFS**
White, R.G. & Walker, J.G.	**NOISE AND VIBRATION**

** In preparation*

PROGRAMMING THE MATRIX ANALYSIS OF SKELETAL STRUCTURES

P. BHATT, B.Eng., M.Eng., Ph.D.
Department of Civil Engineering
University of Glasgow

ELLIS HORWOOD LIMITED
Publishers · Chichester

Halsted Press: a division of
JOHN WILEY & SONS
New York · Chichester · Brisbane · Toronto

First published in 1986 by
ELLIS HORWOOD LIMITED
Market Cross House, Cooper Street, Chichester, West Sussex, PO19 1EB,
England

The publisher's colophon is reproduced from James Gillison's drawing of the ancient Market Cross, Chichester.

Distributors:

Australia and New Zealand:
Jacaranda-Wiley Ltd., Jacaranda Press,
JOHN WILEY & SONS INC.
GPO Box 859, Brisbane, Queensland 4001, Australia

Canada:
JOHN WILEY & SONS CANADA LIMITED
22 Worcester Road, Rexdale, Ontario, Canada

Europe and Africa:
JOHN WILEY & SONS LIMITED
Baffins Lane, Chichester, West Sussex, England

North and South America and the rest of the world:
Halsted Press: a division of
JOHN WILEY & SONS
605 Third Avenue, New York, NY 10158, USA

British Library Cataloguing in Publication Data
Bhatt, P.
Programming the matrix analysis of skeletal structures. —
(Ellis Horwood series in civil engineering)
1. Structures, Theory of — Matrix methods
2. Structures, Theory of — Computer programs
I. Title
624.1′71′01512934   TA647

ISBN 0–85312–994–0 (Ellis Horwood Limited — Library Edn.)
ISBN 0–7458–0068–8 (Ellis Horwood Limited — Student Edn.)
ISBN 0–470–20310–2 (Halsted Press)

Printed in Great Britain by Unwin Brothers of Woking

Table of Contents

To the memory of my father
Kavi Shankara Bhatta (1899?–1983)

Preface

In recent years a large number of textbooks have been published on the matrix analysis of skeletal structures. Many of these books give, almost as an afterthought, some simple versions of computer programs for the analysis of skeletal structures. As a rule these programs do not explain the organization, strategy and other details of the programs. Consequently, these programs are generally unreadable. It is a common experience of teachers of structural analysis that most students experience great difficulty in translating the theory of matrix analysis into 'good' computer programs. This is unfortunate because this difficulty prevents the students from fully appreciating the unified approach and the power of the matrix method of structural analysis.

In recent years, many authorities have warned against the dangers of 'black box' approach to using programs. It has even been suggested that the only solution to prevent this attitude from gaining ground is to train an engineer to read a program in the same way that he reads a bill of quantities or a drawing.

The object of the present book is to remedy this unsatisfactory situation. It was inspired by the excellent text *Programming the Finite Element Method* by E. Hinton and D. R. J. Owen, Academic Press, 1977. Experience in the classroom showed that a textbook such as that of Hinton and Owen would certainly enhance considerably the understanding and appreciation of the matrix methods of structural analysis.

Some of the special features of the book are as follows.

(1) The programs are readable. Towards this end the names of the variables have been chosen with great care to make them immediately meaningful

(2) The watchword is clarity. Apparent computational efficiency has been sacrificed for the sake of clarity. Intermediate variables have been used sparingly.

(3) Each chapter explains only one specific step of the stiffness method of analysis such as the calculation of the element stiffness matrices and the assembling of the structural stiffness matrix. In the same chapter, all the associated subroutines are discussed and explained in detail. Great emphasis is placed on explaining the logic of each subroutine.

(4)   The programs cover more ground than is normal. For example the text includes not only programs for the linear elastic analysis of two-dimensional skeletal structures but also programs for the solution of eigenvalue problems associated with the calculation of load factors to cause elastic instability and natural frequency calculation.

(5)   The programs include not only orthodox prismatic beam elements but also special elements such as beams on an elastic foundation, semi-rigid elements and nonprismatic elements. Similarly the programs include torsion elements with or without the inclusion of warping restraint. This should make the programs attractive for use in courses in design studies.

The author is grateful to the University of Glasgow, Scotland, and in particular Professors Coull and Sutherland for leave of absence and to Royal Military College (RMC) of Canada, Kingston, Ontario, and in particular Professor P. M. Jarrett for a visiting professorship which enabled the author to complete the book. The author is grateful to the following individuals for the assistance rendered: Ms Tina Lahaie and Ms Casey Campbell both from RMC for excellent typing of the draft; Mr Larry Harvey from RMC and Mr Robert Watson and Dr John Ebireri from Glasgow University for the excellent diagrams; Dr Claude Tahiani from RMC for useful comments on the draft; Mr Gordon Irving from Glasgow University for assistance in computational matters; Sheila, Arun and Ranjana for encouragement and moral support.

Glasgow   P. Bhatt
November, 1985

CHAPTER 1   Matrix Analysis of Structures by the Stiffness Method

Structural analysis is concerned with the determination of stresses and displacements in a structure for a given loading so that the designer can provide sufficient resistance to resist the stresses developed. Although the designer has to design many complex structures, most of the structures in practice fall into the following two categories.

(a) *Skeletal structures.* Skeletal structures are easily identified by the presence of a framework or 'skeleton' built from elements such as bars and beams connected at joints. Typical examples of skeletal structures are two-dimensional (2-D) and three-dimensional (3-D) pin-jointed structures (trusses), 2-D and 3-D rigid-jointed structures (plane and space frames), plane grids, etc. In the great majority of such structures, the members are straight although plane grids in bridge structures very often contain curved members.

(b) *Continuum structures.* In this type of structure, the material is not concentrated along certain members as in the skeletal structures but the material is continuously distributed with no clearly identifiable members or joints. Typical examples of continuum structures are plates, shells, walls, dams, etc.

Over the years, innumerable methods for analysing structures have been developed. Many of these methods were geared to the manual analysis of specific structures. However, the ready availability of digital computers has revolutionized especially the analysis and to a lesser extent the design of structures. Digital computers can be programmed to perform extremely complex calculations with a minimum input of basic data. Consequently, many methods of structural analysis which exploit the power of the computers have been developed over the last three decades. However, at the present time, the method known as the stiffness (or displacement) method is the one most widely used. One of the very attractive features of the stiffness method is that it enables the development of a unified approach to the analysis of complex skeletal and continuum structures. This unity of concept enables a program which can analyse pin-jointed structures to be modified with minimum effort to analyse another type of structure such as, say, rigid-jointed structures. In this chapter the analysis of skeletal structures by the stiffness method will be described in detail. The programming aspects will be discussed in the succeeding chapters.

1.1 STRUCTURAL ANALYSIS—BASIC CONDITIONS TO BE SATISFIED

Every method of structural analysis has to satisfy the following basic requirements in order that the solution is 'correct'. The conditions are as follows.

(a) Equilibrium between the internal stresses (or, for skeletal structures, stress resultants such as axial forces, bending moments, shear forces, torques, etc.) and external loads must be maintained throughout the structure.
(b) The displacements and their derivatives must satisfy the basic continuity requirements as required by the specific structure. For example, in a pin-jointed structure, all the members meeting at a joint must experience the same displacement in two orthogonal directions. Similarly, in a 2-D rigid-jointed structure, continuity of displacement requires continuity of both slopes and displacements.
(c) The stresses and strains must satisfy the prescribed material laws.
(d) The stresses and strains must satisfy the prescribed forces and displacements at the boundaries such as supports, etc.

1.2 ELEMENT IN A STRUCTURE—THE BUILDING BLOCK

In a skeletal structure, the basic building block is the member or the element in a structure. The structure is an assembly of elements connected at joints (also called nodes). It is therefore important to understand the derivation of the 'properties' of the element and the way in which the basic requirements of equilibrium, etc., stated in section 1.1 are satisfied. In addition, because the element will be joined to other members only at its ends, attention will be concentrated on deriving the relationship between the forces and the corresponding displacements at the ends of the element. In the following sections, the relationship between the forces and the displacements at the ends of some common elements such as bars, beams and shafts will be derived.

1.3 A BAR ELEMENT—THE BASIC EQUATIONS

Consider the element which is subjected to axial loads only as shown in Fig. 1.1. The following assumptions are made in deriving the properties of the element.

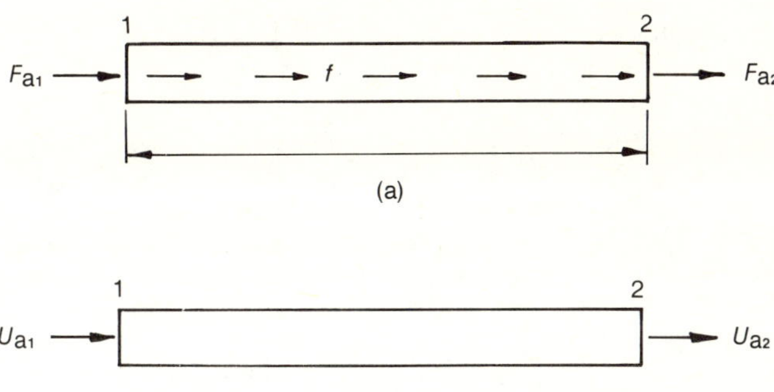

(a)

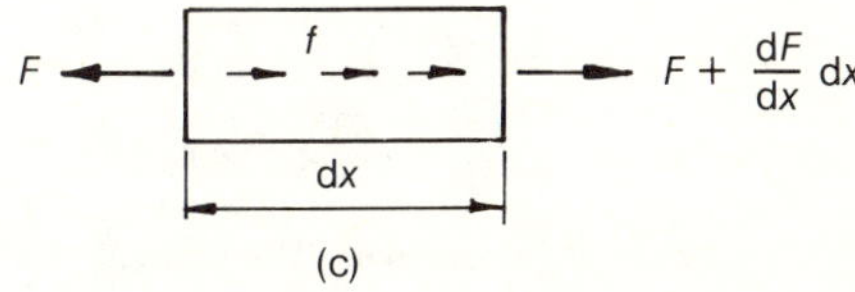

(b)

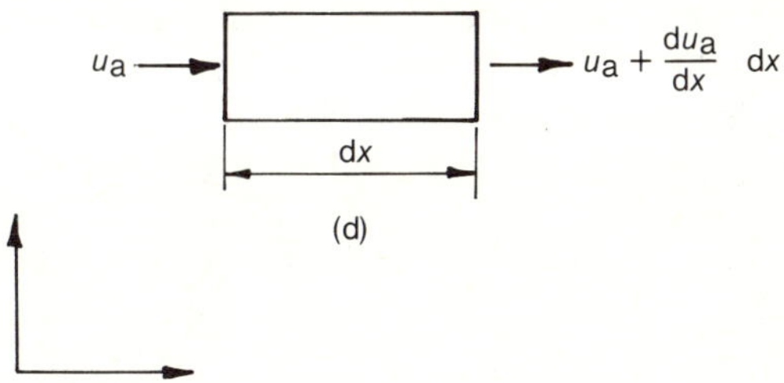

(c)

(d)

Fig. 1.1   An axial element: (a) forces on the element: (b) displacements at the ends; (c) forces on an infinitesimal segment; (d) displacements at the ends of an infinitesimal segment.

(a)   The element has a constant cross-sectional area A.
(b)   The material is elastic with the value of Young's modulus of E.
(c)   The intensity of the distributed load on the element is f per unit length.
(d)   The axial forces at the ends of the element are F_{a1} and F_{a2} at the ends 1 and 2 respectively.
(e)   The axial displacements at the ends 1 and 2 are respectively U_{a1} and U_{a2}.

It is required to derive the relationship between the forces F_{a1} and F_{a2} and the displacements U_{a1} and U_{a2}, ensuring that the basic requirements of equilibrium, etc., are satisfied. This can be done as follows.

(i) *Equilibrium.*   Consider the forces on an infinitesimal length dx of the member. Assuming that the axial tensile force is F at a section, the axial force at a neighbouring section dx from the first is $[F + (dF/dx)\,dx]$. Summing up the forces in the axial direction we have

$$-F + (f\,dx) + \left(F + \frac{dF}{dx}\,dx\right) = 0, \qquad \text{i.e. } \frac{dF}{dx} + f = 0 \qquad (1.1)$$

This is the basic equilibrium equation relating the internal 'stress' F to the external force f on the element. This equation is independent of the material constants and is therefore valid for all materials.

(ii) *Strain–displacement relationship.* Considering the infinitesimal length dx of the element, if the axial displacement at a section is u_a, then the axial displacement at a section dx from the first is $[u_a + (du_a/dx)\,dx]$. The change in length is therefore $(du_a/dx)\,dx$. The axial strain ε_a is given by

$$\varepsilon_a = \frac{du_a}{dx} \qquad (1.2)$$

As in the equilibrium equation, the strain–displacement relationship is independent of material constants. It is a purely geometric relationship.

(iii) *Stress–strain relationship.* If it is assumed that the axial stress σ_a is the only stress present and is uniformly distributed over the cross section, then the axial force $F = A\sigma_a$, where A is the cross-sectional area. If further the material is linearly elastic, then $\sigma_a = E\varepsilon_a$ where E is Young's modulus. Using equations (1.2), the axial force F is given by

$$F = AE\frac{du_a}{dx} \qquad (1.3)$$

Substituting equation (1.3) in equation (1.1), the equilibrium equation can be expressed as

$$AEu_a'' + f = 0 \qquad (1.4)$$

where u_a'' is the second derivative of u_a w.r.t. x.

This is the fundamental differential equation for an elastic bar element with constant properties along its length and subjected to a distributed axial load f along its length. This equation satisfies all the fundamental requirements stated previously and therefore a solution of this equation is the 'correct' solution.

1.4 SOLUTION OF THE BAR EQUATION

Equation (1.4) is easily integrated as follows:

$$u_a = C_1 x + C_2 - \left(\frac{f}{2AE}\right)x^2$$

where C_1 and C_2 are integration constants. The boundary conditions are as follows:

$$\text{at end 1,} \quad u_a = U_{a1}, \quad x = 0$$
$$\text{at end 2,} \quad u_a = U_{a2}, \quad x = L$$

To satisfy the boundary conditions, the constants are given by

$$C_1 = \frac{U_{a2} - U_{a1}}{L} + \frac{fL}{2AE}, \quad C_2 = U_{a1}$$

Using these values for C_1 and C_2,

$$u_a = \left(1 - \frac{x}{L}\right)U_{a1} + \frac{x}{L} U_{a2} + \frac{f}{2AE} x(L - x) \tag{1.5}$$

The above solution was obtained in the 'orthodox' manner. A slightly different approach is useful as it enables one to calculate the final displacement in two convenient stages. In this approach, the final solution is obtained as the sum of two partial solutions as follows. This is valid because the differential equation (1.4) is a linear differential equation.

1.5   FIXED END AND FIXED DISPLACEMENT SOLUTIONS

The two parts of the solution are called as fixed end solution and fixed displacement solution respectively. The solution to each part is obtained as follows.

(i) *Fixed end solution.*   As the name implies, it is assumed that, for this part of the solution, the boundary conditions are

$$\text{at end 1,} \quad u_a = 0, \quad x = 0$$
$$\text{at end 2,} \quad u_a = 0, \quad x = L$$

This requires that

$$C_1 = 0.5fL, \quad C_2 = 0$$

This leads to the fixed end solution as

$$u_a = \frac{f}{2AE} x(L - x) \tag{1.6}$$

(ii) *Fixed displacement solution.*   In this case it is assumed that $f = 0$ and the boundary conditions are

$$\text{at end 1,} \quad x = 0, \quad u_a = U_{a1}$$
$$\text{at end 2,} \quad x = L \quad u_a = U_{a2},$$

This requires that $u_a = C_1 x + C_2$. The integration constants are given by

$$C_1 = \frac{U_{a2} - U_{a1}}{L}, \quad C_2 = U_{a1}.$$

This leads to

$$u_a = \left(1 - \frac{x}{L}\right)U_{a1} + \frac{x}{L}U_{a2} = N_1 U_{a1} + N_2 U_{a2} \qquad (1.7)$$

where $N_1 = 1 - x/L$ and $N_2 = x/L$ are called shape functions because they give the value of u_a due to the unit values U_{a1} and U_{a2} respectively of the end displacements. As can be easily verified, equation (1.5) is the sum of equations (1.6) and (1.7). It is important to keep in mind that the term containing the external distributed load f appears in the fixed end part of the solution only.

1.6 THE ELEMENT STIFFNESS MATRIX—A BAR ELEMENT

Using equations (1.5) and (1.3), the axial tensile force F is given by

$$F = \frac{AE}{L}(U_{a2} - U_{a1}) + 0.5f(L - 2x)$$

At end 1, $F = -F_{a1}$, $x = 0$, and, at end 2, $F = F_{a2}$, $X = L$. Therefore

$$F_{a1} = \frac{AE}{L}(U_{a1} - U_{a2}) - 0.5fL, \qquad F_{a2} = \frac{AE}{L}(U_{a2} - U_{a1}) - 0.5fL$$

The above equations can be expressed in matrix form as

$$\begin{bmatrix} F_{a1} \\ F_{a2} \end{bmatrix} = \frac{AE}{L}\begin{bmatrix} 1 & -1 \\ -1 & 1 \end{bmatrix}\begin{bmatrix} U_{a1} \\ U_{a2} \end{bmatrix} + \begin{bmatrix} -0.5fL \\ -0.5fL \end{bmatrix} \qquad (1.8)$$

The four matrices in equation (1.8) from left to right are generally known as the element joint force vector, the element stiffness matrix, the element joint displacement vector and the fixed end joint force vector respectively. Note that the element stiffness matrix is a symmetric matrix. This is true of all linearly elastic elements and is a consequence of the well-known Maxwell–Betti theorem.

1.7 BEAM ELEMENT—THE BASIC EQUATIONS

Consider the beam element shown in Fig. 1.2. The element is subjected at the ends to forces normal to the axis and couples. In addition to the forces at the ends, a distributed load q per unit length acts on the element. The following assumptions are made in deriving the properties of the element.

(a)   The element has a constant second moment of area I.
(b)   The material is elastic with Young's modulus equal to E.
(c)   The intensity of distributed load is q per unit length.
(d)   The couples (assumed positive clockwise) at the ends 1 and 2 of the element are M_1 and M_2 respectively.
(e)   The rotations (assumed positive clockwise) at the ends 1 and 2 of the element are θ_1 and θ_2 respectively.

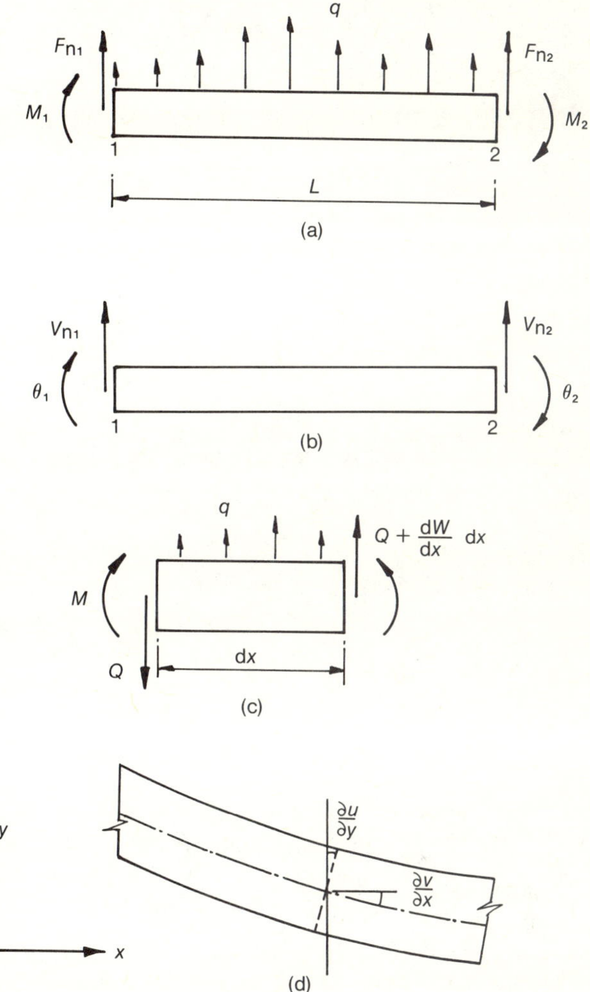

Fig. 1.2   A bending element: (a) forces on the element; (b) displacements at the ends; (c) forces on an infinitesimal segment; (d) deformation of the cross section.

(f)   The forces normal to the axis at the ends 1 and 2 are respectively F_{n1} and F_{n2}.

(g)   The translations normal to the axis at the ends 1 and 2 are respectively V_{n1} and V_{n2}.

It is required to derive the relationship between the forces at the ends, i.e. M_1, F_{n1}, M_2 and F_{n2}, and the corresponding displacements, i.e. θ_1, V_{n1}, θ_2 and V_{n2}, ensuring that all the basic requirements of equilibrium, etc., are satisfied. This can be done as follows.

(i) *Equilibrium.*   Consider the forces acting on an infinitesimal length dx of the element. Assuming that the positive bending moment causes tension on the bottom fibre and that an 'anticlockwise' shear force is positive, the forces acting on the infinitesimal length is as shown is Fig. 1.2(c). Summing up the forces in the vertical direction, we have

$$-Q + \left(Q + \frac{\mathrm{d}Q}{\mathrm{d}x}\,\mathrm{d}x\right) + q\,\mathrm{d}x = 0$$

This leads to

$$\frac{\mathrm{d}Q}{\mathrm{d}x} + q = 0$$

Taking moment about the right-hand end, we have

$$M - \left(M + \frac{\mathrm{d}M}{\mathrm{d}x}\,\mathrm{d}x\right) - Q\,\mathrm{d}x + q\,\mathrm{d}x\,(0.5\,\mathrm{d}x) = 0$$

Ignoring terms of second order and simplifying, we have

$$\frac{\mathrm{d}M}{\mathrm{d}x} + Q = 0$$

The equations of equilibrium are therefore

$$\frac{\mathrm{d}Q}{\mathrm{d}x} + q = 0, \qquad \frac{\mathrm{d}M}{\mathrm{d}x} + Q = 0$$

Eliminating Q from the above two equations, we have

$$M'' - q = 0 \tag{1.9}$$

where M'' represents the second derivative of M w.r.t. x.

This is the fundamental equation of equilibrium relating the internal 'stress' M and the external load q. As for the bar element, the equation is independent of the material properties of the element.

(ii) *Strain–Displacement relationship.*   In general, any point in a two dimensional space can have two independent displacements in two orthogonal directions. However, depending on the constraints present, the two displacements will not be independent. In the simplified engineer's theory of beam bending, it is assumed that plane sections remain plane 'before and after bending'. Assuming that the beam axis coincides with the x axis and that the origin for the y axis is at the centroid of the cross section, if the plane sections assumption is made, then the displacement in the axial direction is given by

$$u = a(x) + b(x)y \tag{1.10}$$

where $a(x)$ and $b(x)$ are unknown functions of x only.

In the engineer's theory of bending, two further assumptions are normally made. These assumptions are as follows.

(a)   Shear strains are negligible. It should be emphasized that only shear strains and not shear stresses are neglected. This is equal to assuming that the shear modulus is infinitely large.

(b)   The strain in the direction of the depth is negligible.

Assumption (a) leads to the condition that $\partial u/\partial y + \partial v/\partial x = 0$ as shown in Fig. 1.2(d) where v is the displacement in the depth direction and u is the displacement in the axial direction. From equation (1.10), $\partial u/\partial y = b(x)$ and, because shear strain is zero, $b(x) = -\partial v/\partial x$. Therefore the displacement u in the axial direction is given by $u = a(x) - (\partial v/\partial x)y$. Since the variation in displacement v in the depth direction is negligible, it is permissible to replace $\partial v/\partial x$ by $\mathrm{d}v/\mathrm{d}x$. Therefore the axial displacement u is given by

$$u = a(x) - \frac{\mathrm{d}v}{\mathrm{d}x}\,y \tag{1.11}$$

The axial strain $\varepsilon_x = \partial u/\partial x = \mathrm{d}a(x)/\mathrm{d}x - v''y$, where v'' is the second derivative of v w.r.t. x.

(iii) *Stress–strain relationship.*   Assuming that the material is linearly elastic, then the axial stress σ_x is given by

$$\sigma_x = E\varepsilon_x = E\left[\frac{\mathrm{d}a(x)}{\mathrm{d}x} - v''y\right] \tag{1.12}$$

The stress resultants, i.e. the axial force F and the bending moment M, are given by

$$\text{axial force } F = \int \sigma_x \, \mathrm{d}A$$

$$\text{bending moment } M = -\int \sigma_x y \, \mathrm{d}A$$

where $\mathrm{d}A$ is the element of area. Since the origin for y is at the centroid of the section,

$$A = \int \mathrm{d}A, \quad 0 = \int y \, \mathrm{d}A, \quad I = \int y^2 \, \mathrm{d}A$$

Therefore, equation (1.12) can be written as

$$\sigma_x = \frac{F}{A} - \frac{M}{I}\,y$$

where $M = EIv''$ and $F = AE[\mathrm{d}a(x)/\mathrm{d}x]$. The equation

$$M = EIv'' \tag{1.13}$$

is the fundamental 'stress–strain' relationship for the beam element and is generally known as the moment–curvature relationship. It is not necessarily a general relationship for all beam elements because of the many assumptions made in its derivation. However, it is generally assumed to be reasonably correct for shallow beams (i.e. beams where the span-to-depth ratio is large) away from regions of concentrated loads and reactions.

Combining equations (1.9) and (1.13), the equation of equilibrium in terms of the displacement v is given by

$$EIv'''' = q \tag{1.14}$$

This is the fundamental equation for the solution of beam problems based on the assumptions of the engineer's theory of bending. It is valid only if flexural rigidity EI is constant along the length of the beam. It satisfies all the basic requirements of equilibrium , etc., stated in section 1.1 and therefore any solution of this equation will be the 'correct' solution.

1.8 SOLUTION OF THE BEAM EQUATION

The solution will be obtained in two parts as explained previously.

(i) *Fixed end solution.*   The differential equation is easily integrated and the solution to v is given by

$$v = \frac{C_1}{6} x^3 + \frac{C_2}{2} x^2 + C_3 x + C_4 + \frac{q}{24EI} x^4$$

Because we are considering the fixed end condition, the boundary conditions are

$$\text{at end 1,} \quad v = \frac{dv}{dx} = 0, \quad x = 0$$

$$\text{at end 2,} \quad v = \frac{dv}{dx} = 0, \quad x = L$$

The integration constants are given by

$$C_1 = -0.5qL, \quad C_2 = \frac{q}{12} L^2, \quad C_3 = C_4 = 0, \quad v = \frac{q}{24EI} (x^4 - 2Lx^3 + L^2 x^2)$$

$$(1.15)$$

(ii) *Fixed displacement solution.*   The solution to v is obtained assuming $q = 0$. Therefore the solution to v is given by

$$v = \frac{C_1}{6} x^3 + \frac{C_2}{2} x^2 + C_3 x + C_4$$

The boundary conditions are given by

$$\text{at end 1,} \quad v = V_{n1}, \quad \frac{dv}{dx} = -\theta_1, \quad x = 0$$

$$\text{at end 2,} \quad v = V_{n2}, \quad \frac{dv}{dx} = -\theta_2, \quad x = L$$

Note that, for the coordinate axes chosen, dv/dx is positive if anticlockwise while the sign convention for θ is positive if clockwise. Solving for the integration constants we have

$$C_1 = \frac{12}{L^3}(V_{n1} - V_{n2}) - \frac{6}{L^2}(\theta_1 + \theta_2)$$

$$C_2 = \frac{-6}{L^2}(V_{n1} - V_{n2}) + \frac{\theta_1 + 2\theta_2}{L}$$

$$C_3 = -\theta_1$$

$$C_4 = V_{n1}$$

The solution to v is expressed in terms of the shape functions N_1 to N_4 as

$$v = N_1 V_{n1} + N_2 \theta_1 + N_3 V_{n2} + N_4 \theta_2 \tag{1.16}$$

where

$$N_1 = 1 - 3\alpha^2 + 2\alpha^3, \qquad N_2 = L(-\alpha + 2\alpha^2 - \alpha^3)$$

$$N_3 = 3\alpha^2 - 2\alpha^3, \qquad N_4 = L(\alpha^2 - \alpha^3)$$

$$\alpha = \frac{x}{L}$$

As explained before, the shape functions describe the displacement of the beam due to unit values of displacements at the end of the beam.

(iii) *The complete solution.*   The complete solution is the sum of the two solutions given by equations (1.15) and (1.16). Therefore,

$$v = N_1 V_{n1} + N_2 \theta_1 + N_3 V_{n2} + N_4 \theta_2 + \frac{q}{24EI}(x^4 - 2Lx^3 + L^2 x^2) \tag{1.17}$$

1.9   THE ELEMENT STIFFNESS MATRIX FOR A BEAM ELEMENT

From the moment–curvature relationship $M = EIv''$ and from the equation of equilibrium shear force $Q = -dM/dx = -EIv'''$. Using equation (1.17) for v, the bending moment M and shear force Q are given, using $\alpha = x/L$, by

$$M = \frac{EI}{L^2}[6(2\alpha - 1)(V_{n1} - V_{n2}) + (4 - 6\alpha)L\theta_1 + (2 - 6\alpha)L\theta_2]$$

$$+ \frac{qL^2}{12}(1 - 6\alpha + 6\alpha^2)$$

$$Q = \frac{EI}{L^3}[12(V_{n1} - V_{n2}) - 6L(\theta_1 + \theta_2)] + 0.5qL(2\alpha - 1)$$

Since the bending moment is positive if it causes tension on the bottom face and the shear force Q is positive if it is 'anticlockwise', the forces at the ends of the element are as follows:

$$\text{at end 1,}\quad M = M_1, \quad Q = -F_{n1}, \quad x = 0$$

$$\text{at end 2,}\quad M = -M, \quad Q = F_{n2}, \quad x = L$$

The substitution of the boundary conditions lead to the following equations:

$$F_{n1} = \frac{6EI}{L^3}[2(V_{n1} - V_{n2}) - L(\theta_1 + \theta_2)] - 0.5qL$$

$$M_1 = \frac{EI}{L^2}[-6(V_{n1} - V_{n2}) + 4L\theta_1 + 2L\theta_2] + \frac{qL^2}{12}$$

$$F_{n2} = \frac{6EI}{L^3}[-2(V_{n1} - V_{n2}) + L(\theta_1 + \theta_2)] - 0.5qL$$

$$M_2 = \frac{EI}{L^2}[-6(V_{n1} - V_{n2}) + 2L\theta_1 + 4L\theta_2] - \frac{qL^2}{12}$$

The above four equations can be expressed in matrix form as

$$
\begin{bmatrix} F_{n1} \\ M_1 \\ F_{n2} \\ M_2 \end{bmatrix}
= \frac{EI}{L}
\begin{bmatrix}
\dfrac{12}{L^2} & \dfrac{-6}{L} & \dfrac{-12}{L^2} & \dfrac{-6}{L} \\
 & 4 & \dfrac{6}{L} & 2 \\
 & & \dfrac{12}{L^2} & \dfrac{6}{L} \\
\text{symmetric} & & & 4
\end{bmatrix}
\begin{bmatrix} V_{n1} \\ \theta_1 \\ V_{n2} \\ \theta_2 \end{bmatrix}
+
\begin{bmatrix} -0.5qL \\ \dfrac{qL^2}{12} \\ 0.5qL \\ \dfrac{-qL^2}{12} \end{bmatrix}
$$

As can be observed, the element stiffness matrix is once again symmetric.

1.10 THE EFFECT OF SHEARING DEFORMATIONS

In the derivation of the moment–curvature relationship in section 1.7, it was assumed that the beam was made of material which was infinitely stiff in shear. However, real materials have a finite shear modulus and, by ignoring the deformation due to shear, one underestimates the true deformations of the structure. In general, compared with the deformation due to bending stresses, the deformation due to shear stresses is small, except for 'stocky' beams which have a low span-to-depth ratio. However, when computers are used for structural analysis, it is important to ensure that all beams, irrespective of the span-to-depth ratio, are properly analysed. It is therefore important to modify the element stiffness matrix to include the effect of shearing deformations. Consider an infinitesimal element subjected to shear

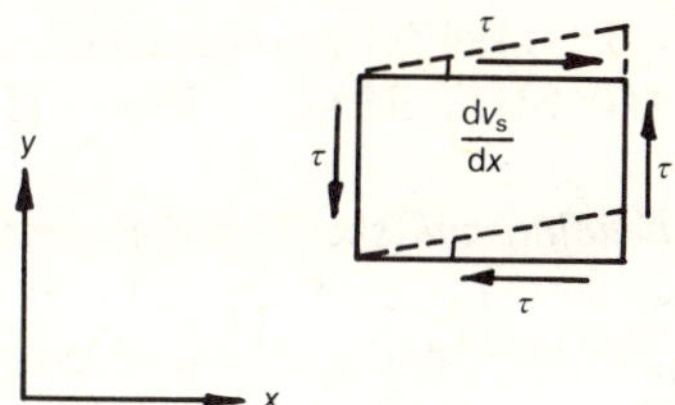

Fig. 1.3   Deformation due to shear stresses.

stress as shown in Fig. 1.3. Assuming that the material is elastic, the shear strain is given by

$$\frac{dv_s}{dx} = \frac{\tau}{G}$$

where G is the shear modulus, τ the shear stress and v_s the displacement due to shear. In a beam cross section, the shear stress is not constant. For example, for a beam of rectangular cross section, the shear stress varies parabolically across the depth. Therefore, for the sake of simplicity, it is generally assumed that a representative value of τ is given by

$$\tau = \frac{Q}{A_s}$$

where Q is the shear force and A_s is the effective cross section for shear (which is equal to the area of cross section divided by the area shear factor). For rectangular sections the area shear factor is 1.2. For I sections bending about the major axis, because the shear force is carried mainly by the web, it is generally assumed that A_s = web area/1.2. Therefore the general 'stress–strain' relationship is given by

$$\frac{dv_s}{dx} = \frac{Q}{GA_s}$$

Assuming that Q is constant, the above equation can be integrated as

$$v_s = \frac{Q}{GA_s} x + C_1$$

where C_1 is a constant of integration. It should be noted that dv_s/dx is the slope of the beam axis to the horizontal while the vertical sections remain undeformed as shown in Fig. 1.3. If v_s is zero at, say, $x = x_1$, then v_s is given by

$$v_s = \frac{Q}{GA_s}(x - x_1) \tag{1.18}$$

In the next section, equation (1.18) will be used to modify the element stiffness matrix previously derived, to include the effects of shearing deformations.

1.11 THE BEAM STIFFNESS MATRIX INCLUDING THE EFFECT OF SHEARING DEFORMATIONS

(i) *The effect of bending and shearing deformation on the translation.* Consider the beam element shown in Fig. 1.4. Let the element be subjected to a translation of V_{n1} only at end 1. If the shearing deformations are ignored, then, from the stiffness matrix previously derived, the shear force at end 1 is $Q = -F_{n1}$ and this shear force is constant throughout the beam. Assuming that v_s is zero at $x = L$, then the shear deformation is given, from equation (1.18), by

$$v_s = -\frac{F_{n1}}{GA_s}(x - L)$$

where $F_{n1} = 12(EI/L^3)V_{n1}$. At $x = 0$,

$$v_s = \frac{F_{n1}L}{GA_s} = \beta V_{n1}$$

where $\beta = 12EI/GA_s L^2$.

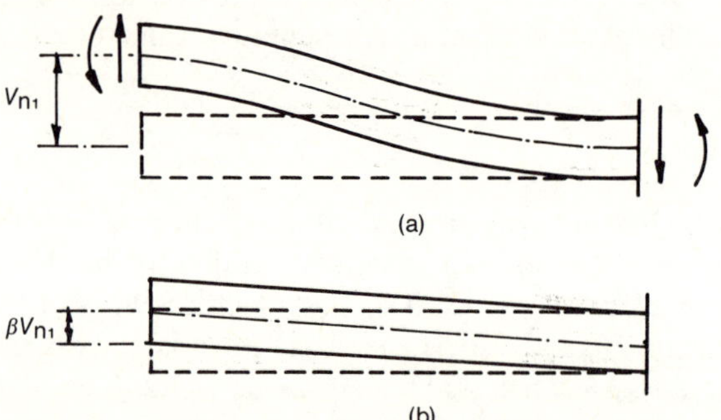

Fig. 1.4   The effect of bending and shearing deformation on the translation at end 1: (a) deformation due to bending action only; (b) additional deformation due to shearing action.

Because, during the shearing deformations, the vertical sections of the beam do not rotate, shearing deformations will affect only the translation at end 1 but does not affect the translation at end 2 and rotations at the ends of the beam. The total translation at end 1 is calculated as follows:

$$\begin{Bmatrix} \text{translation} \\ \text{at} \\ \text{end 1} \end{Bmatrix} = \begin{Bmatrix} \text{translation due to} \\ \text{bending deformation} \\ \text{only } (V_{n1}) \end{Bmatrix} + \begin{Bmatrix} \text{translation due to} \\ \text{shearing deformation} \\ \text{only } (\beta V_{n1}) \end{Bmatrix}$$

$$= (1 + \beta)V_{n1}$$

Therefore the inclusion of the effects of shearing deformation has increased the net translation from V_{n1} to $(1 + \beta)V_{n1}$. Similar arguments apply to translation at end 2 only of V_{n2}. Therefore, to correct the stiffness matrix for the inclusion of shearing deformations, it is necessary to divide the

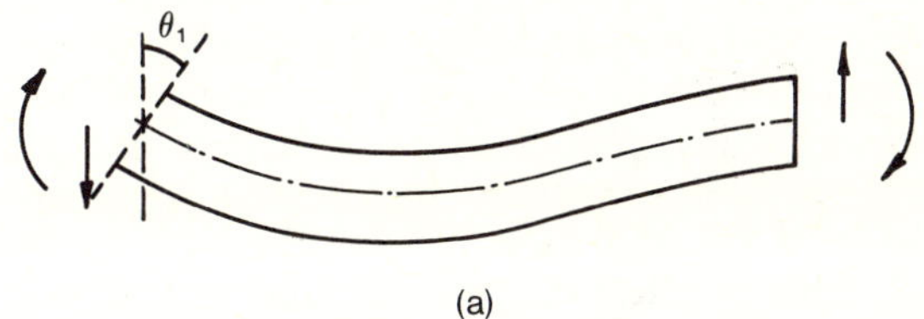

(a)

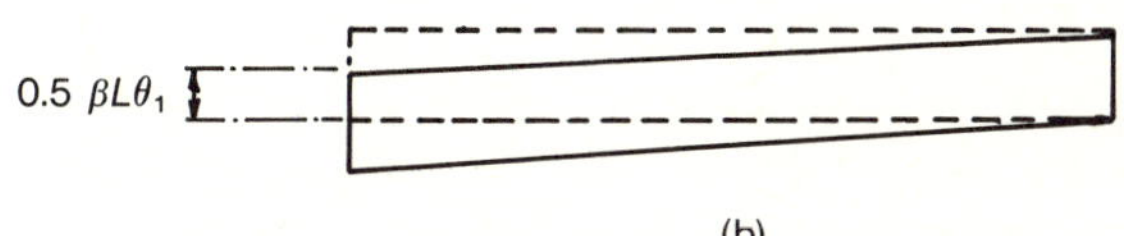

$0.5\,\beta L\theta_1$

(b)

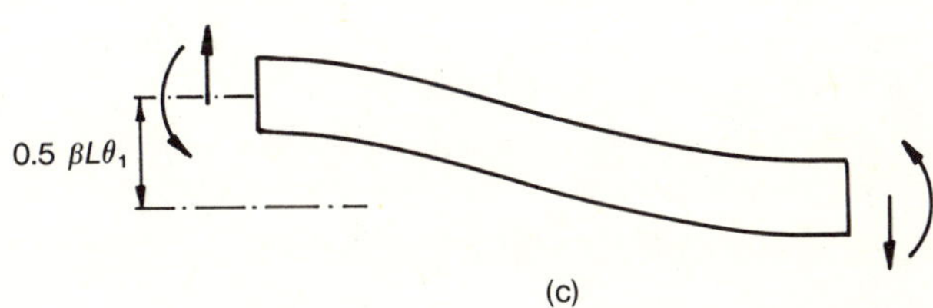

$0.5\,\beta L\theta_1$

(c)

Fig. 1.5   The effect of bending and shearing deformation on the rotation at end 1:
(a) deformation due to bending action only; (b) shearing deformation caused by forces
developed in (a); (c) forces developed when deformation in (b) is compensated by
including bending and shearing effects.

coefficients of the first and third columns of the stiffness matrix by $1 + \beta$ so
that the net translation is V_{n1} and V_{n2} only at ends 1 and 2 respectively.

(ii) *The effect of bending and shearing deformation on the rotation.*   Consider
the beam subjected to a rotation of θ_1 only at end 1 while all other
displacements at the ends are suppressed as shown in Fig. 1.5. Considering
the effects of bending deformation only, the shear force induced is calculated
from the stiffness matrix derived previously and is given by

$$Q = -F_{n1} = 6\frac{EI}{L^2}\theta_1$$

The shear deformation v_s is given, if v_s is assumed to be zero at end 2 (i.e.
$x = L$), by

$$v_s = -\frac{F_{n1}}{GA_s}(x - L)$$

where $F_{n1} = -(6EI/L^2)\theta_1$.

Therefore, at end 1 (i.e. $x = 0$), $v_s = -0.5\beta L\theta_1$, where $\beta = 12EI/GA_sL^2$.

In other words, the shearing deformation has 'disturbed' the zero
translation condition at end 1. However, because we know how to calculate

the forces induced as a result of translation at end 1 (or end 2) only when the effects of both bending and shearing deformation effects are included, the above disturbance in translation can be easily corrected as follows.

(a) The forces induced at the ends due to a rotation of θ_1 at end 1 when shearing deformations are ignored, are

$$F_{n1} = -F_{n2} = -\frac{6EI}{L^2}\theta_1, \quad M_1 = \frac{4EI}{L}\theta_1, \quad M_2 = \frac{2EI}{L}\theta_2$$

(b) If the effects of the shearing deformation due to the shear force given above is included, then a net translation of $-0.5\beta L\theta_1$ results at end 1 but no additional forces are induced at the ends.

(c) If the beam is subjected to a translation at end 1 only of $0.5\beta L\theta_1$ (without disturbing the other displacements at the ends), then the forces induced at the ends are given, when both the bending and the shearing deformation effects are included, by

$$F_{n1} = -F_{n2} = \frac{12EI}{L^3}\frac{V_{n1}}{1+\beta}, \quad M_1 = M_2 = -\frac{6EI}{L^2}\frac{V_{n1}}{1+\beta}$$

where $V_{n1} = 0.5\beta L\theta_1$.

Summing up the forces from (a) to (c) above we have the forces induced as a result of a rotation θ_1 only when the effects of both bending and shearing deformations are included:

$$F_{n1} = -F_{n2} = -\frac{6EI/L^2}{1+\beta}\theta_1$$

$$M_1 = \frac{(4+\beta)EI}{(1+\beta)L}\theta_1, \quad M_2 = \frac{(2-\beta)EI}{(1+\beta)L}\theta_1$$

Since similar corrections apply to rotation θ_2 as well, the modified beam stiffness matrix which includes the effects of both bending and shearing deformations is given by

$$
\begin{bmatrix} F_{n1} \\ M_1 \\ F_{n1} \\ M_2 \end{bmatrix} = \frac{EI}{L(1+\beta)}
\begin{bmatrix}
\dfrac{12}{L^2} & -\dfrac{6}{L} & -\dfrac{12}{L^2} & -\dfrac{6}{L} \\[2mm]
 & 4+\beta & \dfrac{6}{L} & 2-\beta \\[2mm]
 & & \dfrac{12}{L^2} & \dfrac{6}{L} \\[2mm]
\text{symmetric} & & & 4+\beta
\end{bmatrix}
\begin{bmatrix} V_{n1} \\ \theta_1 \\ V_{n2} \\ \theta_2 \end{bmatrix}
+
\begin{bmatrix} -0.5qL \\[2mm] \dfrac{qL^2}{12} \\[2mm] -0.5qL \\[2mm] \dfrac{qL^2}{12} \end{bmatrix}
$$

$$(1.19)$$

1.12 THE EFFECT OF SHEARING DEFORMATIONS ON THE FIXED END REACTIONS

If the load on a beam is symmetrically distributed, then the effects of shearing deformations do not disturb the end displacements, as shown in Fig. 1.6. Therefore the fixed end reactions remain unaffected. In other words, for symmetric loading, the fixed end reactions can be calculated by considering bending deformations only. However, if the loading is unsymmetric, then the shearing deformation upsets the conditions at the ends but the stiffness matrix derived above can be used to restore the fixed end conditions and hence to calculate the corrections to the fixed end forces. The details are left as an exercise to the reader.

1.13 A TORSION ELEMENT—THE BASIC EQUATIONS

Consider the torsional element shown in Fig. 1.7. The element is subjected to torques at the ends in addition to a distributed torque t per unit length. Note that in Fig. 1.7 the torques are represented by 'double-headed' arrows. The convention used is that of the right-hand rule, i.e. the direction of rotation of the torque is given by the direction in which the fingers of the right-hand wrap when the thumb is directed along the arrowhead. The following assumptions are made in deriving the properties of the element.

(a)   The element has constant properties along its length.
(b)   The material is elastic.
(c)   The torques are T_1 and T_2 at the ends 1 and 2 respectively.
(d)   The twists are ψ_1 and ψ_2 at the ends 1 and 2 respectively.

Before deriving the stiffness matrix for the element, it is necessary to discuss some important points in connection with the torsion problem. When a beam is subjected to torsional forces, in addition to the twisting of the cross section, displacements along the axis of the beam occur as shown in Fig. 1.8. The axial displacements are called warping displacements. For 'solid' sections such as rectangular beams, the warping displacements are generally small. However, for 'open' sections such as thin-walled rolled steel beams of I and channel shapes, the warping displacements tend to be large. If the axial displacements are restrained, then the flanges act as beams bending in their own plane and are able to resist torsion in the 'bending' mode. This is sometimes called 'nonuniform' or restrained warping torsion or even bending torsion. In this section the effects of warping restraint are ignored and it is assumed that one is considering the so-called Saint Venant's torsion where the torsion is resisted by the shear stresses in the cross section and the beam is permitted unrestrained warping displacements. Nonuniform torsion is discussed in Appendix 4.

The element stiffness matrix expresses the relationship between the forces at the ends, i.e. T_1 and T_2 and the corresponding displacements ψ_1 and ψ_2. It

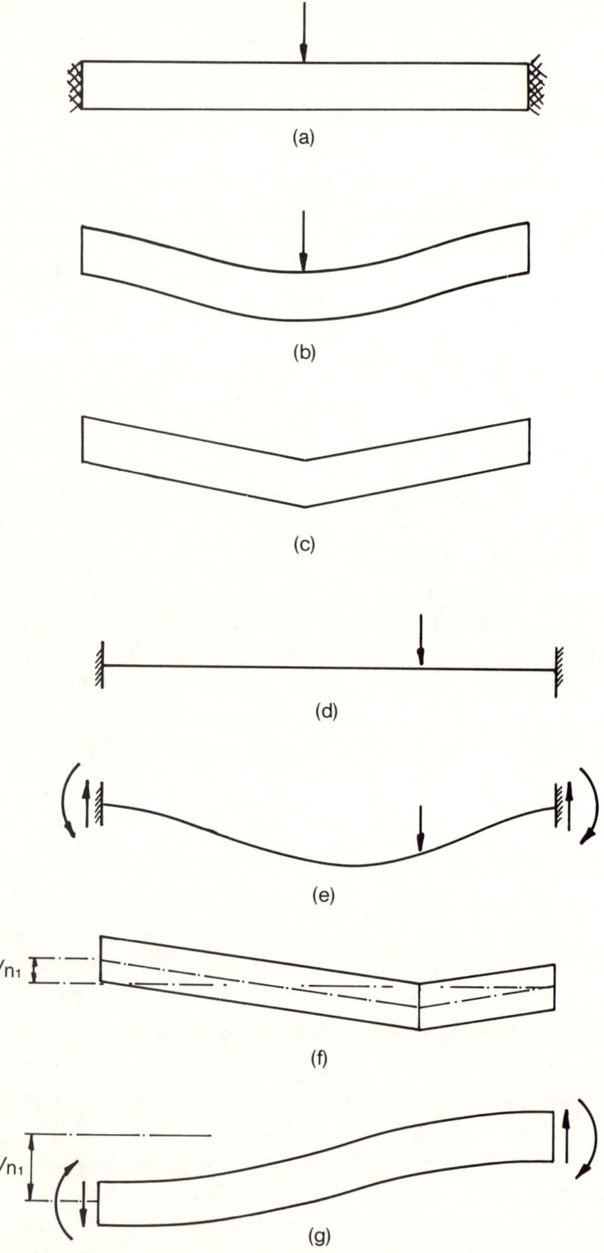

Fig. 1.6 The effect of including bending and shearing deformation on the fixed end forces in a symmetrically loaded beam: (a) beam under a symmetrical lateral load; (b) deformation due to bending action only; (c) additional symmetrical deformation due to shearing action; (d) unsymmetrically loaded beam; (e) deformation due to bending action; (f) unsymmetrical deformation due to shearing action which disturbs the boundary condition; (g) forces developed due to restoring the boundary condition.

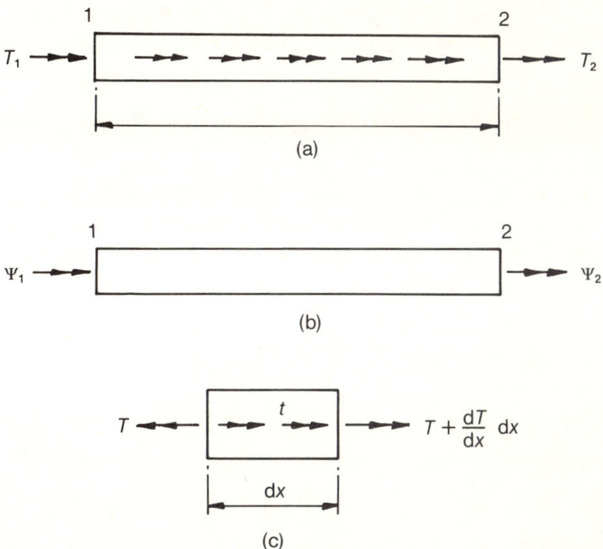

Fig. 1.7   A torsion element: (a) forces on the element; (b) twists at the ends; (c) forces on an infinitesimal segment.

is necessary that all the basic requirements stated in section 1.1 must be satisfied. This can be done as follows.

(i) *Equilibrium.*   Considering the forces acting on an infinitesimal length dx of the element, the equilibrium requires that

$$-T + t\,dx + \left(T + \frac{dT}{dx}dx\right) = 0, \quad \text{i.e.} \ \frac{dT}{dx} + t = 0 \qquad (1.20)$$

This is the equation of equilibrium in terms of 'stress' T and is independent of material properties.

(ii) *Strain–displacement relationship.*   In the classical (or Saint Venant's) torsion problem it is shown that the displacements consist of the twist about the axis and warping of the cross section such that

(a)   The rate of twisting along the axis is the same for all cross sections.

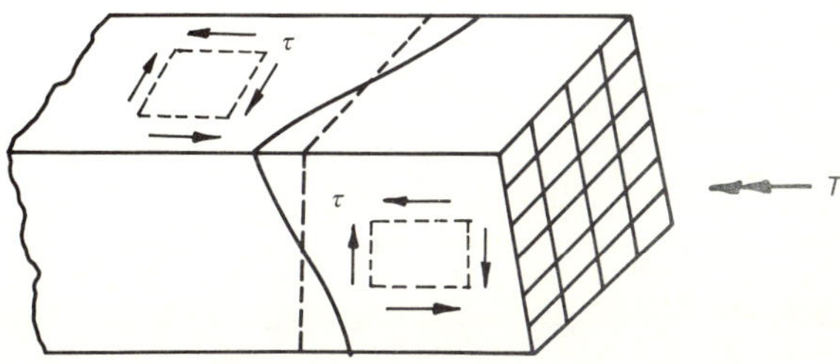

Fig. 1.8   Warping displacements due to torsion.

(b)   The rate of warping along the axis is the same for all cross sections. It is of course not necessary for all points in a cross section to have the same value of warping displacement.

The above assumptions lead to the conclusion that in a cross section there are only two shearing strains and a normal strain present.

(iii) *Stress–strain relationship.*   Because of the assumption of free warping there are only two shear stresses present in the cross section which produce a resultant of a torque T which is related to the twist by

$$T = GJ \frac{\mathrm{d}\psi}{\mathrm{d}x} \tag{1.21}$$

where G is the shear modulus and $J =$ the torsion 'inertia' known as Saint Venant's torsion constant. J is a property of the cross section and, for a rectangular beam of dimension $b \times d$, an approximate expression for J is given by

$$J = kb^3 d, \quad k = \frac{0.3}{1 + (b/d)^2}, \quad b < d$$

For other types of 'open' and 'closed' sections, expressions for J are given in many standard works on the theory of elasticity.

Using the above 'stress–strain' relationship, the equation of equilibrium (1.21) can be expressed as

$$GJ\psi'' + t = 0 \tag{1.22}$$

In the above, ψ'' is the second derivative of ψ w.r.t. x. This is the fundamental differential equation for the elastic torsion element. This equation satisfies all the basic requirements of equilibrium, etc., and therefore a solution of the equation leads to the correct solution.

1.14   SOLUTION OF THE TORSION EQUATION

A comparison of equation (1.22) with the corresponding equation (1.4) for the bar element, and in a similar fashion a comparison of equation (1.21) with equation (1.3), shows that the axial force and torsion problems are identical with u_a replaced by ψ, AE replaced by GJ, and F replaced by T. Therefore the following stiffness matrix for the torsion element can be derived without much ado:

$$\psi = \left(1 - \frac{x}{L}\right)\psi_1 + \frac{x}{L}\psi_2 + \frac{t}{2GJ}x(L - x)$$

$$T = \frac{GJ}{L}(\psi_2 - \psi_1) + 0.5t(L - 2x)$$

$$\begin{bmatrix} T_1 \\ T_2 \end{bmatrix} = \frac{GJ}{L}\begin{bmatrix} 1 & -1 \\ -1 & 1 \end{bmatrix}\begin{bmatrix} \psi_1 \\ \psi_2 \end{bmatrix} + \begin{bmatrix} -0.5tL \\ -0.5tL \end{bmatrix} \tag{1.23}$$

1.15 FIXED END REACTIONS DUE TO THERMAL RESTRAINT

In the previous sections, the fixed end forces due to a uniformly distributed load on the element was calculated. In many cases, in practice, it is necessary to consider the forces which arise because of the prevention of free thermal expansion. In this section the fixed end forces which arise as a result of thermal restraint will be derived. Consider a beam element subjected to a temperature change which is linear with depth as shown in Fig. 1.9. It is further assumed that the change in temperature is constant throughout the length of the element. Let the temperature change in the upper and lower surfaces be T_u and T_b respectively. If the origin for the y axis is at the centroid of the cross section, then the temperature change at any depth y from the centroid is given by

$$T = T_b + \frac{T_u - T_b}{d}(Y_b + y) \tag{1.24}$$

where $d\ (= Y_u + Y_b)$, is the depth of the beam, Y_u is the distance from the centroid to the top face and Y_b is the distance from the centroid to the bottom face. Equation (1.24) can also be expressed as

$$T = \frac{T_u Y_b + T_b Y_u}{d} + \frac{(T_u - T_b)y}{d}$$

If the coefficient of linear thermal expansion is α_t, then the strain in the axial direction is given by $\alpha_t T$. However, if the thermal expansion is restrained, then axial forces and bending moments develop at the ends. If the

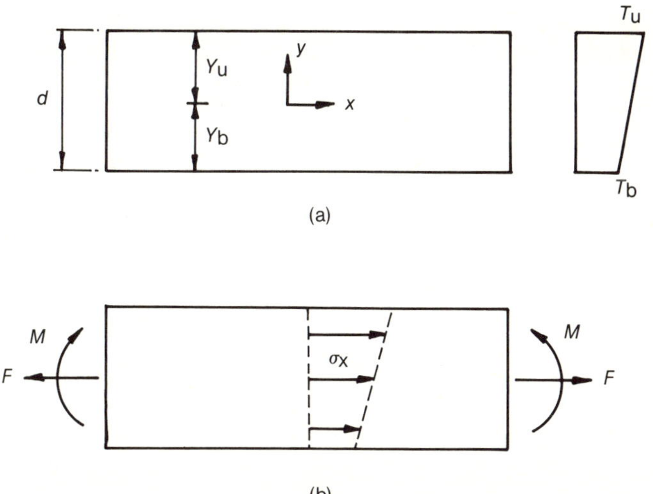

Fig. 1.9 Restraining forces due to temperature changes: (a) beam element with a linear temperature change; (b) restraining forces developed.

axial restraint force is F ($+$ is tensile) and a bending moment M ($+$ produces tension on the bottom face), then the axial strain due to these forces is given, as explained in section 1.7, by

$$\varepsilon_x = \frac{F/A - My/I}{E}$$

If the thermal expansion is fully restrained, then the total strain due to restraint forces at the ends and thermal strain must be zero, i.e. $\alpha_t T + \varepsilon_x = 0$. Equating the terms in y to zero we have

$$F = \frac{-AE\alpha_t(T_u Y_b + T_b Y_u)}{d}, \quad M = \frac{EI\alpha_t(T_u - T_b)}{d} \tag{1.25}$$

For a bar element it is generally assumed that $T_u = T_b = T$ and $F = -AE\alpha_t T$.

1.16 THE ELEMENT STIFFNESS MATRIX WITH FORCES AND DISPLACEMENTS AT JOINTS IN GENERAL COORDINATE DIRECTIONS

In the previous sections, the stiffness matrix was derived on the assumption that the element is horizontal and the displacement and forces are either along the axis or normal to the axis. This assumption made the derivation reasonably simple but it is not general enough for practical use. The reason for this is that, in general, the elements in a structure are inclined to the coordinate axes. In this section the stiffness matrix derived previously will be 'converted' to a more usable form.

(i) *A bar element.* Consider the bar element shown in Fig. 1.10. The displacement along the x axis and y axis at the ends are (u_1, v_1) and (u_2, v_2) at ends 1 and 2 respectively. Resolving the displacements along the axial direction we have

$$U_{a1} = u_1 l + v_1 m, \quad U_{a2} = u_2 l + v_2 m$$

where $l = \cos \phi$, $m = \cos(90 - \phi)$ are the direction cosines of the vector directed along end 1 to end 2, and ϕ is the inclination of the vector to the x axis.

Similarly, if the forces at the ends of the element are (F_{x1}, F_{y1}) and (F_{x2}, F_{y2}) at ends 1 and 2 respectively, the relationship between the axial forces and the forces along the coordinate axes are given by

$$F_{x1} = F_{a1}l, \quad F_{y1} = F_{a1}m, \quad F_{x2} = F_{a2}l, \quad F_{y2} = F_{a2}m$$

From the element stiffness relationship given by equation (1.8) we have

$$F_{a1} = \frac{AE(U_{a1} - U_{a2})}{L} - 0.5fL, \quad F_{a2} = \frac{AE(U_{a2} - U_{a1})}{L} - 0.5fL$$

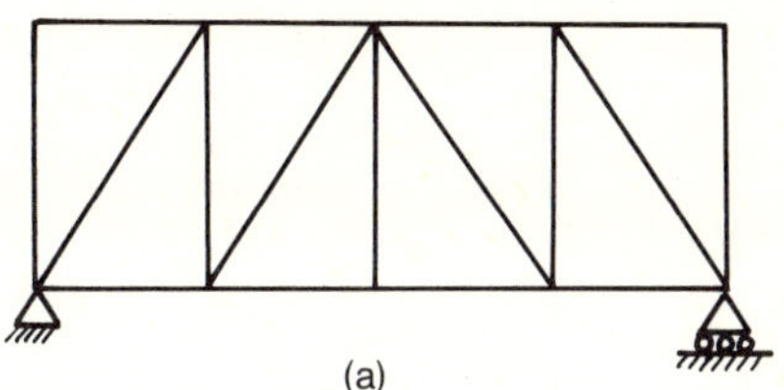

(a)

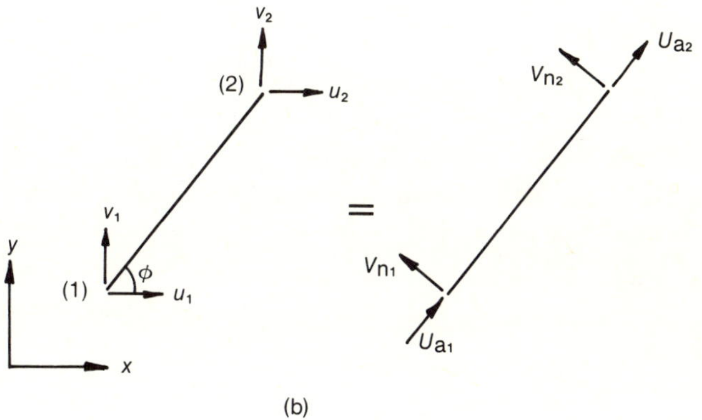

(b)

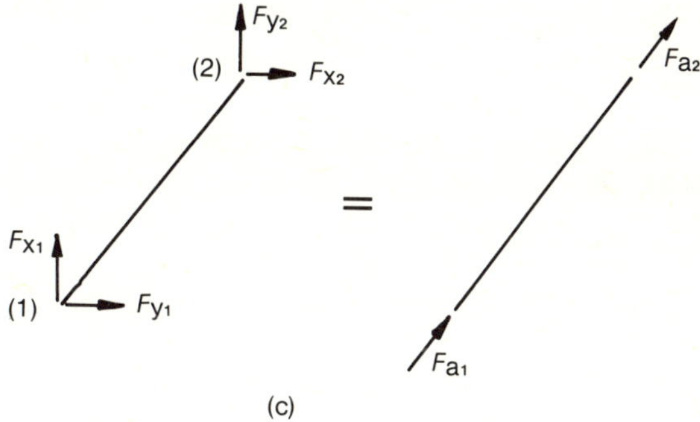

(c)

Fig. 1.10   Forces and displacements at the ends of a 2-D truss element: (a) 2-D truss; (b) displacements at the ends of an element; (c) forces at the ends of an element.

Substituting for U_{a1} and U_{a2} in terms of (u_1, v_1) and (u_2, v_2) we have

$$F_{a1} = \frac{AE[(u_1 - u_2)l + (v_1 - v_2)m]}{L} - 0.5fL$$

$$F_{a2} = \frac{AE[(u_2 - u_1)l + (v_2 - v_1)m]}{L} - 0.5fL$$

Having obtained F_{a1} and F_{a2}, the forces (F_{x1}, F_{y1}) and (F_{x2}, F_{y2}) can be easily calculated:

$$F_{x1} = F_{a1}l = \frac{AE[(u_1 - u_2)l^2 + (v_1 - v_2)lm]}{L} - 0.5fLl$$

$$F_{y1} = F_{a1}m = \frac{AE[(u_1 - u_2)lm + (v_1 - v_2)m^2]}{L} - 0.5fLm$$

$$F_{x2} = F_{a2}l = \frac{AE[(u_2 - u_1)l^2 + (v_2 - v_1)lm]}{L} - 0.5fLl$$

$$F_{y2} = F_{a2}m = \frac{AE[(u_2 - u_1)lm + (v_2 - v_1)m^2]}{L} - 0.5fLm$$

The above four equations can be expressed in matrix form as shown in Table 1.1, resulting in an element stiffness matrix which is in a more general form than equation (1.8). As can be seen, the resulting element stiffness matrix is symmetric.

Table 1.1

The element stiffness relationship for a 2-D bar element.

$$\begin{bmatrix} F_{x1} \\ F_{y1} \\ F_{x2} \\ F_{y2} \end{bmatrix} = \frac{AE}{L} \begin{bmatrix} l^2 & lm & -l^2 & -lm \\ & m^2 & -lm & -m^2 \\ & & l^2 & lm \\ & \text{symmetric} & & l^2 \end{bmatrix} \begin{bmatrix} u_1 \\ v_1 \\ u_2 \\ v_2 \end{bmatrix} - 0.5fL \begin{bmatrix} l \\ m \\ l \\ m \end{bmatrix}$$

(ii) *A beam element.* A beam element occurs in structures such as continuous beams, as members of 2-D rigid-jointed structure (plane frames), etc. In continuous beams, the element is subjected to bending forces only (i.e. bending moment and shear force). However, in rigid-jointed structures, the element is subjected not only to bending forces but also to axial forces. In order to be able to develop the stiffness matrix for a general bending element, in the following section both bending as well as axial forces will be included. Consider the general 'bending' element shown in Fig. 1.11. The displacements (which include both translations as well as rotations) at end 1 are (u_1, v_1, θ_1) and at end 2 are (u_2, v_2, θ_2). Similarly the forces at the ends are (F_{x1}, F_{y1}, M_1) at end 1 and (F_{x2}, F_{y2}, M_2) at end 2. The object is to establish a relationship between the forces and displacements at the ends of the element.

Resolving the displacements along the axial and normal directions, we have

$$\text{at end 1,} \quad U_{a1} = u_1l + v_1m, \quad V_{n1} = -u_1m + v_1l$$

$$\text{at end 2,} \quad U_{a2} = u_2l + v_2m, \quad V_{n2} = -u_2m + v_2l$$

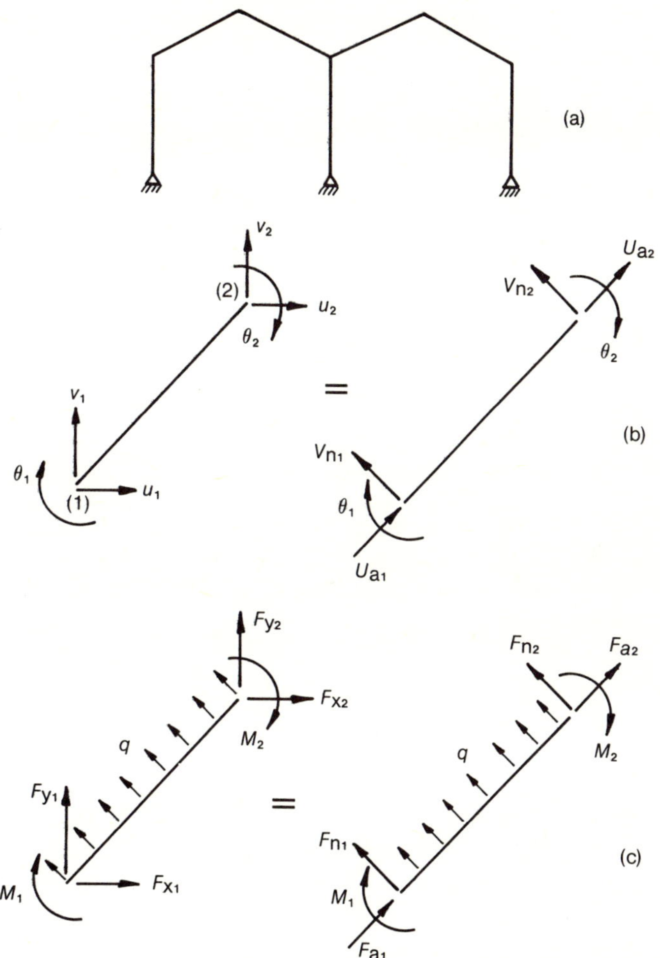

Fig. 1.11   Forces and displacements at the ends of a 2-D rigid-jointed frame element: (a) 2-D rigid-jointed frame; (b) displacements at the ends of the element; (c) forces on the element.

Note that, because θ is a vector about the z axis (assumed normal to the plane of the paper), it remains unaltered. Similarly, resolving the axial and the normal forces at the ends along the x and y axes, we have

$$\text{at end 1,} \quad F_{x1} = -F_{n1}m + F_{a1}l, \quad F_{y1} = F_{n1}l + F_{a1}m$$

$$\text{at end 2,} \quad F_{x2} = -F_{n2}m + F_{a2}l, \quad F_{y2} = F_{n2}l + F_{a2}m$$

The final stiffness matrix is obtained by expressing the axial and normal displacements in terms of the displacements along the coordinate axes directions and calculating the resulting bending (equation (1.19)) and axial (equation (1.8)) forces. The next step is to calculate the forces in the coordinate axes directions in terms of the normal and axial forces. Table 1.2 shows the resulting stiffness relationship.

Table 1.2

The stiffness relationship for an element in a 2-D rigid-jointed structure.

$$
\begin{bmatrix} F_{x1} \\ F_{y1} \\ M_1 \\ F_{x2} \\ F_{y2} \\ M_2 \end{bmatrix}
= \frac{EI}{L(1+\beta)}
\begin{bmatrix}
\frac{12}{L^2}(m^2+\alpha l^2) & \frac{-12}{L^2}(1-\alpha)lm & \frac{6m}{L} & \frac{-12}{L^2}(m^2+\alpha l^2) & \frac{12}{L^2}(1-\alpha)lm & \frac{6m}{L} \\[2mm]
 & \frac{12}{L^2}(l^2+\alpha m^2) & \frac{-6l}{L} & \frac{12}{L^2}(1-\alpha)lm & \frac{-12}{L^2}(l^2+\alpha m^2) & \frac{-6l}{L} \\[2mm]
 & & 4+\beta & \frac{-6m}{L} & \frac{6l}{L} & 2-\beta \\[2mm]
 & & & \frac{12}{L^2}(m^2+\alpha l^2) & \frac{-12}{L^2}(1-\alpha)lm & \frac{-6m}{L} \\[2mm]
 & \text{symmetric} & & & \frac{12}{L^2}(l^2+\alpha m^2) & \frac{6}{L}l \\[2mm]
 & & & & & 4+\beta
\end{bmatrix}
\begin{bmatrix} u_1 \\ v_1 \\ \theta_1 \\ u_2 \\ v_2 \\ \theta_2 \end{bmatrix}
+
\begin{bmatrix}
\frac{-fLl}{2}+\frac{qL}{2}m \\[2mm]
\frac{-fLm}{2}-\frac{qLl}{2} \\[2mm]
\frac{-qL^2}{12} \\[2mm]
\frac{-fLl}{2}+\frac{qL}{2}m \\[2mm]
\frac{-fL}{2}m-\frac{qL}{2}l \\[2mm]
\frac{qL^2}{12}
\end{bmatrix}
$$

$$
\beta = \frac{12EI}{GA_{s}L^2}, \qquad \alpha = \frac{AL^2}{12I}(1+\beta)
$$

(iii) *An element in a plane grid structure.* Plane grid structures are 2-D structures in which the external loads are applied normal to the plane of the structure. Because of this, the direction of the shear force is normal to the plane of the structure and the element in general is subjected to not only bending forces but also twisting forces. Fig. 1.12(a) shows a typical grid structure. As for elements discussed previously, it is convenient to develop the stiffness matrix with the forces and displacements along the coordinate axes. For a plane grid element it is convenient to have the x and y axes in the plane of the grid and the applied load along the z axis. The forces at the ends of the element are M_x which is the couple about the x axis, M_y which is the couple about the y axis and F_z which is the force along the z axis. The corresponding displacements are θ_x which is the rotation about the x axis, θ_y which is the rotation about the y axis and w which is the translation along the z axis.

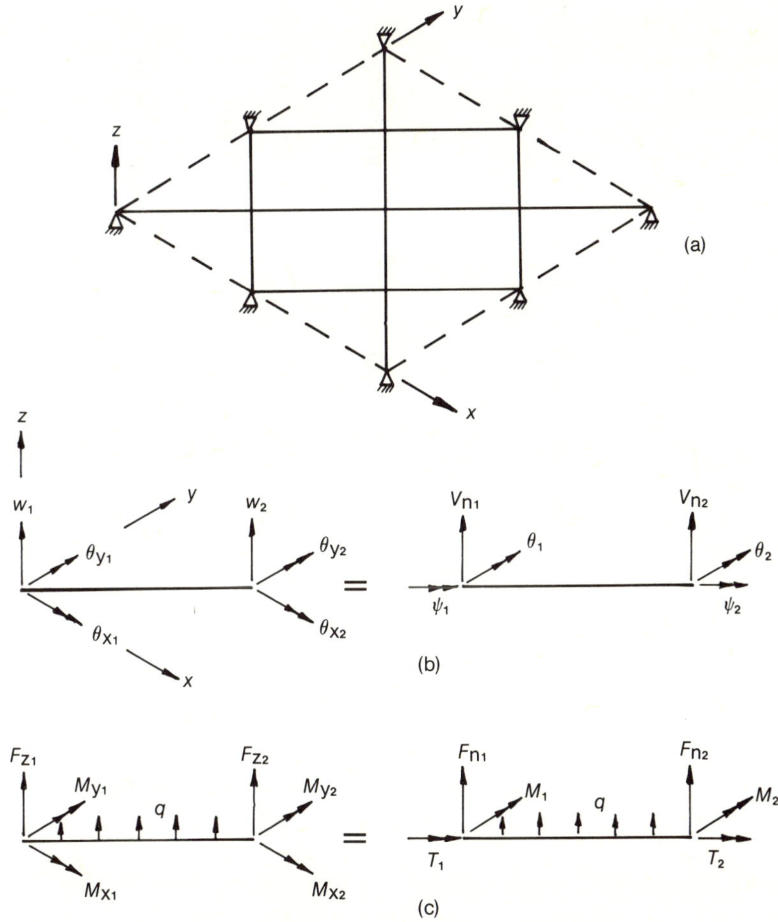

Fig. 1.12   Forces and displacements at the ends of a plane grid element: (a) plane grid; (b) displacements at the ends of the element; (c) forces on the element.

Table 1.3

The stiffness relationship for an element in a plane grid structure.

$$
\begin{bmatrix} F_{z1} \\ M_{x1} \\ M_{y1} \\ F_{z2} \\ M_{x1} \\ M_{y2} \end{bmatrix} = \frac{EI}{L(1+\beta)}
\begin{bmatrix}
\dfrac{12}{L^2} & \dfrac{6m}{L} & \dfrac{-6l}{L} & \dfrac{-12}{L^2} & \dfrac{6m}{L} & \dfrac{-6l}{L} \\[2mm]
 & (4+\beta)m^2 + \alpha l^2 & -(4+\beta-\alpha)lm & \dfrac{-6m}{L} & (2-\beta)m^2 - \alpha l^2 & -(2-\beta+\alpha)lm \\[2mm]
 & & (4+\beta)l^2 + \alpha m^2 & \dfrac{6l}{L} & -(2-\beta+\alpha)lm & (2-\beta)l^2 - \alpha m^2 \\[2mm]
 & & & \dfrac{12}{L^2} & \dfrac{-6m}{L} & \dfrac{6l}{L} \\[2mm]
 & \text{symmetric} & & & (4+\beta)m^2 + \alpha l^2 & -(4+\beta-\alpha)lm \\[2mm]
 & & & & & (4+\beta)l^2 + \alpha m^2
\end{bmatrix}
\begin{bmatrix} w_1 \\ \theta_{x1} \\ \theta_{y1} \\ w_2 \\ \theta_{x2} \\ \theta_{x2} \end{bmatrix}
+
\begin{bmatrix}
\dfrac{-qL}{2} \\[3mm]
\dfrac{-qL^2}{12}m - \dfrac{tL}{2}l \\[3mm]
\dfrac{qL^2}{12}l - \dfrac{tL}{2}m \\[3mm]
\dfrac{-qL}{L} \\[3mm]
\dfrac{qL^2}{12}m - \dfrac{tL}{2}l \\[3mm]
\dfrac{-qL^2}{12}l - \dfrac{tL}{2}m
\end{bmatrix}
$$

$$
\beta = \frac{12EI}{GA_s L^2}, \qquad \alpha = \frac{GJ}{EI}(1+\beta)
$$

Resolving the displacements into bending rotations (θ_1, θ_2) and torsional twists (ψ_1, ψ_2) we have

$$\text{at end 1,} \quad \theta_1 = -\theta_{x1}m + \theta_{y1}l, \quad \psi_1 = \theta_{x1}l + \theta_{y1}m$$
$$\text{at end 2,} \quad \theta_2 = -\theta_{x2}m + \theta_{y2}l, \quad \psi_2 = \theta_{x2}l + \theta_{y2}m$$

Similarly the moment M and torque T can be resolved into couples along the x and y axes as follows:

$$\text{at end 1,} \quad M_{x1} = T_1 l - M_1 m, \quad M_{y1} = T_1 m + M_1 l, \quad F_{z1} = F_{n1}$$
$$\text{at end 2,} \quad M_{x2} = T_2 l - M_2 m, \quad M_{y2} = T_2 m + M_2 l, \quad F_{z2} = F_{n2}$$

However, using equations (1.19) for bending forces and (1.23) for torsional forces, the final stiffness relationship can be calculated. Table 1.3 gives the final stiffness relationship.

1.17 ANALYSIS OF SKELETAL STRUCTURES BY THE STIFFNESS METHOD

In the previous sections, the element stiffness matrices for various types of element were derived. As was emphasized, these stiffness matrices were derived on the basis of ensuring the satisfaction of equilibrium, strain-displacement relationship, material laws, etc., over the length of the element. Since structures are assembled using elements, in order to ensure the satisfaction of equilibrium, etc., over the whole structure, all that is necessary is to ensure that equilibrium and compatibility are satisfied at the joints as well. In the following sections, attention will be focused on the consideration of equilibrium and compatibility at the joints.

1.18 COMPATIBILITY AT THE JOINTS—THE NODE FREEDOM ARRAY

Compatibility at joints in general means that all the members meeting at a joint have the same displacements. Considering the 2-D pin-jointed structure shown in Fig. 1.13, there are 10 joints and, since in general each joint can

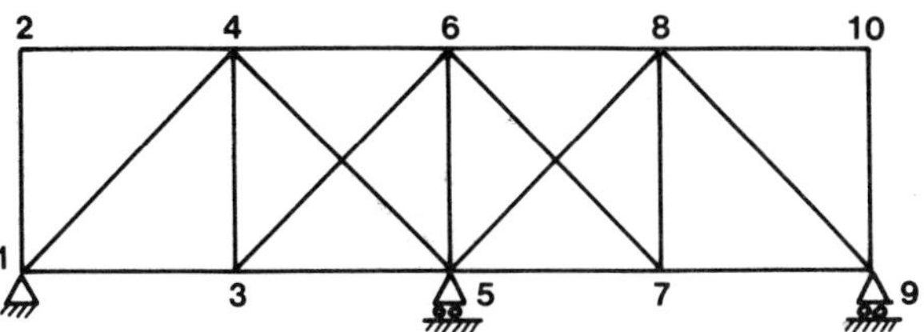

Fig. 1.13   A 2-D pin-jointed truss.

move in both x and y directions, there are 20 possible joint displacements. However, four displacements are known to be zero because of the fact that joint 1 is a pinned support where displacements in both the x and the y directions are restrained and similarly, as both joint 5 and joint 9 are roller supports, they have the displacements in the y direction suppressed. The number of unknown displacements is thus $10 \times 2 - 4 = 16$. If the unknown joint displacements are numbered consecutively from 1 to 16, the unknown displacement number at any joint can be inferred from the Table 1.4. Such a table is known as a node freedom table.

In a similar manner for a 2-D rigid-jointed structure, in general at each joint there are three unknown displacements, i.e. translations u and v in the x and y directions and a rotation θ. For a plane grid structure again at a joint, there are three displacements, i.e. rotations θ_x and θ_y about the x and y axes and a translation w about the z axis. Tables 1.5 and 1.6 show the node freedom table for the 2-D rigid-jointed structure (Fig. 1.14) and plane grid (Fig. 1.15) respectively.

Great care is needed in determining the node freedom numbers. As an example, consider the diagramatic representation of a 'balanced cantilever' bridge shown in Fig. 1.16(a). Because joints 3 and 4 are intermediate hinges, member 3–2 and member 3–4 do not rotate by the same amount at joint 3. Similar conditions hold good for joint 4. One simple way to enforce this condition is to number the joints as shown in Fig. 1.16(b). The compalibility condition demands that at joints 3 and 4 it is necessary that the translations in x and y directions are the same. Similar conditions are true at joints 5 and 6 as well. Table 1.7 shows the node freedom table.

Table 1.4

The node freedom table for the pin-jointed structure in Fig. 1.13.

| | Unknown displacement number in | |
Node or joint number	x direction u	y direction v
1	0	0
2	1	2
3	3	4
4	5	6
5	7	0
6	8	9
7	10	11
8	12	13
9	14	0
10	15	16

Table 1.5

The node freedom table for a 2-D rigid-jointed structure (Fig. 1.14).

Joint number	Unknown joint displacement number in		
	x-direction u	y-direction v	rotation θ
1	0	0	0
2	1	2	3
3	4	5	6
4	0	0	7
5	8	9	10
6	11	12	13
7	14	0	15
8	16	17	18

Note that, at a fixed support, $u = v = \theta = 0$ and, at a hinged support, $u = v = 0$ but $\theta \neq 0$ since rotation is possible and, at a roller support, only one translation is suppressed.

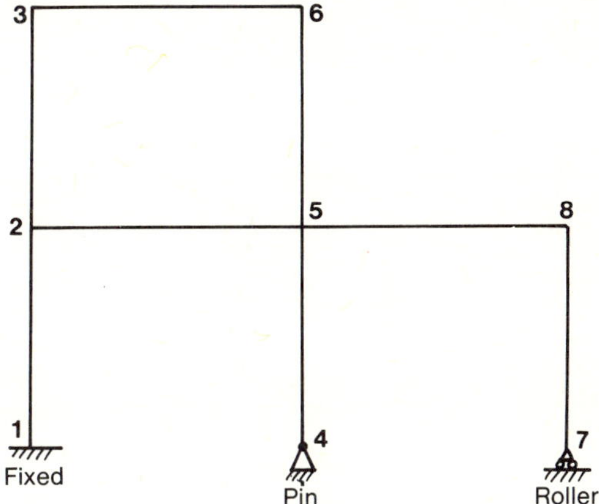

Fig. 1.14   A 2-D rigid-jointed frame.

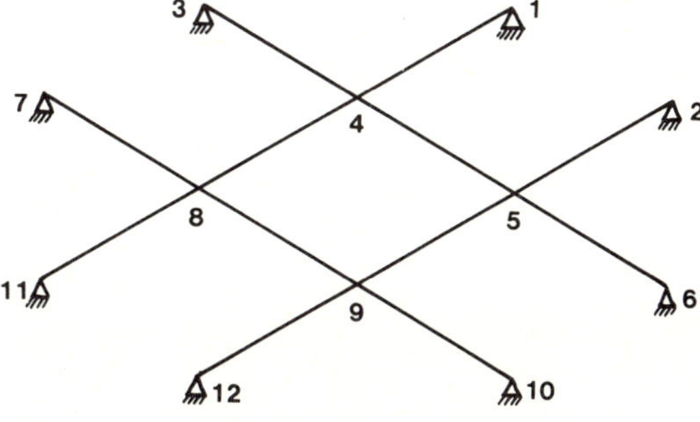

Fig. 1.15   A plane grid.

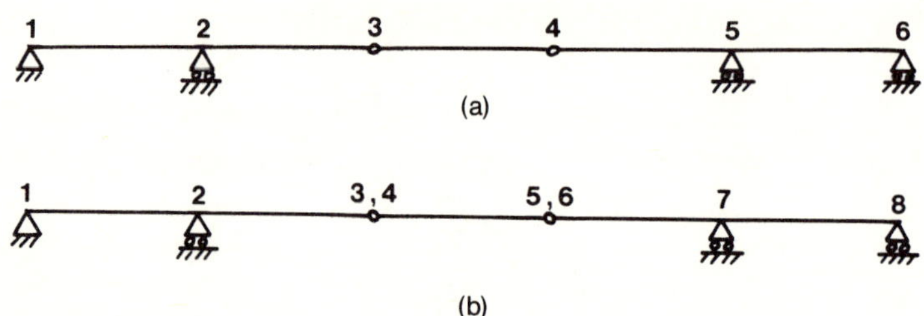

Fig. 1.16   A balanced cantilever bridge: (a) conventional joint numbering; (b) special joint numbering to ensure discontinuity of rotation at the intermediate hinges.

Table 1.6

The node freedom table for a plane grid (Fig. 1.15).

Joint number	Unknown joint displacement number in		
	w	θ_x	θ_y
1	0	1	2
2	0	3	4
3	0	5	6
4	7	8	9
5	10	11	12
6	0	13	14
7	0	15	16
8	17	18	19
9	20	21	22
10	0	23	24
11	0	25	26
12	0	27	28

Table 1.7

The node freedom table for balanced cantilever bridge (Fig. 1.16(b)).

Node number	Unknown joint displacement number in		
	x-direction u	y-direction v	rotation θ
1	0	0	1
2	2	0	3
3	4	5	6
4	4	5	7
5	8	9	10
6	8	9	11
7	12	0	13
8	14	0	15

1.19 EQUILIBRIUM AT THE JOINTS

Consider the pin-jointed structure shown in Fig. 1.13. To illustrate the conditions of equilibrium at a joint, choose joint 4 as shown in Fig. 1.17. Evidently laws of statics demand that

$$\sum (\text{horizontal components of forces in members meeting at joint 4}) = W_{x4}$$

$$\sum (\text{vertical components of forces in members meeting at joint 4}) = W_{y4}$$

where W_{x4} and W_{y4} are the external forces at joint 4 in the x and y directions respectively.

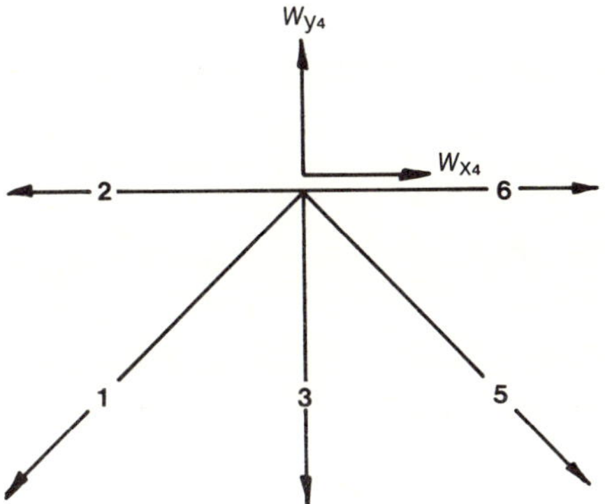

Fig. 1.17   Forces at a joint.

The element stiffness matrix gives the forces at the ends of the member in the x and y directions in terms of the displacements at the ends of the members. As an example, consider the member 4–5. If end 1 is considered to be at joint 4 and end 2 at joint 5 (it is quite arbitrary which end is considered as end 1 and which is end 2 as long as the appropriate direction cosines are used), then from the node freedom table, Table 1.4, the displacements at joints 4 and 5 are as follows:

$$\text{end } 1 = \text{joint } 4, \quad u_1 = \text{displacement } 5, \quad v_1 = \text{displacement } 6$$
$$\text{end } 2 = \text{joint } 5, \quad u_2 = \text{displacement } 7, \quad v_2 = 0$$

Therefore the F_{x1} and F_{y1} forces from member 4–5 at joint 4 will be a function of unknown displacements 5, 6 and 7. In a similar fashion the F_x and F_y forces at joint 4 from member 2–4 will be a function of unknown displacements at joints 2 and 4, i.e. displacements 1, 3, 5 and 6. In other words the equations of equilibrium at joint 4 will be a function of the unknown displacements at the ends of all the members meeting at joint 4. Since we have to ensure that joint equilibrium is established at every joint in two directions in which external loads are applied, one is assured of as many equations of equilibrium as there are unknown joint displacements. It should be noted that it is not necessary to consider joints or directions in which reactions can develop because, by definition, reaction develops so as to maintain equilibrium. The designer has to ensure that these reactions can be safely transmitted to the foundations. The equations of joint equilibrium thus establish the relationship between the external loads on the structure and the unknown displacements of the joints of the structure. By analogy with the element stiffness matrix, the relationship between the external loads on the structure and the displacements of the joints of the structure is known as the structural stiffness matrix. The size of the structural matrix is the same as the number of unknown joint displacements considering all the joints of the structure. The structural stiffness matrix is a mathematical expression of the equations of equilibrium at the joints of the structure.

1.20   ASSEMBLING THE STRUCTURAL STIFFNESS MATRIX

As was explained in the previous section, the structural stiffness matrix is formed by considering the equations of equilibrium of the joints of the structure. In terms of computational procedure the following two options are open.

(a)   Completely form the equation of equilibrium at a particular joint in a particular direction before considering the equation of equilibrium at another joint.

(b)   Determine the contributions from one member to the forces at the joints to which it is connected and repeat it for all the members.

Generally the second procedure is adopted because it is computationally more efficient. In order to explain the basic procedure, consider member 4–5 shown in Fig. 1.18. Assuming that end 1 = joint 4 and end 2 = joint 5, we have from the node freedom numbers in Table 1.4 that

$$\text{at end 1,} \quad u_1 = \text{displacement } 5 = d_5, \quad v_1 = \text{displacement } 6 = d_6$$
$$\text{at end 2,} \quad u_2 = \text{displacement } 7 = d_7, \quad v_2 = \text{zero displacement} = 0$$

Since the length L, area of cross section A, Young's modulus E and the direction cosines l and m are known, using the expression for the element

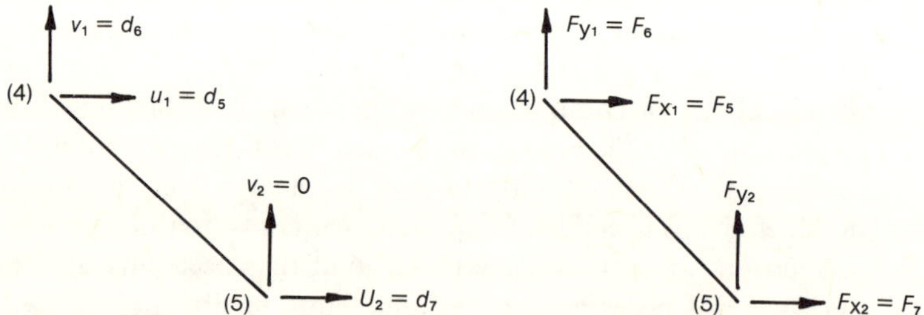

Fig. 1.18   Forces and displacements at the ends of element 4–5.

stiffness matrix given in Table 1.1, the following matrix for the element 4–5 can be calculated:

$$
\begin{bmatrix} F_{x1} \\ F_{y1} \\ F_{x2} \\ F_{y2} \end{bmatrix} = \begin{bmatrix} k_{11} & k_{12} & k_{13} & k_{14} \\ k_{21} & k_{22} & k_{23} & k_{24} \\ k_{31} & k_{32} & k_{33} & k_{34} \\ k_{41} & k_{42} & k_{43} & k_{44} \end{bmatrix} \begin{bmatrix} u_1 = d_5 \\ v_1 = d_6 \\ u_2 = d_7 \\ v_2 = 0 \end{bmatrix}
$$

where the stiffness coefficients k_{ij}, $i = 1$ to 4 and $j = 1$ to 4, are numbers which depend on the properties and orientation of the member. As can be seen from Fig. 1.18, F_{x1} contributes to the force in the x direction at joint 4. The x direction at joint 4 is the same as the direction of d_5. Therefore we can say that F_{x1} is one component of force in the d_5 direction. Similarly, F_{y1} is one component of force in the d_6 direction while F_{x2} is one component of force in the d_7 direction and F_{y2} is one component of force in the direction of vertical reaction at joint 5.

Since, in the entire structure, there are 16 unknown displacements, there are 16 equations of equilibrium to be considered. Member 4–5 contributes forces to only three of these equations, i.e. equations (5), (6) and (7). Other members of the structure contribute to forces in other directions depending on which joints they are connected to. Considering joint 4, it is connected to joints 1, 2, 3, 5 and 6. From the node freedom numbers in Table 1.4, it is clear that the equation of equilibrium in the direction of d_5 and d_6 will be a function of d_1, d_2 (at joint 2), d_3, d_4 (at joint 3), d_5, d_6 (at joint 4), d_7 (at joint 5) and d_8, d_9 (at joint 6). Consider the structural matrix as yet to be formulated as shown in Fig. 1.19. From the discussion of the element stiffness matrix for member 4–5, we have

$$
\begin{aligned}
F_{x1} &= \text{contribution to } F_5 = k_{11}d_5 + k_{21}d_6 + k_{13}d_7 + k_{14}0 \\
F_{y1} &= \text{contribution to } F_6 = k_{21}d_5 + k_{22}d_6 + k_{23}d_7 + k_{24}0 \\
F_{x2} &= \text{contribution to } F_7 = k_{31}d_5 + k_{32}d_6 + k_{33}d_7 + k_{34}0 \\
F_{y2} &= \text{reaction}
\end{aligned}
$$

Therefore member 4–5 contributes to the fifth row of the structural stiffness matrix k_{11}, k_{12} and k_{13} in the fifth, sixth and seventh columns respectively. Similarly there are contributions in the fifth, sixth and seventh columns of rows 6 and 7. The contributions are shown in Fig. 1.19.

In a similar fashion, if we consider member 3–6, we have end 1 = joint 3, $u_1 = d_3$, $v_1 = d_4$, end 2 = joint 6, $u_2 = d_8$ and $v_2 = d_9$. If the element stiffness matrix for member 3–6 is given by

$$
\begin{bmatrix} F_{x1} = F_3 \\ F_{y1} = F_4 \\ F_{x2} = F_8 \\ F_{y2} = F_9 \end{bmatrix} = \begin{bmatrix} p_{11} & p_{12} & p_{13} & p_{14} \\ p_{21} & p_{22} & p_{23} & p_{24} \\ p_{31} & p_{32} & p_{33} & p_{34} \\ p_{41} & p_{42} & p_{43} & p_{44} \end{bmatrix} \begin{bmatrix} u_1 = d_3 \\ v_1 = d_4 \\ u_2 = d_8 \\ v_2 = d_9 \end{bmatrix}
$$

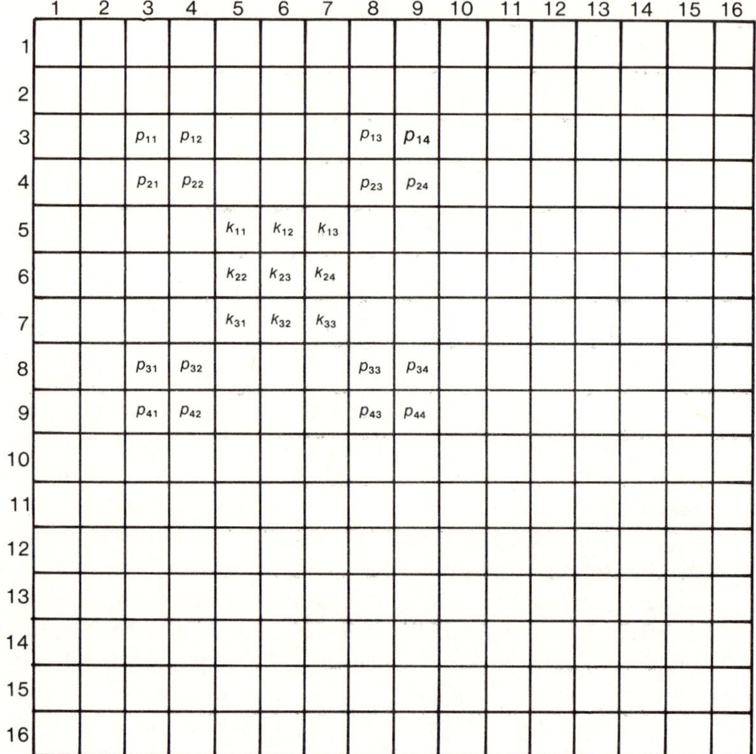

Fig. 1.19   Partially assembled structural stiffness matrix.

Therefore the element 3–6 contributes to the third, fourth, eighth and ninth rows in column positions 3, 4, 8 and 9.

Fig. 1.19 shows the contributions from members 4–5 and 3–6. In a similar manner, other members also contribute to the equations of equilibrium. When the contributions from all members of a structure are added, the structural stiffness matrix represents the equations of equilibrium at the joints of the structure. As can be seen from the above description, the rule for assembling the structural stiffness matrix can be formulated as follows.

Let a temporary vector MEMDIS(NEVAB) contain the information about the unknown displacement numbers pertaining to the joints at the ends of the member. In this connection, NEVAB is equal to the total number of independent displacements at the two ends of the element. As an example, for the element 4–5, the vector MEMDIS contains the following information:

MEMDIS(1) = 5, MEMDIS(2) = 6, MEMDIS(3) = 7, MEMDIS(4) = 0

Similarly, for member 3–6, the vector MEMDIS will contain the following information:

MEMDIS(1) = 3, MEMDIS(2) = 4, MEMDIS(3) = 8, MEMDIS(4) = 9

The contribution from an element is added to the structural stiffness matrix as follows.

The element $ESTIF(i, j)$ of the element stiffness matrix is added to the element $GSTIF(MEMDIS(i), MEMDIS(j))$ of the structural stiffness matrix GSTIF. If $MEMDIS(i)$ is zero, the addition is ignored because one is establishing the equations of equilibrium w.r.t. the forces in the 'direction' of nonzero joint displacements. Similarly, if $MEMDIS(j)$ is zero, then the addition is ignored because a zero displacement does not contribute any forces due to joint displacements. The variables i and j assume a maximum value of NEVAB which in 2-D pin-jointed structures is equal to 4. In 2-D rigid-jointed structures and plane grids, $NEVAB = 6$ because at each joint there are three unknown displacements so that for a member with two ends there are six possible joint displacements.

1.21   THE JOINT FORCE VECTOR

When developing the expressions for the element stiffness matrix, it was shown that for an element, the following relationship is valid:

$$\begin{Bmatrix} \text{element} \\ \text{joint force} \\ \text{vector} \end{Bmatrix} = \begin{Bmatrix} \text{element} \\ \text{stiffness} \\ \text{matrix} \end{Bmatrix} \begin{Bmatrix} \text{element} \\ \text{joint displacement} \\ \text{vector} \end{Bmatrix}$$

$$+ \begin{Bmatrix} \text{fixed end load vector} \\ \text{due to the distributed loads} \\ \text{on the elements} \end{Bmatrix}$$

When the elements are assembled to form the structure, taking due note of the compatibility condition, the above equation can be written, because of equilibrium as

$$\begin{Bmatrix} \text{vector of loads} \\ \text{at the joints} \\ \text{of the structure} \end{Bmatrix} = \begin{Bmatrix} \text{assembled} \\ \text{structural} \\ \text{stiffness matrix} \end{Bmatrix} \begin{Bmatrix} \text{unknown joint} \\ \text{displacement} \\ \text{vector} \end{Bmatrix}$$

$$+ \begin{Bmatrix} \text{assembled fixed end force vector} \\ \text{due to the distributed loads} \\ \text{on all the elements of the structure} \end{Bmatrix}$$

Transferring all the load vectors on to one side, we have

$$\begin{Bmatrix} \text{structural} \\ \text{stiffness} \\ \text{matrix} \end{Bmatrix} \begin{Bmatrix} \text{unknown} \\ \text{joint displacement} \\ \text{vector} \end{Bmatrix} = \begin{Bmatrix} \text{vector of loads} \\ \text{at the joints of} \\ \text{the structure} \end{Bmatrix}$$

$$- \begin{Bmatrix} \text{assembled fixed end force vector} \\ \text{due to the distributed loads} \\ \text{on the elements of the structure} \end{Bmatrix}$$

The assembling of the fixed end force vector is carried out in a manner similar to the assembling of the structural stiffness matrix. For example for a pin-jointed element, the fixed end force vector for an element is given by a temporary vector ELOAD(NEVAB) where

$$ELOAD(1) = -0.5fLl, \quad ELOAD(2) = -0.5fLm$$
$$ELOAD(3) = -0.5fLl, \quad ELOAD(4) = -0.5fLm$$

then the element ELOAD(i) is added to the assembled load vector GLOAD-((MEMDIS(i)), where MEMDIS(NEVAB) contains the information about the unknown joint displacement numbers corresponding to the joints which are joined by the element under consideration.

1.22   DETERMINATION OF THE UNKNOWN JOINT DISPLACEMENTS

In the relationship

$$\begin{Bmatrix} \text{structural} \\ \text{stiffness} \\ \text{matrix} \end{Bmatrix} \begin{Bmatrix} \text{unknown} \\ \text{joint displacement} \\ \text{vector} \end{Bmatrix} = \begin{Bmatrix} \text{joint} \\ \text{force} \\ \text{vector} \end{Bmatrix}$$

where

$$\begin{Bmatrix} \text{joint} \\ \text{force} \\ \text{vector} \end{Bmatrix} = \begin{Bmatrix} \text{applied concentrated} \\ \text{load vector} \\ \text{at the joints} \end{Bmatrix} - \begin{Bmatrix} \text{assembled} \\ \text{fixed end} \\ \text{force vector} \end{Bmatrix}$$

the only unknown is the vector of joint displacements. This is determined by solving the simultaneous equations implied in the equations of equilibrium at the joints of the structure. Gaussian elimination or one of its several variants is used for solving the simultaneous equations. The details will be discussed in Chapter 5.

1.23   DETERMINATION OF FORCES IN MEMBERS

For the purposes of design, it is necessary to determine the stress resultants such as axial forces, bending moments, etc., in the elements of a structure. After the displacements at the joints are calculated, the forces in the various types of members are as follows.

(i) *A 2-D pin-jointed element.*   The axial forces at the ends are given by equation (1.8). The axial displacements U_{a1} and U_{a2} are given by

$$U_{a1} = u_1l + v_1m, \quad U_{a2} = u_2l + v_2m$$

where the values of u_1, u_2, v_1 and v_2 are known because the displacements of the joints of the structure have been determined and the correspondence between the displacements at the ends of a member and the displacements of the joints of the structure is known from the node freedom table.

(ii) *A 2-D rigid-jointed frame member.*   In this case the axial forces are determined from equation (1.8) and the bending forces from equation (1.19). The axial and normal displacements at end *i* are given by

$$U_{ai} = u_i l + v_i m, \quad V_{ni} = -u_i m + v_i m, \quad i = 1, 2$$

(iii) *A plane grid member.*   In this case the bending forces are determined using equation (1.19) and the torsional forces from equation (1.23). The bending and torsional rotations at end *i* are determined as follows:

$$\theta_i = -\theta_{xi} m + \theta_{yi} l, \quad \psi_i = \theta_{xi} l + \theta_{yi} m, \quad i = 1, 2$$

1.24   THE INFLUENCE OF THE AXIAL LOAD ON THE BENDING STIFFNESS COEFFICIENTS

It was explained in section 1.16 that, in general for 2-D rigid-jointed structures, the members are subjected not only to bending forces but also to axial forces. In developing the stiffness matrix, the two effects of bending and axial forces were treated independently with no interaction between them. This assumption is valid only if the axial force in a member is very much smaller than its elastic buckling load.

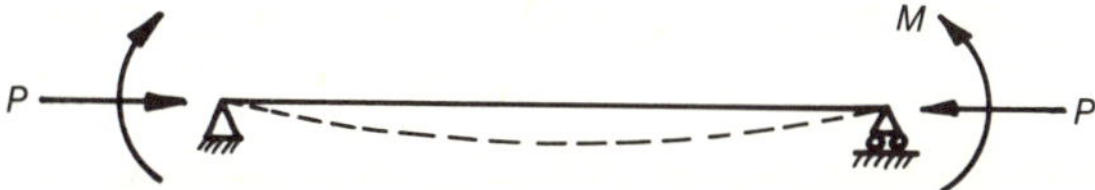

Fig. 1.20   Interaction between the axial and bending forces.

Consider the beam-column shown in Fig. 1.20. Evidently the axial force at the end is eccentric with respect to the displaced position of the structure and therefore induces a bending moment equal to the axial load multiplied by the eccentricity. This interaction of the bending and axial forces has to be taken into consideration especially when studying the stability of rigid-jointed structures.

Consider the beam-column element shown in Fig. 1.21. It is assumed that the axial force *P* is positive if tensile and that the axial force retains its original

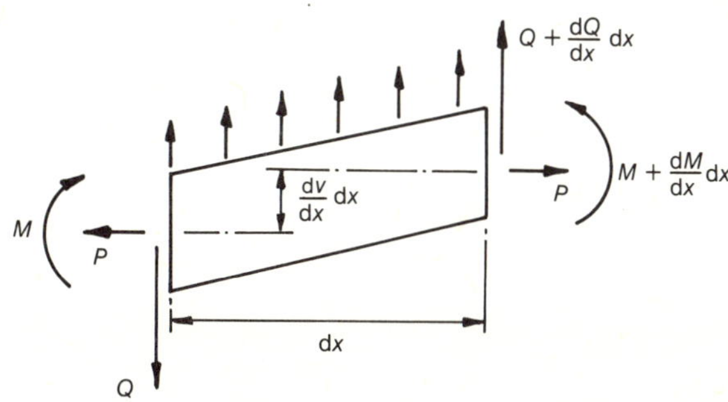

Fig. 1.21   An infinitesimal segment of a beam-column element.

orientation in the deformed position of the structure. As in section 1.9, the equation of equilibrium in the vertical direction is given by

$$\frac{dQ}{dx} + q = 0$$

In considering the moment equilibrium it is important to include the bending moment induced by the axial force. The moment equilibrium equation is given by

$$M - \left(M + \frac{dM}{dx}\,dx\right) - Q\,dx + P\left(\frac{dv}{dx}\,dx\right) + q\,dx\,(0.5\,dx) = 0$$

Ignoring terms of second order, we have

$$\frac{dM}{dx} + Q - P\frac{dv}{dx} = 0$$

Eliminating Q from the above equation, we have

$$M'' - Pv'' - q = 0$$

where M'' and v'' denote the second derivative of M and v respectively w.r.t. x. If the effects of shearing deformation are ignored, then the moment–curvature relationship is given by $M = EIv''$. Expressing M in terms of the curvature, the equation of equilibrium can be expressed as the following so called beam-column equation:

$$EIv'''' - Pv'' - q = 0 \tag{1.26}$$

1.25   SOLUTION OF THE BEAM-COLUMN EQUATION

The solution is obtained in the usual manner as follows.

(i) *Fixed end solution.*   The solution is given by

$$v = C_1 \cos kx + C_2 \sin kx + C_3 kx + C_4 + \frac{0.5qx^2}{P}, \quad k^2 = -\frac{P}{EI}$$

As the definition of k^2 makes clear, the above solution is valid only if P is negative, i.e. compressive. If P is tensile, then the solution is given by

$$v = C_1 \cosh kx + C_2 \sinh kx + C_3 kx + C_4 - \frac{0.5qx^2}{P}, \quad k^2 = \frac{P}{EI}$$

The integration constants are determined so as to satisfy the boundary conditions which are

$$\text{at end 1,} \quad v = \frac{dv}{dx} = 0 \text{ at } x = 0$$

$$\text{at end 2,} \quad v = \frac{dv}{dx} = 0 \text{ at } x = L$$

The integration constants are given as follows.

(a) *Compressive P.*   If P is compressive, then

$$C_1 = \frac{0.5qL^2}{PkL} \frac{2\sin kL - kL(\cos kL + 1)}{D}$$

$$C_2 = -\frac{0.5qL^2}{PkL} \frac{kL\sin kL + 2(\cos kL - 1)}{D}$$

$$C_3 = -C_2$$

$$C_4 = -C_1,$$

where

$$D = 2(1 - \cos kL) - kL\sin kL.$$

(b) *Tensile P.*   If P is tensile, the equations will be the same as for P compressive except that $\cos kx$ and $\sin kx$ are replaced by $\cosh kx$ and $\sinh kx$ respectively.

(ii) *Fixed displacement case.*   The solution in this case is given by

$$v = \overline{C_1}\cos kx + \overline{C_2}\sin kx + \overline{C_3}kx + \overline{C_4}$$

if P is compressive and by

$$v = \overline{C_1}\cosh kx + \overline{C_2}\sinh kx + \overline{C_3}kx + \overline{C_4}$$

if P is tensile. The boundary conditions are given by

$$\text{at end 1,}\quad x = 0,\quad v = V_{n1},\quad \frac{dv}{dx} = -\theta_1$$

$$\text{at end 2,}\quad x = L,\quad v = V_{n2},\quad \frac{dv}{dx} = -\theta_2$$

The integration constants are given as follows.

(a) *Compressive P.*   If P is compressive,

$$\overline{C_1} = \frac{(V_{n1} - V_{n2})(1 - c) + \theta_1 L(c - s/kL) + \theta_2 L(s/kL - 1)}{D}$$

$$\overline{C_2} = \frac{(V_{n2} - V_{n1})s + \theta_1 L[s - (1 - c)/kL] + \theta_2 L(1 - c)/kL}{D}$$

$$\overline{C_3} = \frac{(V_{n1} - V_{n2})s - \theta_1 L[(1 - c)/kL] - \theta_2 L(1 - c)/kL}{D}$$

$$\overline{C_4} = \frac{V_{n1}(1 - c - skL) + V_{n2}(1 - c) + \theta_1 L(s/kL - c) + \theta_2 L(1 - s/kL)}{D}$$

where $c = \cos kL$, $s = \sin kL$ and $D = 2(1 - c) - skL$.

(b) *Tensile P*. If P is tensile, the equations given for P compressive are valid except that $\cos kL$ and $\sin kL$ are replaced by $\cosh kL$ and $\sinh kL$ respectively.

Using the expressions above, the displacement v is expressed as

$$v = N_1 V_{n1} + N_2 \theta_1 + N_3 V_{n2} + N_4 \theta_2$$

where the shape functions N_1 to N_4 are given by

$$N_1 = \frac{(1 - c)(1 + \cos kx) - s(\sin kx - kx + kL)}{D}$$

$$N_2 = \frac{L\{(s/kL - c)(1 - \cos kx) + [s - (1 - c)/kL]\sin kx - (1 - c)x/L\}}{D}$$

$$N_3 = \frac{(1 - c)(1 - \cos kx) + s(\sin kx - kx)}{D}$$

$$N_4 = \frac{L\{[(1 - s)/kL](1 - \cos kx) + [(1 - c)/kL](\sin kx - kx)\}}{D}$$

where $D = 2(1 - c) - skL$, $s = \sin kL$ and $c = \cos kL$.

Note that, if P is tensile, $s = S = \sinh kL$ and $c = C = \cosh kL$ should be used and $\sin kx$ and $\cos kx$ should be replaced by $\sinh kx$ and $\cosh kx$ respectively.

(c) *Complete solution*.   The complete solution is the sum of solutions (a) and (b) above. Therefore

$$v = C_1 \cos kx + C_2 \sin kx + C_3 kx + C_4$$

$$+ N_1 V_{n1} + N_2 \theta_1 + N_3 V_{n2} + N_4 \theta_2 + 0.5 \frac{qx^2}{P}$$

1.26   THE ELEMENT STIFFNESS MATRIX FOR A BEAM-COLUMN ELEMENT

The bending moment M and shear force Q at a section can be determined from displacement v as follows:

$$M = EIv'', \quad Q = -\frac{dM}{dx} + P\frac{dv}{dx} = -EIv''' + P\frac{dv}{dx}$$

The forces at the ends of the element are given by

$$\text{at end 1,} \quad x = 0, \quad F_{n1} = -Q, \quad M_1 = M$$

$$\text{at end 2,} \quad x = L, \quad F_{n2} = Q, \quad M_2 = -M$$

Calculating M and Q in terms of the end displacements and the lateral load on the element, the following stiffness relationship can be derived:

$$
\begin{bmatrix} F_{n1} \\ M_1 \\ F_{n2} \\ M_2 \end{bmatrix} = \frac{EI}{L} \begin{bmatrix} \dfrac{T_{11}}{L^2} & \dfrac{-Q_{11}}{L} & \dfrac{-T_{11}}{L^2} & \dfrac{-Q_{11}}{L} \\[2mm] & S_{11} & \dfrac{Q_{11}}{L} & S_{12} \\[2mm] & & \dfrac{T_{11}}{L^2} & \dfrac{Q_{11}}{L} \\[2mm] \text{symmetric} & & & S_{11} \end{bmatrix} \begin{bmatrix} V_{n1} \\ \theta_1 \\ V_{n2} \\ \theta_2 \end{bmatrix} + \begin{bmatrix} -0.5qL \\[2mm] \dfrac{\mu qL^2}{12} \\[2mm] -0.5qL \\[2mm] \dfrac{-\mu qL^2}{12} \end{bmatrix} \quad (1.27)
$$

where

$$S_{11} = S, \; S_{12} = SC$$

$$Q_{11} = S_{11} + S_{12}$$

$$T_{11} = 2Q_{11} - (kL)^2$$

$$S = \frac{kL[1 - kL \cot kL]}{2 \tan(0.5kL) - kL}$$

$$C = \frac{kL - \sin kL}{\sin kL - kL \cos kL}$$

$$\mu = -12EI \frac{0.5kL \cot(0.5kL) - 1}{PL^2}$$

The functions S and C are generally known as stability functions. Note that, if P is a tensile force, it is important to replace all trigonometric functions by their hyperbolic equivalents.

1.27   THE EFFECT OF SHEARING DEFORMATIONS ON THE STIFFNESS MATRIX OF A BEAM-COLUMN ELEMENT

In the previous section, the element stiffness matrix of a beam-column was derived, taking into account the effect of bending deformations only. In general it is not important to consider the effect of shearing deformations in studying the elastic stability of beam-columns occurring in 2-D rigid-jointed structures because, if the members are 'stocky', then buckling is unlikely to be a problem. However, in 'built-up' columns it is important to consider the effect of shearing deformations. The corrections to equation (1.27) can be carried out following the procedure presented in section 1.11 for a beam element. The details are given in Appendix 5. Only the final equations are given below. In order to include the effects of shearing deformations, in equations (1.27) the unbarred quantities are replaced by their barred equivalents as follows:

$$\overline{Q_{11}} = Q_{11}/(1 + D)$$

$$\overline{T_{11}} = 2\overline{Q_{11}} - (kL)^2$$

$$\overline{S_{11}} = S_{11} - 0.5\overline{Q_{11}}D,$$

$$\overline{S_{12}} = S_{12} - 0.5\overline{Q_{11}}D,$$

where

$D = \beta Q_{11}/6$, $\beta = 12EI/GA_sL^2$ and GA_s is the effective shear rigidity.

1.28 THE CONCEPT OF ELASTIC STABILITY

In ordinary parlance, a system is said to be stable if a small disturbance from the 'equilibrium' position produces only a small response. However, if a small disturbance from the equilibrium position produces a large response, then the system is said to be unstable. This is applicable whether the system one is referring to is the economy of a country, a structure or even a human being. In other words, there are degrees of stability (or instability). In connection with structures, one generally refers not so much to degrees of stability but rather to the limit of stability when almost zero disturbing force is required to produce 'infinite' displacements. In order to keep the calculations simple, one important assumption is normally made, i.e. that the material remains elastic irrespective of the stress level. In practical structures, when the displacements (or rather strains) are large, then in general the material does not remain elastic but plasticity sets in. However, in elastic stability calculations, it is assumed that, irrespective of the strains, the material remains elastic. This is a convenient rather than a realistic assumption. In other words the load at which the structure becomes elastically unstable is, in most structures, purely a mathematical concept and does not really have a realistic meaning. However, designers use this concept frequently and it is therefore useful to be able to calculate the load at which the structure becomes unstable.

1.29 WHY STRUCTURES BECOME UNSTABLE

Consider the beam shown in Fig. 1.22(a). From equation (1.27), it is clear that the relationship between the moment M_1 and the rotation θ_1 for this beam is given by

$$M_1 = \frac{EI}{L}S_{11}\theta_1$$

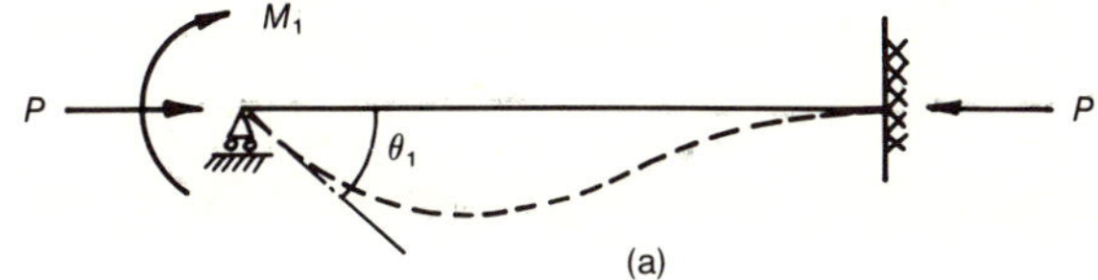

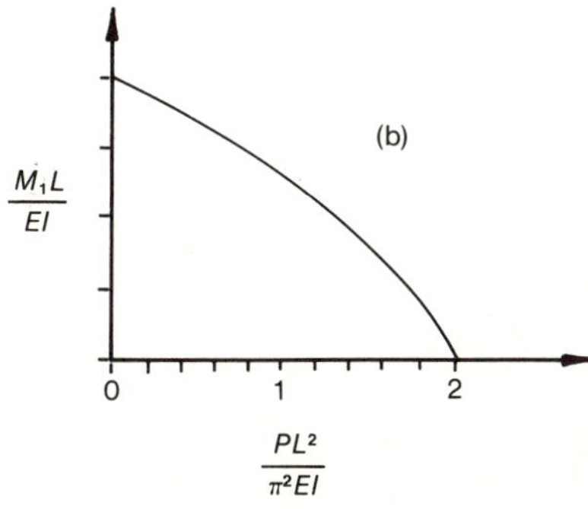

Fig. 1.22   The behaviour of a beam-column element: (a) beam-column element; (b) variation in rotational stiffness with axial compressive load.

Figure 1.22(b) shows the variation in M_1 to cause a rotation of θ_1 equal to unity when the axial compressive force P is gradually increased. As can be seen, the increase in P leads to a reduction in M_1. In other words the structure is losing its stiffness. Evidently, at a certain value of the axial load, the M_1 value required to produce unit rotation θ_1 is almost zero. In other words, if M_1 is viewed as a disturbing force, then at a certain value of P even a very small value of M_1 leads to an infinitely large value of rotation θ_1. At this stage, one would say that the beam has reached the limit of its elastic stability.

This simple example shows that the axial compressive forces, if they are sufficiently large, reduce the stiffness of a structure, leading to the elastic instability of the structure.

1.30   THE BEAM-COLUMN ELEMENT STIFFNESS MATRIX WITH FORCES AND DISPLACEMENTS ALONG GENERAL COORDINATE DIRECTIONS

Equation (1.27) is equivalent to equation (1.19) when the effect of the axial load on the bending stiffness coefficients is ignored. Following the procedures outlined in section 1.16, the general beam-column element stiffness matrix can be derived. Table 1.8 shows the general beam-column element stiffness matrix. This matrix is sometimes also referred to as the 2-D rigid-jointed stability matrix.

Table 1.8

The stability stiffness relationship for an element in a 2-D rigid-jointed structure.

$$
\begin{bmatrix} F_{x1} \\ F_{y1} \\ M_1 \\ F_{x2} \\ F_{y2} \\ M_2 \end{bmatrix} = \frac{EI}{L}
\begin{bmatrix}
\dfrac{T_{11}m^2 + \alpha l^2}{L^2} & \dfrac{-(T_{11} - \alpha)lm}{L^2} & \dfrac{Q_{11}}{L}m & \dfrac{-(T_{11}m^2 + \alpha l^2)}{L^2} & \dfrac{(T_{11} - \alpha)lm}{L^2} & \dfrac{Q_{11}}{L}m \\[2ex]
 & \dfrac{T_{11}l^2 + \alpha m^2}{L^2} & \dfrac{-Q_{11}l}{L} & \dfrac{(T_{11} - \alpha)lm}{L^2} & \dfrac{-(T_{11}l^2 + \alpha m^2)}{L^2} & \dfrac{-Q_{11}l}{L} \\[2ex]
 & & S_{11} & \dfrac{-Q_{11}m}{L} & \dfrac{Q_{11}l}{L} & S_{12} \\[2ex]
 & & & \dfrac{T_{11}m + \alpha l^2}{L^2} & \dfrac{-(T_{11} - \alpha)lm}{L^2} & \dfrac{-Q_{11}m}{L} \\[2ex]
 & & & & \dfrac{T_{11}l^2 + \alpha m^2}{L^2} & \dfrac{Q_{11}l}{L} \\[2ex]
\text{symmetric} & & & & & S_{11}
\end{bmatrix}
\begin{bmatrix} u_1 \\ v_1 \\ \theta_1 \\ u_2 \\ v_2 \\ \theta_2 \end{bmatrix}
+
\begin{bmatrix}
\dfrac{-fL}{2}l + \dfrac{qL}{2}m \\[2ex]
\dfrac{-fL}{2}m - \dfrac{qL}{2}l \\[2ex]
-\mu\dfrac{qL^2}{12} \\[2ex]
\dfrac{-fL}{2}l + \dfrac{qL}{2}m \\[2ex]
\dfrac{-fL}{2}m - \dfrac{qL}{2}l \\[2ex]
\mu\dfrac{qL^2}{12}
\end{bmatrix}
$$

$$\alpha = \frac{AL^2}{I}$$

1.31 DETERMINATION OF THE LOAD FACTOR TO CAUSE THE ELASTIC INSTABILITY OF 2-D RIGID-JOINTED STRUCTURES

It was explained in section 1.19 that the consideration of equilibrium at the joints of the structure leads to the concept of the structural stiffness matrix relating the forces at the joints and the corresponding displacements. In a 2-D rigid-jointed structure, the external loads induce axial forces in the members of the structure in addition to the bending moments and shear forces. Assuming that at a certain load factor the axial loads in the members are known, then the structural stiffness matrix can be easily assembled using the stability stiffness matrix given in Table 1.8. All that remains is to study whether the displacements tend to infinity at that load factor. If the displacements are well within the 'limits', then the load factor is increased and the element stiffness matrix is calculated using the new axial load, and the structural stiffness matrix is assembled using the new element stiffness matrices. The displacements are calculated and the procedure is repeated until the smallest load factor at which the structure becomes unstable is determined.

As the above description makes abundantly clear, the calculations are iterative and a good initial estimation of the load factor at which the structure becomes unstable vastly reduces the number of iterations and thus reduces the computational cost.

1.32 THE CRITERION USED TO DETERMINE THE LIMIT OF ELASTIC STABILITY

It was indicated in the previous section that, for determining the load factor to cause elastic instability, one needs to determine the smallest load factor at which the displacements tend to infinity. It is well known (it can be proved using Cramer's rule) that the displacements tend to infinity if the determinant of the structural stiffness matrix tends to zero. This is the criterion used in structural stability computations. Further details will be discussed in Chapter 7.

1.33 ANALYSIS OF STRUCTURES SUBJECTED TO VIBRATION

Structures in general are subjected not only to static forces but also to forces which set up vibration in structures. The 'vibratory' forces arise from fluctuating forces due to wind, earthquake, traffic, vibrating machinery, etc. If the structure is sufficiently 'flexible', then very large forces can be developed as a result of apparently 'small' vibratory forces. Generally the proneness of a structure to the vibratory force is assessed by comparing the natural frequencies of the structure with the frequencies of the vibrating force. In the analysis of structures subjected to static forces only, the forces acting on the structure are in general independent of the deformations of the structure provided that the displacements are small. In a vibrating structure, the dynamic force arises from the vibration of the mass of the elements of the structure. Since, according to Newton's second law of motion, force acting on a mass is defined as mass multiplied by the acceleration of the mass, following D'Alembert's principle, the analysis of a structure which is vibrating can be treated as an equivalent structure subjected to 'static' external loads called inertial forces given by

$$\text{inertial force} = -\text{mass} \cdot \text{acceleration}$$

Structures are subjected not only to inertial forces as explained above but also to forces called damping forces which, as the name implies, damp out the vibration. In structures, damping arises naturally because of relative motion across a crack in a reinforced concrete structure, friction at a bolted joint in a steel structure, viscous damping due to the surrounding fluid such as wind or water, etc. However, in most structures, such damping forces are small and can be ignored especially when calculating the natural frequencies of a structure.

1.34 ANALYSIS OF UNDAMPED FREE VIBRATIONS

As was explained in the previous sections, the majority of 'design checks' for structures subjected to vibration involve calculating the natural frequency of a structure and comparing it with the frequencies of the vibratory force acting on the structure. In other words the calculation is mainly confined to the study of a structure which has been disturbed from its equilibrium position and is vibrating without either applied external forces or damping forces. Such motion is generally known as undamped free (i.e. no applied external vibratory forces are acting) vibration. Because motion is undamped, the motion is always simple harmonic. In other words, the motion of a particle is represented as

$$\text{displacement as a function of time} = \text{amplitude of displacement} \cdot \cos \omega t$$

where t is the time and ω the circular frequency of vibration. Since acceleration is a second differential of displacement w.r.t. time, we have

$$\left\{ \begin{matrix} \text{acceleration as} \\ \text{a function of time} \end{matrix} \right\} = -\omega^2 \cdot \text{amplitude of displacement} \cdot \cos \omega t$$

$$\left\{ \begin{matrix} \text{inertial force as a} \\ \text{function of time} \end{matrix} \right\} = \text{mass} \cdot \omega^2 \cdot \text{amplitude of displacement} \cdot \cos \omega t$$

As can be seen from the above equations, when the motion is free and undamped, then displacements, strains and inertial forces all vary w.r.t. time as $\cos \omega t$. It is therefore convenient to ignore the time dimension altogether and to treat equilibrium, etc., only in terms of amplitudes of forces, displacements, etc. This approach will be adopted in the following sections.

1.35 THE DYNAMIC STIFFNESS MATRIX—A BAR ELEMENT

For a bar element, the basic differential equation is given by equation (1.4), i.e. $AEu_a'' + f = 0$. If the free undamped vibration problem as shown in Fig. 1.23 is being analysed, then $u_a = U_a$ is the amplitude of axial displacement, $f = m\omega^2 U_a$ is the amplitude of the inertial force and m is the uniformly distributed mass. The differential equation is therefore

$$AEU''_a + m\omega^2 U_a = 0 \qquad (1.28)$$

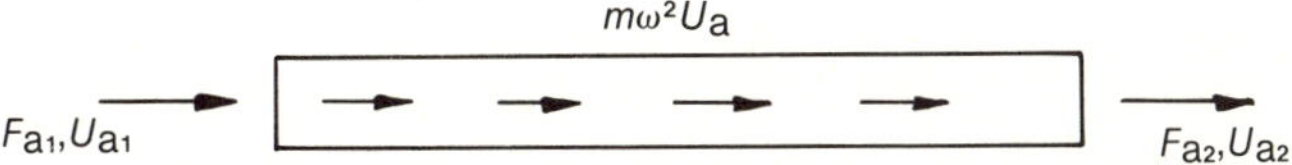

Fig. 1.23   Amplitudes of the dynamic forces of a bar element.

The solution is given by

$$U_a = C_1 \cos kx + C_2 \sin kx$$

where $k^2 = m\omega^2/AE$. The boundary conditions are

$$\text{at end 1,} \quad x = 0, \quad U_a = U_{a1}$$

$$\text{at end 2,} \quad x = L, \quad U_a = U_{a2}$$

The constants of integration are given by

$$C_1 = U_{a1}, \quad C_2 = (U_{a2} - U_{a1} \cos kL) \operatorname{cosec} kL$$

The expression of U_a is given by

$$U_a = U_{a1}(\cos kx - \cot kL \sin kx) + U_{a2} \frac{\sin kx}{\sin kL}$$

The axial tensile force $F = AE\, dU_a/dx$. Therefore,

$$F = AEk\left[-U_{a1}(\sin kx + \cot kL \cos kx) + U_{a2} \frac{\cos kx}{\sin kL} \right]$$

The boundary conditions for F are

$$\text{at end 1,} \quad x = 0, \quad F = -F_{a1}$$

$$\text{at end 2,} \quad x = L, \quad F = F_{a2}$$

The expressions for the end forces are

$$F_{a1} = \frac{AE}{L}(A_1 U_{a1} + A_2 U_{a2}), \quad F_{a2} = \frac{AE}{L}(A_2 U_{a1} + A_1 U_{a2})$$

where $A_1 = kL \cot kL$ and $A_2 = -kL \operatorname{cosec} kL$. The dynamic element stiffness matrix is given by

$$\begin{bmatrix} F_{a1} \\ F_{a2} \end{bmatrix} = \frac{AE}{L} \begin{bmatrix} A_1 & A_2 \\ A_2 & A_1 \end{bmatrix} \begin{bmatrix} U_{a1} \\ U_{a2} \end{bmatrix} \qquad (1.29)$$

Comparing equation (1.29) with equation (1.8), one can see that, for a 'static' matrix, $A_1 = 1.0$ and $A_2 = -1.0$ while, for a dynamic matrix (i.e. equation (1.29)), A_1 and A_2 are functions of the mass and the frequency of vibration.

1.36 THE DYNAMIC ELEMENT STIFFNESS MATRIX— A BEAM ELEMENT

For a beam element the basic differential equation is equation (1.14), i.e.

$$EIv'''' = q$$

When the free undamped vibration problem as shown in Fig. 1.24 is being analysed, $v = V$ is the amplitude of normal displacement, $q = m\omega_2 V$ is the amplitude of inertial force and m is the uniformly distributed mass. The differential equation is given by

$$EIV'''' - m\omega^2 V = 0 \tag{1.30}$$

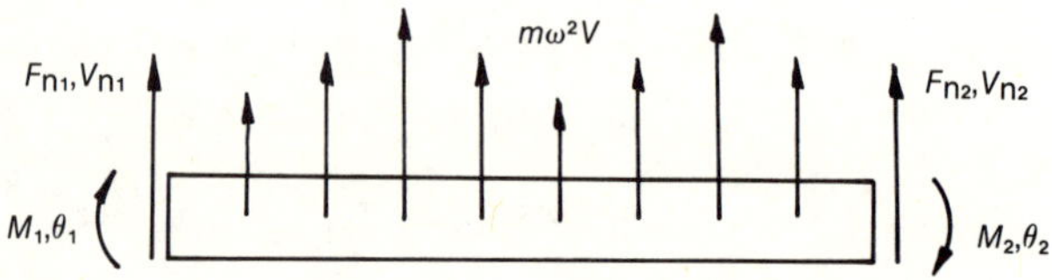

Fig. 1.24   Amplitudes of the dynamic forces of a bending element.

The solution to (1.30) is given by

$$V = C_1 \sin kx + C_2 \cos kx + C_3 \sinh kx + C_4 \cosh kx, \quad k^4 = \frac{m\omega^2}{EI}$$

The boundary conditions are as usual:

$$\text{at end 1,} \quad x = 0, \quad V = V_{n1}, \quad \frac{\mathrm{d}V}{\mathrm{d}x} = -\theta_1$$

$$\text{at end 2,} \quad x = L, \quad V = V_{n2}, \quad \frac{\mathrm{d}V}{\mathrm{d}x} = -\theta_2$$

The integration constants C_1 to C_4 are given by

$$C_1 = \frac{-V_{n1}(\mathrm{Sc} + \mathrm{Cs}) + V_{n2}(\mathrm{S} + \mathrm{s}) + \theta_1 L(\mathrm{Ss} + \mathrm{Cc} - 1)/kL + \theta_2\{\mathrm{C} - \mathrm{c}\}/kL}{D}$$

$$C_2 = \frac{-V_{n1}(\mathrm{Ss} - \mathrm{Cc} + 1) - V_{n2}(\mathrm{C} - \mathrm{c}) - \theta_1 L(\mathrm{Cs} - \mathrm{Sc})/kL - \theta_2 L(\mathrm{S} - \mathrm{s})/kL}{D}$$

$$C_3 = \frac{V_{n1}(\mathrm{Sc} + \mathrm{Cs}) - V_{n2}(\mathrm{S} + \mathrm{s}) - \theta_1 L(\mathrm{Ss} - \mathrm{Cc} + 1)/kL - \theta_2 L(\mathrm{C} - \mathrm{c})/kL}{D}$$

$$C_4 = \frac{V_{n1}(1 - \mathrm{Ss} - \mathrm{Cc}) + V_{n2}(\mathrm{C} - \mathrm{c}) + \theta_1 L(\mathrm{Cs} - \mathrm{Sc})/kL + \theta_2 L(\mathrm{S} - \mathrm{s})/kL}{D}$$

where $D = 2(1 - \mathrm{Cc})$, $\mathrm{S} = \sinh kL$, $\mathrm{C} = \cosh kL$, $\mathrm{s} = \sin kL$ and $\mathrm{c} = \cos kL$.

Using the above values for the constants of integration, V can be expressed in terms of the shape functions N_1 to N_4 as

$$V = N_1 V_{n1} + N_2 \theta_1 + N_3 V_{n2} + N_4 \theta_2$$

The expressions for the shape functions N_1 to N_4 are given by

$$N_1 = \frac{(Sc + Cs)(\sinh kx - \sin kx) - Ss(\cosh kx - \cos kx) + (1 - Cc)(\cosh kx + \cos kx)}{D}$$

$$N_2 = \frac{-Ss(\sinh kx - \sin kx) - (1 - Cc)(\sinh kx + \sin kx) + (Cs - Sc)(\cosh kx - \cos kx)}{DkL}$$

$$N_3 = \frac{-(S + s)(\sinh kx - \sin kx) + (C - c)(\cosh kx - \cos kx)}{D}$$

$$N_4 = \frac{-(C - c)(\sinh kx - \sin kx) + (S - s)(\cosh kx - \cos kx)}{DkL}$$

The amplitudes of the bending moment M and the shear force Q are given by

$$M = EIV'', \quad Q = -EIV'''$$

The amplitudes of forces at the ends are given by

$$\text{at end 1,} \quad x = 0, \quad M = M_1, \quad Q = -F_{n1}$$

$$\text{at end 2,} \quad x = L, \quad M = -M_2, \quad Q = F_{n2}$$

After the forces at the ends have been circulated, the relationship between the amplitude of forces at the ends and the corresponding amplitudes of displacements are given by

$$\begin{bmatrix} F_{n1} \\ M_1 \\ F_{n2} \\ M_2 \end{bmatrix} = \frac{EI}{L} \begin{bmatrix} \dfrac{T_{11}}{L^2} & \dfrac{-Q_{11}}{L} & \dfrac{-T_{12}}{L^2} & \dfrac{-Q_{12}}{L} \\[2ex] & S_{11} & \dfrac{Q_{12}}{L} & S_{12} \\[2ex] & & \dfrac{T_{11}}{L^2} & \dfrac{Q_{11}}{L} \\[2ex] \text{symmetric} & & & S_{11} \end{bmatrix} \begin{bmatrix} V_{n1} \\ \theta_1 \\ V_{n2} \\ \theta_2 \end{bmatrix} \quad (1.31)$$

$$S_{11} = B(Cs - Sc), \qquad S_{12} = B(S - s)$$

$$Q_{11} = BkL\,Ss, \qquad Q_{12} = BkL(C - c)$$

$$T_{11} = B(kL)^2(Cs + Sc), \quad T_{12} = B(kL)^2(S + s)$$

$$B = \frac{kL}{1 - Cc}$$

where $S = \sinh kL$, $C = \cosh kL$, $s = \sin kL$, $c = \cos kL$ and $k^4 = m\omega^2/EI$.

It should be noted that the distributed loading on the element arises from the inertial forces. The inertial forces themselves are functions of displacement V but they can be expressed through shape functions as functions of end displacements (V_{n1}, θ_1, V_{n2}, θ_2). In other words, it is not possible to separate the fixed end forces and the fixed displacement forces in the dynamic analysis of structures.

1.37   THE EFFECT OF SHEARING DEFORMATIONS

In most practical structures the effects of shearing deformation does not substantially affect the natural frequency of a structure. It becomes important only at higher modes when, say, a continuous beam vibrates with many nodes in a span. In general, in most commonplace situations, higher frequencies are of very little interest. Therefore this aspect of the problem will not be discussed further but the interested reader should refer to Cheng [5] [b].

1.38   THE DYNAMIC ELEMENT STIFFNESS MATRIX— A TORSION ELEMENT

For a torsion element (considering free warping torsion) the fundamental differential equation is given by equation (1.23), i.e.

$$GJ\psi'' + t = 0$$

where ψ is the amplitude of twist and t the amplitude of the inertial torsional force.

The inertial force t is calculated as follows. Let m be the mass per unit length. If ψ is the amplitude of twist, then the amplitude of tangential displacement is ψr, where r is the radial distance to the point under consideration. The amplitude of tangential acceleration is $-\omega^2\psi r$. Considering an infinitesimal element of area dA as shown in Fig. 1.25, the mass of the infinitesimal element of area is $(m/A)\,dA$. The inertial tangential force is $-(m/A)\,dA\,(-\omega^2 r\psi)$. The torque due to the inertial force on the infinitesimal area is given by

$$\text{tangential force}\cdot r = \frac{m}{A}\,dA\,\{\omega^2 r\psi\}r$$

$$= \frac{m}{A}\,\omega^2 r^2\psi\,dA$$

The total torque t is equal to

$$t = \int \frac{m}{A}\,\omega^2 r^2\psi\,dA$$

$$= \frac{m}{A}\,\omega^2\psi \int r^2\,dA$$

$$= \frac{m}{A}\,\omega^2 I_{\mathrm{p}}\psi$$

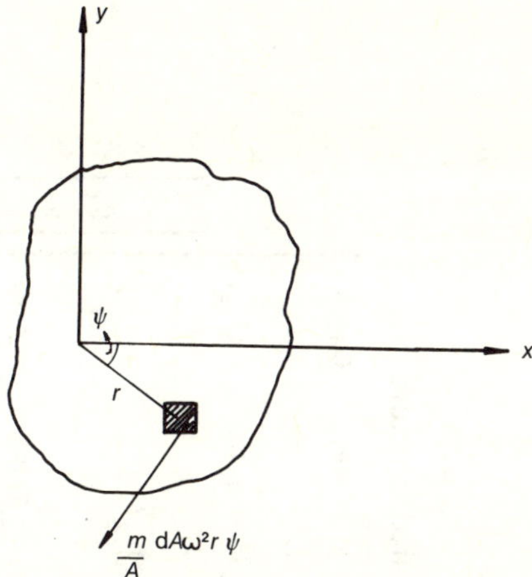

Fig. 1.25   Amplitudes of the dynamic forces on the cross section of a torsion element.

where

$$I_\mathrm{p} = \int r^2 \, \mathrm{d}A \text{ is the polar second moment of area.}$$

The differential equation for the dynamic case is given by

$$GJ\psi'' + \frac{m}{A} I_\mathrm{p}\omega^2\psi = 0 \tag{1.32}$$

Equation (1.32) is similar to equation (1.28). Following the derivation of equation (1.28), the stiffness matrix for the torsion element is given by

$$\begin{bmatrix} T_1 \\ T_2 \end{bmatrix} = \frac{GJ}{L} \begin{bmatrix} A_1 & A_2 \\ A_2 & A_1 \end{bmatrix} \begin{bmatrix} \psi_1 \\ \psi_2 \end{bmatrix} \tag{1.33}$$

where the definitions of A_1 and A_2 are as in equation (1.29) and $k^2 = m\omega^2 I_\mathrm{p}/GJA$.

1.39 DYNAMIC ELEMENT STIFFNESS MATRIX WITH FORCES AND DISPLACEMENTS IN GENERAL COORDINATE DIRECTIONS

In the previous sections, the element stiffness matrix was established for longitudinal, flexural and torsional undamped free vibration. As was explained in section 1.16, the displacements in a member of a 2-D rigid-jointed structure are a combination of axial and bending displacements. Similarly, for a member in a plane grid, the displacements are a combination of bending and torsional displacements. Following the arguments in section 1.16, the dynamic element stiffness matrix for the 2-D rigid-jointed element and the plane grid element can be derived and the corresponding stiffness relationships are given in Tables 1.9 and 1.10 respectively.

Table 1.9

The dynamic stiffness relationship for an element in a 2-D rigid-jointed structure.

$$
\begin{bmatrix} F_{x1} \\ F_{y1} \\ M_1 \\ F_{x2} \\ F_{y2} \\ M_2 \end{bmatrix} = \frac{EI}{L}
\begin{bmatrix}
\dfrac{T_{11}m^2 + \alpha_1 l^2}{L^2} & \dfrac{(-T_{11} + \alpha_1)lm}{L^2} & \dfrac{Q_{11}m}{L} & \dfrac{-(T_{12}m^2 + \alpha_2 l^2)}{L^2} & \dfrac{(T_{12} + \alpha_2)lm}{L^2} & \dfrac{Q_{12}m}{L} \\[2mm]
 & \dfrac{T_{11}l^2 + \alpha_1 m^2}{L^2} & \dfrac{-Q_1 l}{L} & \dfrac{(T_{12} + \alpha_2)lm}{L^2} & \dfrac{(-T_{12}l^2 + \alpha_2)lm}{L^2} & \dfrac{-Q_{12}l}{L} \\[2mm]
 & & S_{11} & \dfrac{-Q_{12}m}{L} & \dfrac{Q_{12}l}{L} & S_{12} \\[2mm]
 & & & \dfrac{T_{11}m^2 + \alpha_{11}l^2}{L^2} & \dfrac{(-T_{11} + \alpha_1)lm}{L^2} & \dfrac{-Q_{11}m}{L} \\[2mm]
 & & & & \dfrac{T_{11}l^2 + \alpha_1 m^2}{L^2} & \dfrac{Q_{11}l}{L} \\[2mm]
\text{symmetric} & & & & & S_{11}
\end{bmatrix}
\begin{bmatrix} u_1 \\ v_1 \\ \theta_1 \\ u_2 \\ v_2 \\ \theta_2 \end{bmatrix}
$$

$$
\alpha_1 = \frac{AL^2}{I} A_1, \qquad \alpha_2 = \frac{AL^2}{I} A_2
$$

Table 1.10

The dynamic stiffness relationship for an element in a plane grid structure.

$$
\begin{bmatrix} F_{z1} \\ M_{x1} \\ M_{y1} \\ F_{z2} \\ M_{x2} \\ M_{y2} \end{bmatrix} = \frac{EI}{L}
\begin{bmatrix}
\dfrac{T_{11}}{L^2} & \dfrac{Q_{11}m}{L} & \dfrac{-Q_{11}l}{L} & \dfrac{-T_{12}}{L^2} & \dfrac{Q_{12}m}{L} & \dfrac{-Q_{12}l}{L} \\[2ex]
 & S_{11}m^2 + \alpha_1 l^2 & -(S_{11} - \alpha_1)lm & \dfrac{-Q_{12}m}{L} & S_{12}m^2 + \alpha_2 l^2 & -(S_{12} - \alpha_2)lm \\[2ex]
 & & S_{11}l^2 + \alpha_1 m^2 & \dfrac{Q_{12}l}{L} & -(S_{12} - \alpha_2)lm & S_{12}l^2 + \alpha_2 m^2 \\[2ex]
 & & & \dfrac{T_{11}}{L^2} & \dfrac{-Q_{11}m}{L} & \dfrac{Q_{11}l}{L} \\[2ex]
 & & & & S_{11}m^2 + \alpha_1 l^2 & -(S_{12} - \alpha_1)lm \\[2ex]
 & & & & & S_{11}l^2 + \alpha_1 m^2
\end{bmatrix}
\begin{bmatrix} w_1 \\ \theta_{x1} \\ \theta_{y1} \\ w_2 \\ \theta_{x2} \\ \theta_{y2} \end{bmatrix}
$$

$$
\alpha_1 = \frac{GJ}{EI} A_1, \qquad \alpha_2 = \frac{GJ}{EI} A_2
$$

1.40   CONCENTRATED MASS AT JOINTS

If a concentrated mass M is at a joint, then the inertial force induced by it is given, when considering undamped free vibration, by

$$\text{inertial force at the joint} = -M \cdot \text{acceleration}$$

$$\text{amplitude of inertial force} = M\omega^2 \cdot \text{amplitude of displacement}$$

In general, one ignores the inertial force due to rotatory motion but includes only the inertial forces due to translational displacements. This is treated as an external load at the joint where the mass is present.

1.41   THE CONCEPT OF RESONANCE

Consider a structure subjected to a vibratory force whose frequency of vibration can be changed. Figure 1.26 shows a qualitative description of the relationship between the amplitude of deformation at a point and the frequency of vibration of the applied force. As can be seen, whenever the frequency of vibration of the force coincides with a natural frequency of the structure, then the displacement tends to infinity. This is the phenomenon of resonance and the structure is said to resonate. In practical structures, the displacement is somewhat controlled because of damping which is always present. Nevertheless, if at a conceptual level one is dealing with a very lightly damped structure (as most structures are), then one can say that the criterion for determining the natural frequency of a structure is that, if it is vibrated by an external force whose frequency is equal to that of a natural frequency of the structure, then the structure resonates.If a structure is resonating, then the displacements tend to infinity and, as already discussed in section 1.30 (in connection with stability analysis), the determinant of the dynamic structural stiffness matrix tends to zero. This is the criterion used for determining the natural frequency of a structure. Further details will be discussed in Chapter 8.

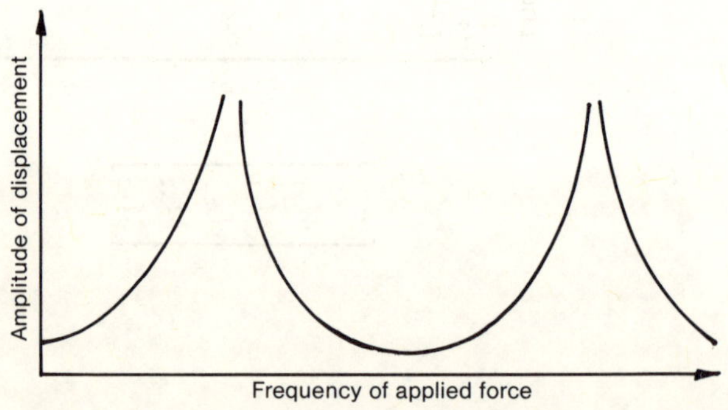

Fig. 1.26   The response of a structure to an external vibratory force.

1.42   THE STIFFNESS METHOD AND COMPUTER PROGRAMMING

The stiffness method discussed in the previous sections involves a great deal of numerical computations. In practice it is therefore meant to be used with the help of a computer program to carry out the complex numerical work. In the following sections, some general aspects of computer programs are discussed. Details about the programs will be discussed in the relevant chapters.

Before going into the computer program aspects, it is useful to keep in mind the various steps involved in the stiffness method. The basic steps can be summarised as follows.

(a)   Inputting of basic information about the coordinates of the nodes, node numbers at the ends of the elements, cross-sectional and material properties of elements, nodes where one or more displacements are restrained, etc.

(b)   Calculations of the element stiffness matrix and adding it as appropriate to the structural stiffness matrix.

(c)   Inputting of data about the external loads and thermal loads on the structure and the calculation of the load vector corresponding to the displacement freedoms of the structure.

(d)   Solution of simultaneous equations to calculate the displacements of the joints.

(e)   Calculation of the forces in the members.

It is useful to remember the above five basic steps during the writing of programs as they enable the program to be divided into a series of convenient blocks.

1.43   WRITING COMPUTER PROGRAMS

Before undertaking the writing of computer programs for the stiffness method of analysis, it is important that the student is familiar with the following.

(a)   A good working knowledge of a computer language such as ALGOL, BASIC, FORTRAN, PASCAL, etc. In this book, all the programs are written in FORTRAN.

(b)   Solution of simultaneous equations by the Gaussian elimination method.

(c)   A thorough knowledge of the stiffness method as applied to the analysis of skeletal structures.

Although the aspects listed above are necessary to undertake the writing of programs, they are not sufficient to ensure that the programs which are written are 'good'. This is especially true if the programs are to be used

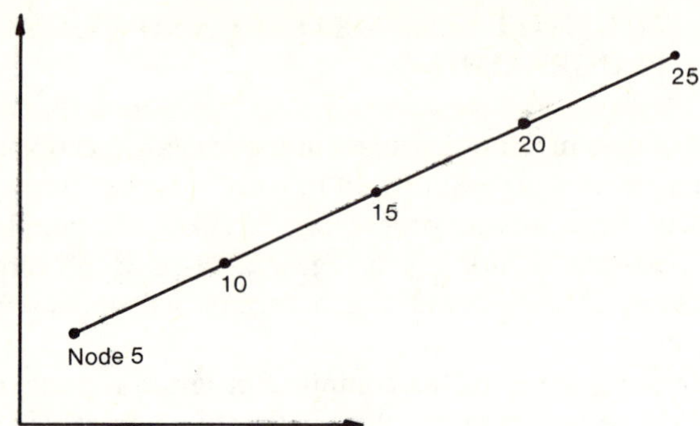

Fig. 1.27 Equally spaced nodes on a line.

'commercially'. Some of the additional points to be kept in mind to aid in the production of 'good' programs are as follows.

(a) It is important to organize the input of basic data such that only the most basic data are input and to allow the computer to calculate the derived data. To illustrate this point, consider a set of equally spaced and equally incremented nodes on a line as shown in Fig. 1.27. If it is required to calculate the coordinates of all the nodes on the line, then the basic data that need to be input are as follows.

    (1) The node number at the beginning of the line and the corresponding coordinates.
    (2) The node number at the end of the line and the corresponding coordinates.
    (3) The number of nodes in between the beginning and the end of the line.

From the above information a properly organized program should be able to calculate the node numbers of all the intermediate nodes and their coordinates. It is important to appreciate this idea of letting the computer do the 'dirty' work because it not only takes more effort to input all the nonbasic data but also creates more opportunities for errors, thus leading to unnecessary increase in the cost of analysis.

(b) In any large structure, there will inevitably be only a small number of groups of elements having different cross sections and material properties. Therefore, in order to reduce data input, it is better to identify each element by the group number of element to which it belongs and to input the large number of cross-sectional and material properties of the elements to only a small number of element groups.

(c) In practice, it is necessary to analyse a structure under a large number of loading cases. It is important that the program is so organized that, after the first loading case is analysed, any additional loading cases are analysed with a minimum number of computations.

(d) In practice, a large number of structures under several loading cases will be analysed in one go. It is important that any data error for a structure or loading case should not prevent the proper analysis of the rest of the structures and/or load cases.

(e) The programmer must be clear from the outset whether the program is meant to analyse, say, only regular multi-storey frames or general 2-D rigid-joint structures. Depending upon what the program is intended to do, the organization of the data input will be correspondingly vastly different.

(f) The output must include not only the data that are input but also all the derived data that are calculated. If there are any errors, the error messages must be such as to assist in the speedy detection of errors.

1.44 IMPROVING THE READABILITY OF PROGRAMS

All programs are inevitably subjected to alteration and improvement. It is, therefore, important to recognize that a program, apart from being able to perform a set of calculations correctly, should also have an easy-to-see logic of the calculations and be readable. The ability to write readable programs is acquired through experience. However, the following points taken from the book by Kernighan and Plauger [7] [a] help to focus on some key points which aid in writing readable programs.

(a) Write clearly—do not be too clever.
(b) Say what you mean, simply and directly.
(c) Avoid temporary variables.
(d) Write clearly—do not sacrifice clarity for 'efficiency'.
(e) Let the machine do the dirty work.
(f) Avoid unnecessary branches.
(g) Use subroutines.
(h) Make the coupling between the modules visible.
(j) Let the data structure the program.
(k) Terminate the input by the end-of-file marker and not by count.
(l) Identify bad input. Recover if possible.
(m) Use names for variables that do mean something.

1.45 NOMENCLATURE OF NAMES OF VARIABLES

One important point that considerably improves the readability of any program is the names chosen for variables. As already suggested, they should mean something—generally some recognizable abbreviation for the name of the variable. The names chosen for all the important variables are listed in Appendix 1. Many of these are taken from the book by Hinton and Owen [7][e]. It is well worth the effort to become very familiar with the names of these variables as they will considerably enhance the ability to follow the program steps easily.

Another point that improves the readability is to choose, following Hinton and Owen, a common root principle. In this book, this principle can be explained as follows.

For example, NPOIN indicates a given number, while IPOIN or JPOIN is a DO loop variable reaching a maximum value of NPOIN. This is illustrated in the following segment of a program.

```
       DO  10  IPOIN = 1,NPOIN
       DO  20  JPOIN = 1,NPOIN
       COORD(IPON,JPOIN) = 0.0
    20 CONTINUE
    10 CONTINUE
```

1.46   SELECTED REFERENCES

Many excellent books are available on the basic aspects of the strength of materials and theory of structures. The following list is basically aimed at providing a list of references on topics which are thought to be at a slightly advanced level.

[1]   Shearing deformations in beams
[a]   S. Timoshenko, *Strength of Materials*, Vol. 1, Van Nostrand, 1955, pp. 170–175.
[b]   A. K. Chugh, Stiffness matrix for a beam element including the transverse shear and axial load effects, *International Journal for Numerical Methods in Engineering*, Vol. 11, 1977, pp. 1681–1697.

[2]   Torsion—basic theory
[a]   S. Timoshenko and J. N. Goodier, *Theory of Elasticity*, McGraw-Hill, 1970, pp. 291–303, 332–336.

[3]   Nonuniform torsion
[a]   S. Timoshenko, *Strength of Materials*, Vol. 2, Van Nostrand, 1956, pp. 255–273.
[b]   J. N. Goodier and M. V. Barton, The effect of web deformations on the torsion of I beams, *Journal of Applied Mechanics*, March 1944, pp. A35–A40.

[4]   Elastic Stability
[a]   M. R. Morne and W. Merchant, *The Stability of Frames*, Pergamon Press, 1965, pp. 1–10, 48–67.

[b]   A. K. Chugh, [1] [b].
[c]   S. Timoshenko and J. M. Gere, *Theory of Elastic Stability*, McGraw-Hill, 1961 pp. 132–142 (for the effect of shearing deformations on the elastic critical load of built-up columns).

[5]   Structural dynamics
[a]   M. Paz, *Structural Dynamics*, Van Nostrand Reinhold, 1980, Chapters 1, 2 and 21.
[b]   F. Y. Cheng, Vibrations of Timoshenko beams and frameworks, *Journal of the Structural Division, Proceedings of the American Society of Civil Engineers*, Vol. 96, ST3, March 1970, pp. 551–570 (this paper discusses the effect of shearing deformations on the natural frequencies of rigid-jointed structures).

[6]   Matrix Analysis of structures
[a]   P. Bhatt, *Problems in Structural Analysis by Matrix Methods*, Construction Press, 1981.
[b]   W. McGuire and R. H. Gallagher, *Matrix Structural Analysis*, Wiley, 1979.
[c]   W. Weaver and J. M. Gere, *Matrix Analysis of Framed Structures*, 2nd edn, Van Nostrand, 1980.

[7]   Computer programming
[a]   B. W. Kernighan and P. J. Plauger, *The Elements of Programming Style*, 2nd edn, McGraw-Hill, 1978.
[b]   B. Meek, P. Heath and N. Rushby, *Guide to Good Programming Practice*, 2nd edn, Ellis Horwood, 1983.
[c]   D. Kirkpatric, Writing clear computer programs, *Civil Engineering* (*American Society of Civil Engineers*), April 1984 pp. 52–54.
[d]   R. A. Pixley and S. A. Ridlon, Checking out computer programs, *Civil Engineering* (*American Society of Civil Engineers*), October 1984 p. 59.
[e]   E. Hinton and D. R. J. Owen, *Programming the Finite Element Method*, Academic Press, 1977.

CHAPTER 2   Data Input

In this chapter, the subroutines used for simplifying the data input are discussed. In addition the major nomenclature used for the names of variables are discussed in detail. It is important to become familiar with the nomenclature in order to be able to study the programs with ease.

2.1   BASIC PARAMETERS

The basic parameters of the structure and its elements are as follows.

2.1.1   Structural parameters

NPOIN   Number of POINts in the structure. Loosely speaking, NPOIN is equal to the number of joints in the structure except in unusual cases such as that discussed in section 1.17 or when concentrated mass acts on a beam in the dynamic analysis of structures when every point in the beam where a concentrated mass acts should be treated as a joint.

NELEM   Number of ELEMents in the structure.

NRESND   Number of REStrained NoDes in the structure. Restrained nodes are where one or more displacements are held at zero value.

NVFIX   Number of nodes where the Value of one or more freedoms have a specified FIXed value. Note the difference between NRESND and NVFIX. In the latter case the displacement freedom is fixed at a nonzero value.

NMATS   Number of different groups of members having different MATerial and cross-sectional geometric propertiesS.

NDIME   Number of DIMEnsions of the structure, i.e. 2-D or 3-D.

NCASE   Number of load CASEs to be considered for a particular structure.

Table 2.1

NDOFN for different elements

Type of element	Displacement	NDOFN
2-D pin jointed	u, v	2
3-D pin jointed	u, v, w	3
2-D rigid jointed	u, v, θ	3
Plane grid[a]	w, θ_x, θ_y	3
Plane grid[b]	$w, \theta_x, \theta_y, \mathrm{d}\psi/\mathrm{d}x$	4

(u, v, w) are translations in the x, y and z directions respectively, $(\theta_x, \theta_y, \theta)$ are the rotations about the x, y and z axes respectively, ψ is the twist at a section and $\mathrm{d}\psi/\mathrm{d}x$ is the rate of twist at the end.
[a] Plane grid analysis ignoring warping restraint.
[b] Plane grid analysis including warping restraint.

2.1.2 Element parameters

NNODE   Number of <u>NODE</u>s defining the element. For skeletal structures, NNODE = 2. For finite elements in the analysis of plate and shell structures, NNODE can be different depending on the type of element being used.

NDOFN   Number of <u>D</u>egrees <u>O</u>f <u>F</u>reedom at the <u>N</u>odes. This depends on the type of element as shown in Table 2.1.

NPROP   Number of basic material and cross-sectional geometric <u>PROP</u>erties required to calculate the element stiffness matrix. This is shown for different types of elements in Table 2.2.

2.1.3 Coordinates of nodes

The coordinates of all NPOIN nodes are stored in an array COORD(NPOIN, NDIME). For any node IPOIN, the data is arranged as follows: COORD(IPOIN,1) is the x coordinate of IPOIN, COORD(IPOIN,2) is the y coordinate of IPOIN, and so on.

Table 2.2

The NPROP properties of elements

Type of element	Elements of array PROPS	NPROP
2-D and 3-D pin jointed	E, A	2
2-D rigid jointed	E, G, I, A, ZASF	5
Plane grid with free warping	E, G, I, J, A, ZASF	6
Plane grid with restrained warping	E, G, I, J, A, ZASF, I_w	7

Note that it is assumed that the bending is about the y_m axis.

2.1.4   The node numbers of elements

The node number of all the NELEM elements are stored in an array
LNODS(NELEM,NNODE). The name LNODS stands for e<u>L</u>ement
<u>NOD</u>e<u>S</u>. For any element IELEM, the node numbers at the two ends are
stored as follows: the node at end 1 is LNODS(IELEM,1); the node at end 2
is LNODS(IELEM,2).

It is immaterial which end is chosen as end 1 and which as end 2 as long
as, once the choice is made, it is consistent throughout the calculations.

2.1.5   Restrained nodes

The information about the restrained nodes is stored in two arrays. The node
numbers of the NRESND nodes which are fully restrained are stored in an
array NDFIX(NRESND). The name NDFIX stands for <u>N</u>o<u>D</u>e freedom
<u>FIX</u>ed. The data about which of the NDOFN freedoms at a node are
restrained are stored in code form in an array IFPRE(NRESND,NDOFN).
The convention used is that 1 means fixed and 0 means free. The name
IFPRE stands for <u>IF</u> <u>PRE</u>scribed. As an example, if the IRESNDth row of
IFPRE has elements the 1, 0, 1, then it means that at the IRESNDth
restrained node the first freedom is fixed, the second freedom is free and the
third freedom is restrained. Table 2.1 shows the order in which the freedoms
at a node are numbered.

2.1.6   Nodes with prescribed nonzero displacements

The information about the nodes with prescribed nonzero values of displacements are stored in a manner similar to the restrained nodes. The node
numbers of all NVFIX nodes where the displacements have a fixed value are
stored in the array NDVFIX(NVFIX). The name NDVFIX stands for <u>N</u>odes
with <u>D</u>isplacement <u>V</u>alue <u>FIX</u>ed. The information about which of the
freedoms have a fixed value and which are free is stored in code form in the
array PRESC(NVFIX,NDOFN). The convention used is that, if the code
about a particular freedom is nonzeo, then it has the value of the displacement shown. As an example, if the IVFIXth row of the array PRESC has
values 0.0, 0.21, 0.03, then it means that the first freedom is free, the second
freedom is fixed at a value of 0.21 and the third freedom is fixed at a value
equal to 0.03. It should be noted that it is perfectly acceptable to have the
same node number appearing both as a restrained node with certain
freedoms fully restrained and as a node with certain freedoms having a
prescribed value.

2.1.7   The freedom number of nodes

The displacement numbers of the displacements at the nodes are stored in the
array NODFRE(NPOIN,NDOFN). If any element of the array is zero, then

that particular displacement is fully restrained. The name NODFRE stands for <u>NOD</u>e <u>FRE</u>edom.

2.1.8 The group number of elements

The group number of an element from the NMATS groups of elements that the NELEM elements belong to are stored in the array MATNO(NELEM,2). The name MATNO stands for <u>MAT</u>erial <u>N</u>umber. The data are arranged as follows.

(a) MATNO(IELEM,1) gives the group number of the IELEMth element from the NMATS group of elements having different cross-sectional geometric and material properties.

(b) MATNO(IELEM,2) shows for beam elements whether a particular element belongs to one of the following four categories according to the following convention.

MATNO(IELEM,2) = 1   Orthodox prismatic beam element.

MATNO(IELEM,2) = 2   Prismatic beam on an elastic foundation.

MATNO(IELEM,2) = 3   Prismatic beam with a semi-rigid connection.

MATNO(IELEM,2) = 4   Nonstandard beam element.

It should be noted that the program calculates the element stiffness matrix for the first three categories of beam elements but for the last category, the element stiffness matrix has to be input as data.

2.1.9 Cross-sectional and material properties

The cross-sectional and material properties are stored in the array PROPS(NMATS,NPROP + ?). The name NPROP stands for <u>N</u>umber of <u>PROP</u>erties and represents the number of cross-sectional and material properties generally required as shown in Table 2.2 for several elements. The + ? sign after NPROP indicates that, for many elements, many more 'properties' than just NPROP will be required. This is discussed later. The nomenclature used is as follows.

YOUNG   Young's modulus E.

SHEARM Shear modulus G.

AREA   Cross-sectional area A.

YINERA Second moment I of area about the y_m axis as shown in Fig. 2.1. It is assumed that y_m is a principal axis.

TINERA St. Venant's torsion inertia J.

WINERA Warping inertia I_w. This is needed when considering the torsion of elements with restrained warping.

ZASF   Area shear factor for shear force along the z_m axis. Note that the effective shear area for shear force along the z_m axis is $A_s =$ AREA/ZASF.

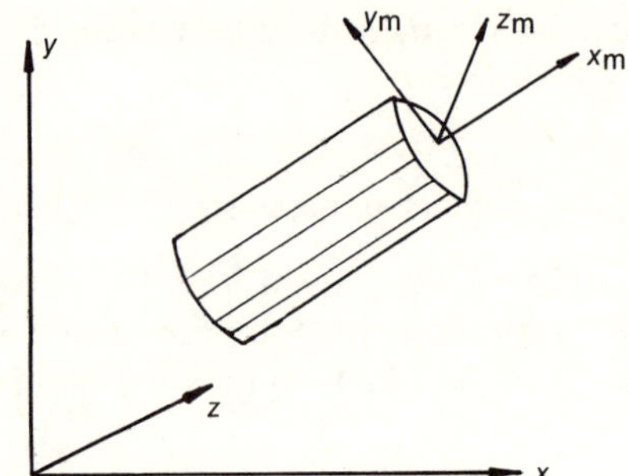

Fig 2.1   Orientation of the principal axes of a member.

2.1.10   Properties additional to NPROP properties

As indicated previously, the array PROPS(NMATS,NPROP + ?) contains properties in addition to the properties discussed in the previous section. As indicated, the programs can handle structures with beams on elastic foundations as shown in Fig. 2.2 and beams with semi-rigid ends to represent the deformation of the joints in bolted steel structures as shown in Fig. 2.3. In addition the programs can handle stress analysis due to a change in temperature. In all these cases, additional 'properties' are needed. For example the beam on an elastic foundation needs the value of subgrade modulus. Similarly the beam with semi-rigid ends needs information about the stiffness of the springs at the ends, and thermal stress analysis requires the value of the coefficient of linear thermal expansion and for beam elements the value of the distance from the neutral axis to the top and bottom surfaces.

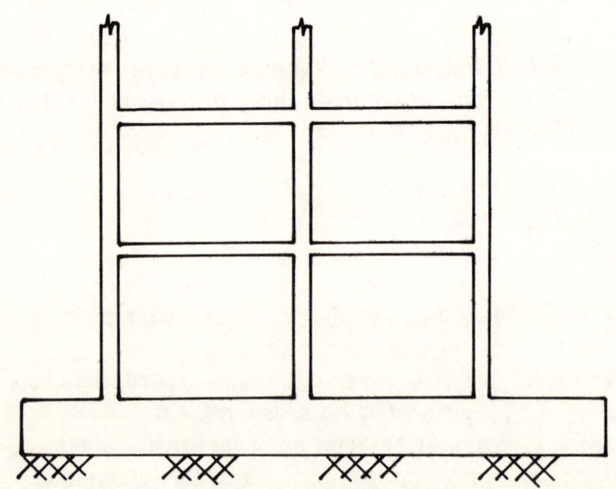

Fig 2.2   A structure with some members supported by an 'elastic' foundation.

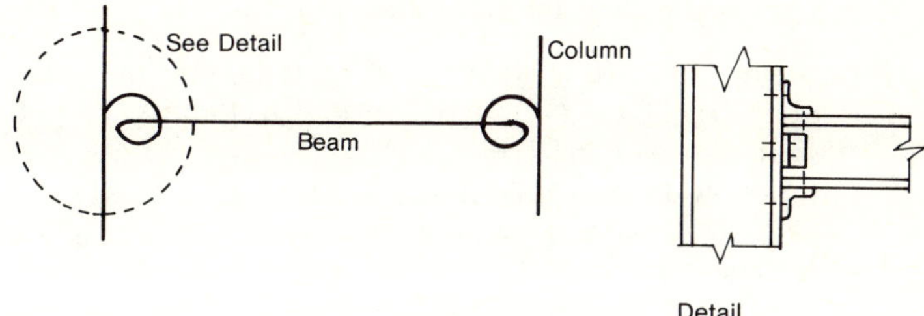

Fig 2.3   A semi-rigid connection.

2.2   SIMPLIFYING THE DATA INPUT

In any large structure there will inevitably be a large number of nodes for which it is necessary to input the data regarding the coordinates of the nodes. Similarly, for a large number of elements it is necessary to input the node numbers of nodes at the ends of the element. This can be a very onerous task, leading to the possibility of errors. However, because most of the data will be fairly regular, it is possible to input data only at the extremities of a regular set of data and to let the computer generate the intermediate data. As an example, consider the set of nodes shown in Fig. 1.27. Evidently, because the intermediate nodes are equally spaced and the node numbers are equally incremented, given the coordinates and the node numbers of the nodes at the two ends of the line and the number of nodes in between, it is possible to generate the coordinates and the node numbers of intermediate nodes. In this book, three subroutines are used to simplify data input. These three subroutines are described below.

2.2.1   Subroutine CORDGEN

This subroutine is used to simplify the data input for the coordinates of nodes. Before the subroutine is described, it is necessary to list the variable names and their meaning.

NODSTR	NODe at the STaRt of a line.
NODEND	NODe at the END of a line.
NSTEP	Number of STEPs to reach from NODSTR to NODEND. For example, in Fig. 1.27, to reach from NODSTR = 5 to NODEND = 25 there are four nodes. Therefore, NSTEP = 4.
NODINC	NODe number INCrement for successive intermediate nodes. NODINC = (NODEND − NODSTR)/NSTEP.
CORINC(NDIME)	The array CORINC stores the COoRdinate INCrement for successive intermediate nodes.

The logic used behind the subroutine is as follows.

(1) Assume that NODEND = 0, say, from a previous calculation.

(2) Set NODSTR = NODEND so that in the first cycle the calculations start with NODSTR = 0.

(3) Input NODEND,NSTEP,COORD(NODEND,NDIME). The value of NSTEP is that between a previously read NODEND and the current NODEND.

(4) If NSTEP = 1, then obviously there are no intermediate nodes to consider. If NODEND = NPOIN, then all the data about node coordinates have been read. If NODEND ≠ NPOIN then go back to step (3) to read more data.

(5) If NSTEP is negative, then obviously there is an error in the data and therefore the calculation will stop by setting the descriminator NFAIL = 1, signalling the end of calculations. In order to assist in error detection, the node number corresponding to the current NODEND will be printed.

(6) If NSTEP > 1, then there are intermediate nodes whose node numbers and coordinates have to be calculated.

(7) The node number of the first intermediate node is (NODSTR + NODINC) and that of the last intermediate node is (NODEND − NODINC).

(8) Between the first and last intermediate nodes the node numbers are successively incremented by NODINC and the IDIMEth coordinate by CORINC(IDIME). Figure 2.4 makes the meaning of the steps more clear.

2.2.2 Subroutine INTGEN

This subroutine is similar in all respects to CORDGEN except for the fact that it operates on dummy integer array IA(NMAX,NVAR). This subroutine is used for generating intermediate data for all regular sets of data in the following arrays.

(a) LNODS(NELEM,NNODE) which stores the node numbers of the nodes at the ends of the NELEM elements.

(b) MATNO(NELEM,2) which stores the group number and, for beam elements, the type number, i.e. an orthodox beam element, a beam on an elastic foundation, etc., as explained in section 2.1.8.

(c) NODFRE(NPOIN,NDOFN) which stores the displacement number of all the nodes. It should be noted that the subroutine INTGEN is used only if one or more freedoms in the structure are forced to move by the same amount, e.g. a multistorey rigid-joint frame in which the horizontal displacements of all the nodes at a storey level are equal to the sway displacement. If this is not the case, then the subroutine NFGEN to be described in section 2.2.3 can be used and it is more efficient. This point is discussed in section 2.2.4.

Table 2.3 shows the relationship between the dummy variables IA(NMAX,NVAR) and the arrays discussed above.

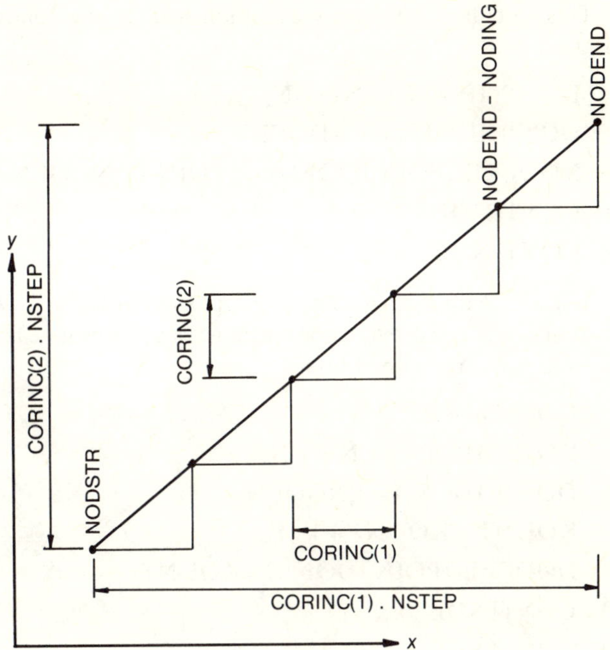

Fig 2.4   Calculation of the intermediate nodal coordinates.

2.2.3   Subroutine NFGEN

This subroutine is used to generate the array NODFRE(NPOIN,NDOFN) when no two (or more) nonzero displacements are constrained to have the same value. The subroutine makes use of the data about which nodes and freedoms are restrained as stored in the arrays NDFIX(NRESND) and IFPRE(NRESND,NDOFN). The logic used for constructing this subroutine can be explained as follows.

Suppose that none of the freedoms is restrained. Then the IDOFNth node freedom at the IPOINth node is given simply by

$$NODFRE(IPOIN,IDOFN) = (IPOIN-1)*NDOFN + IDOFN$$

Table 2.3

The relationship between arrays.

Array	Maximum number of rows	Maximum number of columns
IA	NMAX	NVAR
LNODS	NELEM	NNODE
MATNO	NELEM	1 or 2
NODFRE	NPOIN	NDOFN

This calculation is easily accomplished by the following segment of a program.

```
      DO  10  IPOIN=1,NPOIN
      DO  30  IDOFN=1,NDOFN
      NODFRE(IPOIN,IDOFN)=(IPOIN-1)*NDOFN+IDOFN
   30 CONTINUE
   10 CONTINUE
```

The above task can also be accomplished by the following segment whose advanage over the previous program is that it is more easily modified to take account of the restrained nodes.

```
      KOUNT=0
      DO  10  IPOIN=1,NPOIN
      DO  30  IDOFN=1,NDOFN
      KOUNT=KOUNT+1
      NODFRE(IPOIN,IDOFN)=KOUNT
   30 CONTINUE
   10 CONTINUE
```

Since there will be restrained nodes, the above program can be modified to take account of the restrained nodes as follows.

(1) Choose a point IPOIN as at DO 10 IPOIN=1,NPOIN.
(2) Check IPOIN against the array NDFIX(NRESND) to see whether IPOIN node is a restrained node.
(3) If IPOIN is not a restrained node, then proceed to step DO 30 IDOFN=1,NDOFN and continue the calculations.
(4) If IPOIN is a restrained node, then proceed as follows.
(5) If IPOIN is the IRESNDth restrained node, then from the array IFPRE(IRESND,NDOFN) check which of the NDOFN freedoms are restrained. For those freedoms which are restrained, set the node freedom as zero. For the rest, continue as for step KOUNT= KOUNT+1.

The above ideas are incorporated in the subroutine NFGEN which stands for <u>N</u>ode <u>F</u>reedom <u>GEN</u>eration.

2.2.4   Which to use: INTGEN or NFGEN?

Consider the 2-D rigid-jointed structure shown in Fig. 2.5. Fig. 2.5(a) shows the node freedom numbers when the elements are assumed to be axially rigid while Fig. 2.5(b) shows the corresponding node freedom numbers when the elements are assumed to be axially deformable. As can be seen, in the former case the horizontal displacement at nodes 1, 3 and 5 are all required to be equal to d_2. Therefore in this case the subroutine INTGEN is used for

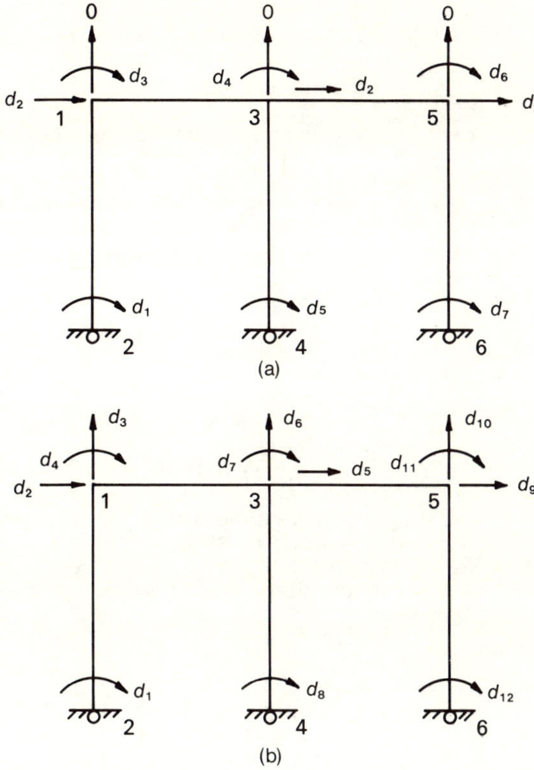

Fig 2.5   Nodal displacements of a 2-D rigid-jointed frame: (a) elements are axially rigid; (b) elements are axially deformable.

calculating the elements of the array NODFRE(NPOIN, NDOFN). In Fig. 2.5(b), since no two nonzero displacements are required to be equal, the array NODFRE(NPOIN,NDOFN) is most efficiently generated by using the subroutine NFGEN.

```
C
C***********************************************
C
      SUBROUTINE CORDGEN
C
C*** THIS SUBROUTINE GENERATES OMITTED DATA FROM
C    THE BLOCK OF DATA AT THE BEGINNING AND END
C    OF A LINE
  *************************************************
  *                                               *
  *      USE APPROPRIATE COMMON BLOCKS HERE        *
  *                                               *
  *************************************************
C
      DIMENSION CORINC(3)
      NODEND=0
    1 NODSTR=NODEND
      READ(5,900) NODEND,NSTEP,(COORD(NODEND,IDIME),
     @IDIME=1,NDIME)
      IF(NSTEP.EQ.1) GOTO 3
      IF(NSTEP.LT.0) GOTO 2
      NODDIF=NODEND-NODSTR
      NODINC=NODDIF/NSTEP
      DO 5 IDIME=1, NDIME
    5 CORINC(IDIME)=(COORD(NODEND,IDIME)-
     @COORD(NODSTR,IDIME))
     /FLOAT(NSTEP)
      NODE1=NODSTR+NODINC
      NODE2=NODEND-NODINC
      DO 6 INODE=NODE1, NODE2, NODINC
      NODE3=INODE-NODINC
      DO 6 IDIME=1, NDIME
    6 COORD(INODE,IDIME)=COORD(NODE3,IDIME)
     @+ CORINC(IDIME)
    3 IF(NODEND.LT.NPOIN) GOTO 1
      IF(NODEND.EQ.NPOIN) GOTO 7
    2 WRITE(6,910)NODEND
      NFAIL=1
  900 FORMAT(I5,I5,3F10.4)
  910 FORMAT(//,' ERROR IN NODE  DATA OF THE
     @NODE',I5)
    7 RETURN
      END
```

```
C
C***********************************************
C
      SUBROUTINE INTGEN(IA,NVAR,NMAX)
C
C***THIS SUBROUTINE GENERATES OMITTED DATA FROM
C    THE  BLOCK OF DATA AT THE BEGINNING AND THE
C    END OF A LINE.
C    THIS WILL BE USED TO GENERATE MATRICES LNODS,
C    NODFRE AND MATNO
C
***********************************************
*                                             *
*      USE APPROPRIATE COMMON BLOCKS HERE      *
*                                             *
***********************************************
```

```
C
      DIMENSION IA(100,3),INCR(3)
      IEND=0
    1 ISTR=IEND
      READ(5,900)IEND,NSTEP,(IA(IEND,JVAR),
     @JVAR=1,NVAR)
      IF(NSTEP.EQ.1) GOTO 3
      IF(NSTEP.LT.0) GOTO 2
      IDIF=IEND-ISTR
      INC=IDIF/NSTEP
      DO 5 IVAR=1,NVAR
    5 INCR(IVAR)=(IA(IEND,IVAR)-IA(ISTR,IVAR))
     @/NSTEP
      I1=ISTR+INC
      I2=IEND-INC
      DO 6 I=I1,I2,INC
      I3=I-INC
      DO 6 IVAR=1,NVAR
      IA(I,IVAR)=IA(I3,IVAR)+INCR(IVAR)
    6 CONTINUE
    3 IF(IEND.LT.NMAX) GOTO 1
      IF(IEND.EQ.NMAX) GOTO 7
    2 IF(NMAX.EQ.NPOIN.AND.NVAR.EQ.NNODE)
     @WRITE(6,910)IEND
      IF(NMAX.EQ.NELEM.AND.NVAR.EQ.NDOFN)
     @WRITE(6,920)IEND
      IF(NMAX.EQ.NELEM.AND.NVAR.LE.2)
     @WRITE(6,930)IEND
      NFAIL=1
  910 FORMAT(//,' ERROR IN ELEMENT CONNECTION
     @DATA OF ELEMENT',I5)
  920 FORMAT(//,' ERROR IN NODE FREEDOM DATA OF
     @NODE',I5)
  930 FORMAT(//,'ERROR IN MATERIAL NO. OF
     @ELEMENT',I5)
  900 FORMAT(5I5)
    7 RETURN
      END
```

```
C
C**************************************************
C
      SUBROUTINE NFGEN(NODFRE)
C
C***THIS SUBROUTINE GENERATES NODE FREEDOM ARRAY
C   FROM THE DATA ABOUT THE NODES WHERE ONE OR
C   MORE FREEDOMS ARE RESTRAINED
C
**************************************************
*                                                *
*     USE APPROPRIATE COMMON BLOCKS HERE         *
*                                                *
**************************************************
C
      DIMENSION NODFRE(100,3)
      KOUNT=0
C
C***CYCLE ON ALL NPOIN POINTS
C
      DO 10 IPOIN=1,NPOIN
```

```fortran
      JRESND=0
C
C***CHECK THROUGH NRESND NODES TO SEE IF IPOIN
C    IS A RESTRAINED NODE
      DO 20 IRESND=1,NRESND
      IF(IPOIN.EQ.NDFFIX(IRESND))JRESND=IRESND
C
C***IF IPOIN IS A RESTRAINED NODE STOP SEARCHING
C
      IF(JRESND.GT.0) GOTO 35
   20 CONTINUE
C
C***CHECK THROUGH NDOFN FREEDOMS TO SEE WHICH
C    FREEDOMS ARE RESTRAINED
C
   35 DO 30 IDOFN=1, NDOFN
      IF(JRESND.EQ.0) GOTO 40
      JDOFN=IFPRE(JRESND,IDOFN)
      IF(JDOFN.EQ.1) NODFRE(IPOIN,IDOFN)=0
      IF(JDOFN.EQ.1) GOTO 30
   40 KOUNT=KOUNT+1
      NODFRE(IPOIN,IDOFN)=KOUNT
   30 CONTINUE
   10 CONTINUE
      RETURN
      END
```

CHAPTER 3   Calculation of Element and Structural Stiffness Matrices

In the stiffness method of analysis, the most important property used is the element stiffness matrix. As shown in Chapter 1, the element stiffness matrix is different for different types of element, and the structural stiffness matrix is assembled from the corresponding element stiffness matrices. In this chapter the subroutines used for the calculation of the element and structural stiffness matrices will be discussed.

3.1 THE SIZE OF THE ELEMENT STIFFNESS MATRIX

The size of the element stiffness matrix is given by NEVAB = NNODE.NDOFN where NNODE is equal to the number of nodes defining the element which for members in a skeletal structure is two and NDOFN is the number of degrees of freedom at a node. Table 3.1 shows NDOFN and NEVAB for several element types.

The element stiffness matrix is denoted by ESTIF(NEVAB,NEVAB) where the name ESTIF stands for <u>E</u>lement <u>STIF</u>fness matrix and NEVAB stands for <u>N</u>umber of <u>E</u>lement <u>VA</u>ria<u>B</u>les.

Table 3.1

NDOFN and NEVAB for several element types.

Type of element	Nodal displacements	NDOFN	NEVAB
2-D pin jointed	u, v	2	4
3-D pin jointed	u, v, w	3	6
2-D rigid jointed	u, v, θ	3	6
Plane grid without warping restraint	w, θ_x, θ_y	3	6
Plane grid with warping restraint	$w, \theta_x, \theta_y, \mathrm{d}\psi/\mathrm{d}x$	4	8

3.2 THE ELEMENT STIFFNESS MATRIX FOR A PIN-JOINTED ELEMENT

The element stiffness matrix for a pin-jointed element was derived in Chapter 1. Tables 3.2 and 3.3 show for easy reference the element stiffness matrices for 2-D and 3-D pin-jointed elements.

Table 3.2

Element stiffness matrix for a 2-D pin-jointed element.

$$
\begin{bmatrix} F_{x1} \\ F_{y1} \\ F_{x2} \\ F_{y2} \end{bmatrix} = \frac{AE}{L} \begin{bmatrix} l^2 & lm & -l^2 & -lm \\ & m^2 & -lm & -m^2 \\ & & l^2 & lm \\ \text{symmetric} & & & m^2 \end{bmatrix} \begin{bmatrix} u_1 \\ v_1 \\ u_2 \\ v_2 \end{bmatrix}
$$

Table 3.3

Element stiffness matrix for a 3-D pin-jointed element.

$$
\begin{bmatrix} F_{x1} \\ F_{y1} \\ F_{z1} \\ F_{x2} \\ F_{y2} \\ F_{z2} \end{bmatrix} = \frac{AE}{L} \begin{bmatrix} l^2 & lm & ln & -l^2 & -lm & -ln \\ & m^2 & mn & -lm & -m^2 & -mn \\ & & n^2 & -ln & -mn & -n^2 \\ & & & l^2 & lm & ln \\ & & & & m^2 & mn \\ \text{symmetric} & & & & & n^2 \end{bmatrix} \begin{bmatrix} u_1 \\ v_1 \\ w_1 \\ u_2 \\ v_2 \\ w_2 \end{bmatrix}
$$

In Table 3.2 and 3.3, AE is the axial rigidity, L is the length of the element, (l, m, n) are the direction cosines of a vector directed from end 1 to end 2, and (u_1, v_1, w_1) and (u_2, v_2, w_2) are the translations in the x, y and z directions at ends 1 and 2 respectively, as shown in Fig. 3.1.

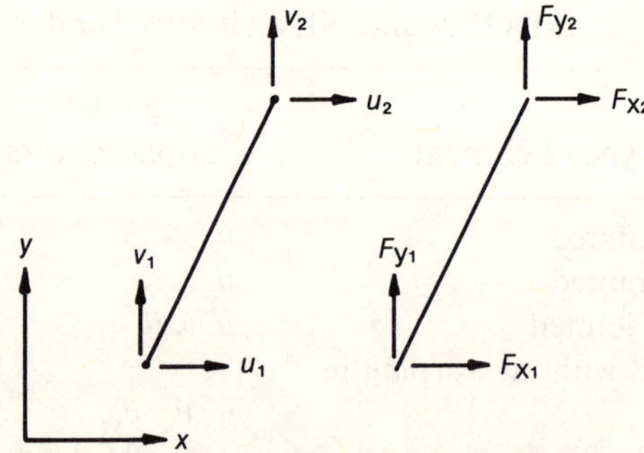

Fig. 3.1   Displacements and forces at the ends of a 2-D pin-jointed element.

3.3 BENDING STIFFNESS COEFFICIENTS FOR A GENERAL BENDING ELEMENT

Figure 3.2 shows the independent bending displacements for a beam element. As can be seen, allowing for the symmetry of stiffness coefficients as required by the Maxwell–Betti law, there are in all a maximum of ten stiffness coefficients as follows.

(a)   Pure rotation coefficients S_{11}, S_{12} and S_{22}.
(b)   Pure translation coefficients T_{11}, T_{12} and T_{22}.
(c)   Cross-rotation–translation coefficients Q_{11} Q_{12}, Q_{21} and Q_{22}.

The value of the stiffness coefficients depends on the type of bending element under consideration. In the programs in this book, only four types of element are considered. The values of the stiffness coefficients for the different types of elements are as follows.

(i) *Prismatic beam element.*   The stiffness coefficients for this type of element was derived in Chapter 1. Because of 'symmetry' of two ends, the properties at the two ends must be same. Therefore, $S_{11} = S_{22}$ and $T_{11} = T_{22}$. In addition, because the normal reactions at the two ends are equal and opposite, the following relationships are valid.

$$Q_{11} = Q_{12} = S_{11} + S_{12}, \quad Q_{22} = Q_{21} = S_{22} + S_{12}$$

$$T_{11} = T_{12} = Q_{11} + Q_{12}, \quad T_{22} = T_{12} = Q_{22} + Q_{21}$$

In other words, all the stiffness coefficients can be expressed in terms of S_{11} and S_{12} given by

$$S_{11} = \frac{4 + \beta}{1 + \beta}, \quad S_{12} = \frac{2 - \beta}{1 + \beta}$$

where $\beta = 12EI/GA_sL^2$, EI is the flexural rigidity and GA_s is the shear rigidity.

(ii) *A Beam on an elastic foundation.*   This type of element often forms part of a structure such as foundation mats, etc., and is particularly valuable when having to deal with 'structure–soil' interaction. In this case, symmetry demands that the properties at the two ends must be same. However, because the foundation provides 'lateral' support, the shear force in the element is not constant. There are therefore six independent stiffness coefficients as the following relationships are valid:

$$S_{11} = S_{22}, \, T_{11} = T_{22}, \, Q_{12} = Q_{21}, \, Q_{11} = Q_{22}$$

The independent stiffness coefficients are quoted below. Detailed derivation can be found in Appendix 2.

$$S_{11} = (SC - sc)D, \qquad S_{12} = (Cs - Sc)D,$$

$$Q_{11} = \lambda L(C^2 - c^2)D, \qquad Q_{12} = \lambda L(2Ss)D$$

$$T_{11} = 2(\lambda L)^2(SC + sc)D, \quad T_{12} = 2(\lambda L)^2(Sc + Cs)D$$

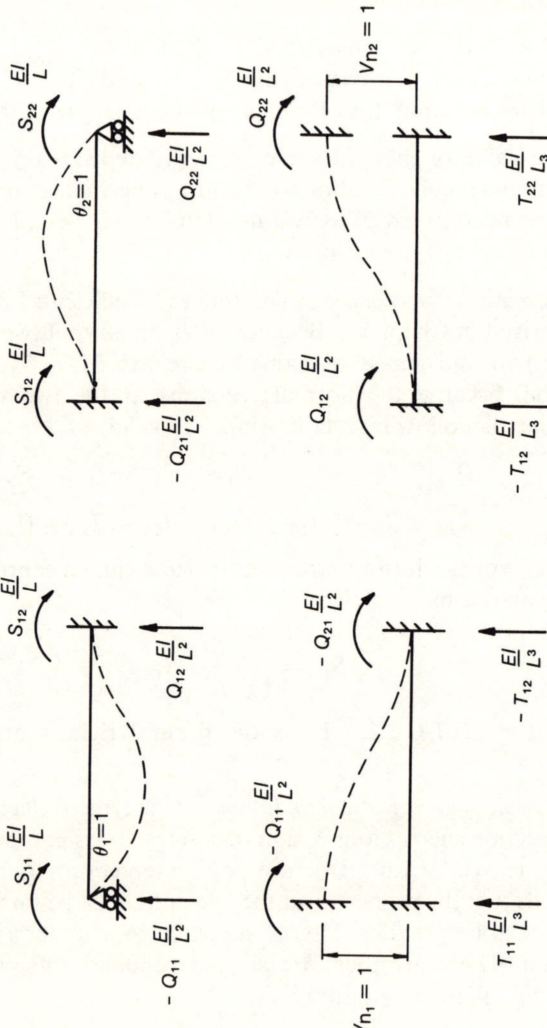

Fig. 3.2   Independent displacements at the ends of a beam element and the associated forces at the ends.

where $D = 2\lambda L/(S^2 - s^2)$, $S = \sinh \lambda L$, $C = \cosh \lambda L$, $s = \sin \lambda L$, $c = \cos \lambda L$, $\lambda = \sqrt{\sqrt{(K/4EI)}}$, K is the subgrade modulus and EI is the flexural rigidity.

(iii) *A prismatic beam with semi-rigid ends.*   This type of element is often used to simulate the deformation at the joints in bolted steel structures. The flexibility of joints is modelled by including springs at the ends of the beam elements. This type of element is also of use when analysing flat slabs using the procedcure of the American Concrete Institute. Because the springs at the two ends may not be same, the element lacks symmetry. However, the shear force in the beam is constant and therefore the following equations need to be satisfied:

$$Q_{11} = Q_{12} = S_{11} + S_{12}, \quad Q_{22} = Q_{21} = S_{22} + S_{12}$$

$$T_{11} = T_{12} = Q_{11} + Q_{12}, \quad T_{12} = T_{22} = Q_{21} + Q_{22}$$

The three independent stiffness coefficients are given below. Detailed derivation is given in Appendix 3.

$$S_{11} = \frac{4 + \beta + 12\alpha_2}{D}, \quad S_{22} = \frac{4 + \beta + 12\alpha_1}{D}, \quad S_{12} = \frac{2 - \beta}{D}$$

where

$$D = [(1 + \beta) + (4 + \beta)(\alpha_1 + \alpha_2) + 12\alpha_1\alpha_2], \quad \alpha_1 = EI/k_1L, \quad \alpha_2 = EI/k_2L,$$

k_1 and k_2 are stiffness of springs at ends 1 and 2 respectively, and EI and β are as defined before.

(iv) *A nonprismatic beam element.*   This is the most general type of element. In the programs in this book it is necessary to input the stiffness coefficients as data. Some useful information on nonprismatic beams can be found in the book by Gere [1] and the publication by the Portland Cement Association, [2].

3.4   THE ELEMENT STIFFNESS MATRIX FOR A TORSION ELEMENT

As discussed in Chapter 1, two types of torsion element are considered in this book, i.e. Saint Venant's or free warping torsion and nonuniform or warping restraint torsion. The stiffness matrix for the free warping case was derived in Chapter 1. The case of restrained warping is discussed and the necessary stiffness coefficients are derived in Appendix 4.

3.5   THE GENERAL ELEMENT STIFFNESS MATRIX FOR A 2-D RIGID-JOINTED STRUCTURE AND PLANE GRIDS

As explained in Chapter 1, the stiffness matrix for an element in a 2-D rigid-jointed structure is a combination of bending and axial stiffness terms while for a plane grid element it is a combination of bending and torsion stiffness

Table 3.4

The element stiffness matrix for a 2-D rigid-jointed member (see Fig. 3.3).

$$
\begin{bmatrix} F_{x1} \\ F_{y1} \\ M_1 \\ F_{x2} \\ F_{y2} \\ M_2 \end{bmatrix} = \frac{EI}{L}
\begin{bmatrix}
\dfrac{T_{11}m^2 + \alpha l^2}{L^2} & \dfrac{(-T_{11} + \alpha)lm}{L^2} & \dfrac{Q_{11}m}{L} & \dfrac{-(T_{12}m^2 + \alpha l^2)}{L^2} & \dfrac{(T_{12} - \alpha)lm}{L^2} & \dfrac{Q_{21}m}{L} \\[2ex]
 & \dfrac{T_{11}l^2 + \alpha m^2}{L^2} & \dfrac{-Q_{11}l}{L} & \dfrac{(T_{12} - \alpha)lm}{L^2} & \dfrac{-(T_{12}l^2 + \alpha m^2)}{L^2} & \dfrac{-Q_{21}l}{L} \\[2ex]
 & & S_{11} & \dfrac{-Q_{12}m}{L} & \dfrac{Q_{12}l}{L} & S_{12} \\[2ex]
 & & & \dfrac{T_{22}m^2 + \alpha l^2}{L^2} & \dfrac{(-T_{22} + \alpha)lm}{L^2} & \dfrac{-Q_{22}m}{L} \\[2ex]
 & & & & \dfrac{T_{22}l^2 + \alpha m^2}{L^2} & \dfrac{Q_{22}l}{L} \\[2ex]
\text{symmetric} & & & & & S_{22}
\end{bmatrix}
\begin{bmatrix} u_1 \\ v_1 \\ \theta_1 \\ u_2 \\ v_2 \\ \theta_2 \end{bmatrix}
$$

$$\alpha = \frac{AE}{EI} L^2$$

Table 3.5

The element stiffness matrix for a plane grid element with warping effects ignored (see Fig. 3.4).

$$
\begin{bmatrix} F_{z1} \\ M_{x1} \\ M_{y1} \\ F_{z2} \\ M_{x2} \\ M_{y2} \end{bmatrix} = \frac{EI}{L}
\begin{bmatrix}
\dfrac{T_{11}}{L^2} & \dfrac{Q_{11}m}{L} & \dfrac{-Q_{11}l}{L} & \dfrac{-T_{12}}{L^2} & \dfrac{Q_{21}m}{L} & \dfrac{-Q_{21}l}{L} \\[2ex]
 & S_{11}m^2 + \alpha l^2 & -(S_{11} - \alpha)lm & \dfrac{-Q_{12}m}{L} & S_{12}m^2 - \alpha l^2 & -(S_{12} + \alpha)lm \\[2ex]
 & & S_{11}l^2 + \alpha m^2 & \dfrac{Q_{12}l}{L} & -(S_{12} + \alpha)lm & S_{12}l^2 - \alpha m^2 \\[2ex]
 & & & \dfrac{T_{22}}{L^2} & \dfrac{-Q_{22}m}{L} & \dfrac{Q_{22}l}{L} \\[2ex]
\text{symmetric} & & & & S_{22}m^2 + \alpha l^2 & -(S_{22} - \alpha)lm \\[2ex]
 & & & & & S_{22}l^2 + \alpha m^2
\end{bmatrix}
\begin{bmatrix} w_1 \\ \theta_{x1} \\ \theta_{y1} \\ w_2 \\ \theta_{x2} \\ \theta_{y2} \end{bmatrix}
$$

$$
\alpha = \frac{GJ}{EI}
$$

Table 3.6

The element stiffness matrix for a plane grid element including the effects of warping restraint.

$$
\begin{bmatrix} F_{z1} \\ M_{x1} \\ M_{y1} \\ B_1 \\ F_{z2} \\ M_{x2} \\ M_{y2} \\ B_2 \end{bmatrix} = \frac{EI}{L}
\begin{bmatrix}
\dfrac{T_{11}}{L^2} & \dfrac{Q_{11}m}{L} & \dfrac{-Q_{11}l}{L} & 0.0 & \dfrac{-T_{12}}{L^2} & \dfrac{Q_{21}m}{L} & \dfrac{-Q_{21}l}{L} & 0.0 \\[2ex]
 & S_{11}m^2 + \alpha_{11}l^2 & (-S_{11} + \alpha_{11})lm & \alpha_{12}l & \dfrac{-Q_{12}m}{L} & S_{12}m^2 - \alpha_{11}l^2 & -(S_{12} + \alpha_{11})lm & \alpha_{12}l \\[2ex]
 & & S_{11}l^2 + \alpha_{11}m^2 & \alpha_{12}m & \dfrac{Q_{12}l}{L} & -(S_{12} + \alpha_{11})lm & S_{12}l^2 - \alpha_{11}m^2 & \alpha_{12}m \\[2ex]
 & & & \alpha_{22} & 0.0 & -\alpha_{22}l & -\alpha_{12}m & \alpha_{24} \\[2ex]
 & & & & \dfrac{T_{22}}{L^2} & \dfrac{-Q_{22}m}{L} & \dfrac{Q_{22}l}{L} & 0.0 \\[2ex]
 & & & & & S_{22}m^2 + \alpha_{11}l^2 & (-S_{22} + \alpha_{12})lm & -\alpha_{12}l \\[2ex]
 & \text{symmetric} & & & & & S_{22}l^2 + \alpha_{11}m^2 & -\alpha_{12}m \\[2ex]
 & & & & & & & \alpha_{22}
\end{bmatrix}
\begin{bmatrix} w_1 \\ \theta_{x1} \\ \theta_{y1} \\ \left(\dfrac{d\psi}{dx}\right)_1 \\ w_2 \\ \theta_{x2} \\ \theta_{y2} \\ \left(\dfrac{d\psi}{dx}\right)_2 \end{bmatrix}
$$

$$\alpha = \frac{GJ}{EI}, \quad \alpha_{11} = \alpha TW_{11}, \quad \alpha_{12} = \alpha TW_{12}, \quad \alpha_{22} = \alpha TW_{22}, \quad \alpha_{24} = \alpha TW_{24} \text{ (see Appendix A4)}$$

where B_1 and B_2 are the bimoments at ends 1 and 2 respectively, and $(d\psi/dx)_1$ and $(d\psi/dx)_2$ are the "rotations" of the flanges at end 1 and end 2 respectively (the rotation about the z axis is given by $\theta_z = (d\psi/dx)$-distance between flanges).

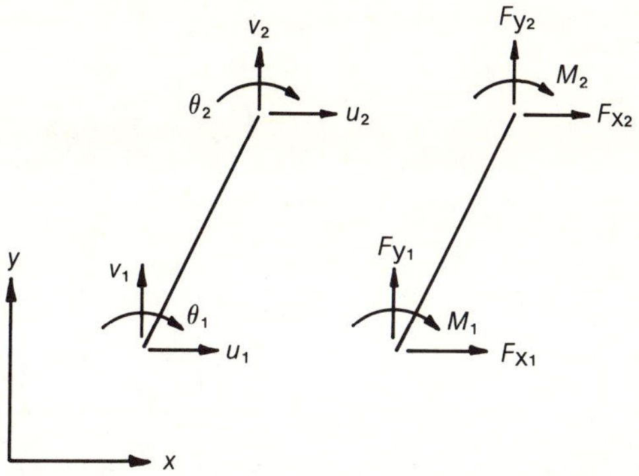

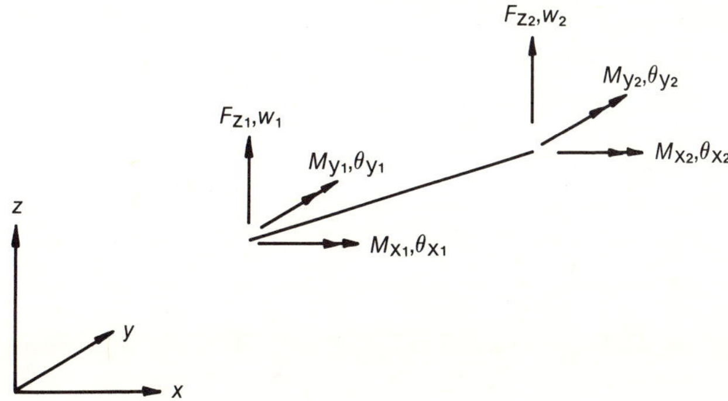

Fig. 3.3   Displacements and forces at the ends of a 2-D rigid-jointed frame element.

coefficients. It should be noted that, when the effects of warping restraint are included in a plane grid element, then the bending terms will remain the same as for the free warping case but the contributions from torsional deformation will be different as explained in Appendix 4. Tables 3.4, 3.5 and 3.6 show the element stiffness matrix for various types of element. Figures 3.3 and 3.4 show the general nodal displacements at the ends of a 2-D rigid-jointed element and a plane grid element respectively.

3.6   ASSEMBLING THE STRUCTURAL STIFFNESS MATRIX

Having calculated the element stiffness matrix, the next step in the stiffness method is to assemble the structural stiffness matrix. As explained in Chapter 1, the assembling of the structural stiffness matrix is a process of building up the equations of equilibrium at the joints of the structure. The structural stiffness matrix is stored in an array GSTIF which stands for <u>G</u>lobal <u>STIF</u>fness matrix. Since the structural stiffness matrix is symmetric, at least for linear and quasi-nonlinear analysis, it is sufficient to store only one symmetrical half of the stiffness matrix. In addition, because the structural stiffness matrix is highly banded as shown in Fig. 3.5(a), the symmetrical half of the stiffness matrix is stored in rectangular form with the width of the

Fig. 3.4   Forces and displacements at the ends of a plane grid element.

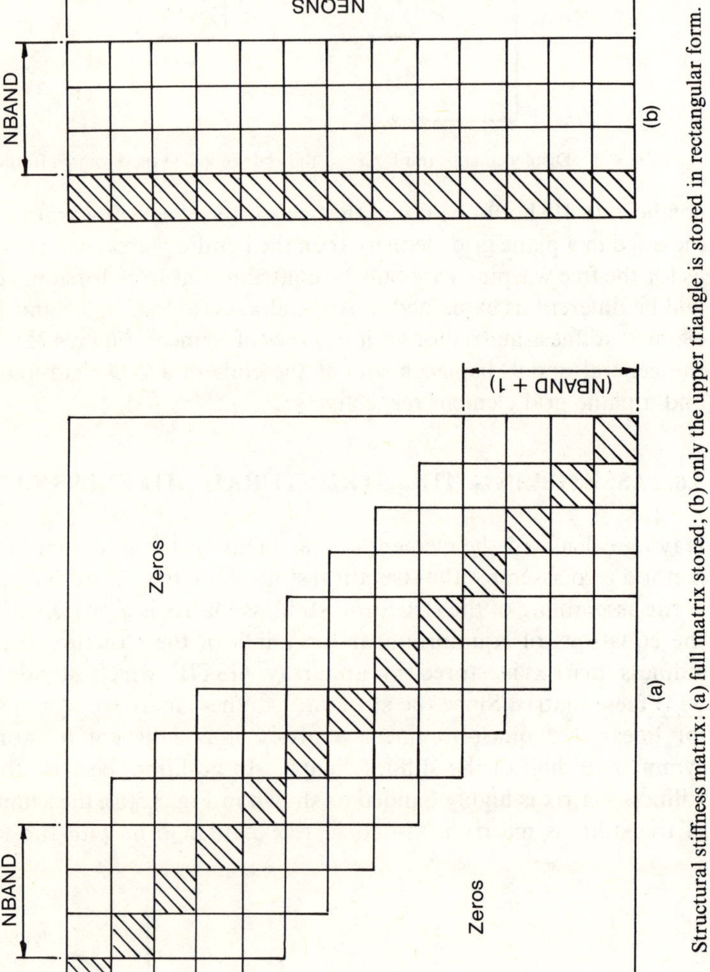

Fig. 3.5   Structural stiffness matrix: (a) full matrix stored; (b) only the upper triangle is stored in rectangular form.

rectangle equal to the half-bandwidth and the diagonal terms in the first column as shown in Fig. 3.5(b). The steps involved in assembling the structural stiffness matrix was explained in Chapter 1 and can be summarized as follows.

(1) Choose an element IELEM. The information about the element is stored in arrays LNODS(NELEM,NNODE) and MATNO-(NELEM,2).

(2) The group number of the element is given by LPROP= MATNO(IELEM,1). The name LPROP stands for e$\underline{L}$ment $\underline{PROP}$erty number. The node numbers at the ends of the element are given by

$$\text{node at end } 1 = \text{IEND} = \text{LNODS(IELEM,1)}$$
$$\text{node at end } 2 = \text{JEND} = \text{LNODS(IELEM,2)}$$

The coordinates of nodes IEND and JEND are given by

$$x_1 = \text{COORD(IEND,1)}, \; y_1 = \text{COORD(IEND,2)}$$
$$x_2 = \text{COORD(JEND,1)}, \; y_2 = \text{COORD(JEND,2)}$$

The length of element is given by

$$L = \sqrt{[(x_2 - x_1)^2 + (y_2 - y_1)^2]}$$

The direction cosines are given by

$$l = \frac{x_2 - x_1}{L}, \quad m = \frac{y_2 - y_1}{L}$$

(3) Since the group number LPROP of the element is known, then its geometrical and elastic properties can be obtained from the array PROPS(NMATS,NPROP + ?). Having calculated the direction cosines (l, m) and the geometrical and elastic properties of the element, the element stiffness matrix can be calculated and stored in the temporary array ESTIF(NEVAB,NEVAB).

(4) Having calculated the array ESTIF, the next step is to add it as appropriate to the structural stiffness matrix GSTIF. In order to do this, the first step is to determine the node freedom numbers corresponding to the nodes IEND and JEND at the ends of the element. The corresponding node freedom numbers are stored in a temporary array MEMDIS(NEVAB). As an example, if we consider the element of a 2-D rigid-jointed structure, then the elements of MEMDIS will be as follows:

$$\text{MEMDIS(1)} = \text{NODFRE(IEND,1)} = \text{node freedom number of } u_1$$
$$\text{MEMDIS(2)} = \text{NODFRE(IEND,2)} = \text{node freedom number of } v_2$$
$$\text{MEMDIS(3)} = \text{NODFRE(IEND,3)} = \text{node freedom number of } \theta_1$$
$$\text{MEMDIS(4)} = \text{NODFRE(JEND,1)} = \text{node freedom number of } u_2$$
$$\text{MEMDIS(5)} = \text{NODFRE(JEND,2)} = \text{node freedom number of } v_2$$
$$\text{MEMDIS(6)} = \text{NODFRE(JEND,3)} = \text{node freedom number of } \theta_2$$

(5)   If the structural stiffness matrix GSTIF is used to store the whole of the structural stiffness matrix, then an element ESTIF(IEVAB,JEVAB) is added to the corresponding element GSTIF(MEMDIS(IEVAB), MEMDIS(JEVAB)) as explained in Chapter 1. However, since GSTIF stores only a symmetrical half of the structural stiffness matrix and that too in a rectangular form, we have to appreciate that all the elements in any row are shifted to the left by (row number − 1) places. Therefore an element ESTIF(IEVAB,JEVAB) is added to the element GSTIF(MEMDIS(IEVAB), (MEMDIS(JEVAB) − (MEMDIS(IEVAB) − 1)). It is understood that, if either MEMDIS(IEVAB) or MEMDIS(JEVAB) is zero, then no addition is done because those displacement freedoms are restrained.

(6)   Repeat the above steps for all the element.

3.7   SUBROUTINES STIFFPJ, STIFFRJ AND STIFFGRID

There are three subroutines used for the calculation of the element stiffness matrix and the assembling of the structural stiffness matrix. The subroutines are discussed briefly below.

(1)   STIFFPJ. As the name implies, it is used for pin-jointed structures.
(2)   STIFFRJ. This is used for rigid-jointed structures. The type of beam element is stored in the second column of the array MATNO-(NELEM,2). The element stiffness matrix depends on the type of bending element, i.e. whether it is an orthodox prismatic element, a beam or an elastic foundation, etc.
(3)   STIFFGRID. This is used for plane grid structures. The discriminator IWARP is used to distinguish between the inclusion or ignoring of warping restraint.

3.8   REFERENCES

[1]   J. M. Gere, *Moment Distribution*, Van Nostrand, 1964.

[2]   *Beam Factors and Moment Coefficients for Members of Variable Section*, Portland Cement Association, Chicago.

```fortran
C
C**********************************************
C
      SUBROUTINE NULLGSTIF
C
C***THIS SUBROUTINE ZEROS THE STRUCTURAL
C   STIFFNESS MATRIX
C
**********************************************
*                                            *
*    USE APPROPRIATE COMMON BLOCKS HERE !!!   *
*                                            *
**********************************************
C
      DO 10 IEQNS=1,NEQNS
      DO 20 JCOLMS=1,NBAND+1
   20 GSTIF(IEQNS,JCOLMS)=0.0
   10 CONTINUE
      RETURN
      END
```

```fortran
C
C**********************************************
C
      SUBROUTINE STIFFPJ(NODFRE,LNODS,MATNO)
C
C***THIS SUBROUTINE COMPUTES THE ELEMENT
C   STIFFNESS MATRIX AND ASSEMBLES THE
C   STRUCTURAL STIFFNESS MATRIX IN BANDED FORMAT
C   WITH THE DIAGONAL TERM IN THE FIRST COLUMN
C   FOR 2-D or 3-D TRUSS
C
      COMMON/CONTRO/NPOIN,NELEM,NNODE,NDOFN,NDIME,
     @NPROP,NMATS,NRESND,NVFIX,NEVAB,NFAIL,NOPTN,
     @NEQNS,NBAND,ITEMP
      COMMON/LGDATA/COORD(100,3),PROPS(10,3)
     @PRESC(80,3),TEMP(100),NDFFIX(80),
     @NDVFIX(80),IFPRE(80,3),ASDIS(400),
     @GSTIF(100,30),GLOAD(100)
      DIMENSION ESTIF(6,6),MEMDIS(6),
     @NODFRE(100,3),LNODS(100,2),MATNO(100,2)
C
C***CYCLE ON ALL THE ELEMENTS. CALCULATE THE
C   ELEMENT STIFFNESS MATRIX AND ADD IT TO THE
C   STRUCTURAL STIFFNESS MATRIX.
C
      DO 10 IELEM=1,NELEM
      LPROP=MATNO(IELEM,1)
C
C*** CALCULATE THE PROPERTIES OF ELEMENTS.
C
      AREA=PROPS(LPROP,2)
      YOUNG=PROPS(LPROP,1)
C
C***CALCULATE THE NODES CORRESPONDING TO THE TWO
C   ENDS OF THE ELEMENT
C
      IEND=LNODS(IELEM,1)
      JEND=LNODS(IELEM,2)
      IF(NDIME.EQ.3) GOTO 15
```

```
C
C***ELEMENT STIFFNESS MATRIX FOR A 2-D PIN-
C    JOINTED ELEMENT.
C
C
C***CALCULATE THE COORDINATES OF THE NODES AT THE
C    ENDS OF THE ELEMENT
C
      XPROJ=COORD(JEND,1)-COORD(IEND,1)
      YPROJ=COORD(JEND,2)-COORD(IEND,2)
      SPAN=SQRT(XPROJ*XPROJ+YPROJ*YPROJ)
      XL=XPROJ/SPAN
      YM=YPROJ/SPAN
      AEBYL=AREA*YOUNG/SPAN
C
C***CALCULATE THE ELEMENTS OF THE UPPER TRIANGLE
C    OF THE STIFFNESS MATRIX FOR A 2-D PIN-
C    JOINTED ELEMENT.
C
      ESTIF(1,1)=AEBYL*XL*XL
      ESTIF(1,2)=AEBYL*XL*YM
      ESTIF(2,2)=AEBYL*YM*YM
      ESTIF(1,3)=-ESTIF(1,1)
      ESTIF(1,4)=-ESTIF(1,2)
      ESTIF(2,3)=-ESTIF(1,2)
      ESTIF(2,4)=-ESTIF(2,2)
      ESTIF(3,3)=ESTIF(1,1)
      ESTIF(3,4)=ESTIF(1,2)
      ESTIF(4,4)=ESTIF(2,2)
      GOTO 25
C
C***ELEMENT STIFFNESS MATRIX FOR A 3-D PIN-
C    JOINTED ELEMENT.
C
   15 XPROJ=COORD(JEND,1)-COORD(IEND,1)
      YPROJ=COORD(JEND,2)-COORD(IEND,2)
      ZPROJ=COORD(JEND,3)-COORD(IEND,3)
      SPAN=SQRT(XPROJ*XPROJ+YPROJ*YPROJ+
     @ZPROJ*ZPROJ)
      XL=XPROJ/SPAN
      YM=YPROJ/SPAN
      ZN=ZPROJ/SPAN
      AEBYL=AREA*YOUNG/SPAN
      ESTIF(1,1)=AEBYL*XL*XL
      ESTIF(1,2)=AEBYL*XL*YM
      ESTIF(1,3)=AEBYL*XL*ZN
      ESTIF(1,4)=-ESTIF(1,1)
      ESTIF(1,5)=-ESTIF(1,2)
      ESTIF(1,6)=-ESTIF(1,3)
      ESTIF(2,2)=AEBYL*YM*YM
      ESTIF(2,3)=AEBYL*YM*ZN
      ESTIF(2,4)=-ESTIF(1,2)
      ESTIF(2,5)=-ESTIF(2,2)
      ESTIF(2,6)=-ESTIF(2,3)
      ESTIF(3,3)=AEBYL*ZN*ZN
      ESTIF(3,4)=-ESTIF(1,3)
      ESTIF(3,5)=-ESTIF(2,3)
      ESTIF(3,6)=-ESTIF(3,3)
      ESTIF(4,4)=ESTIF(1,1)
      ESTIF(4,5)=ESTIF(1,2)
```

```
            ESTIF(4,6)=ESTIF(1,3)
            ESTIF(5,5)=ESTIF(2,2)
            ESTIF(5,6)=ESTIF(2,3)
            ESTIF(6,6)=ESTIF(3,3)
C
C***FILL UP THE SYMMETRICAL LOWER HALF OF THE
C   ESTIF MATRIX
C
   25 DO 20 IEVAB=2,NEVAB
      IEVAB1=IEVAB-1
      DO 20 JEVAB=1,IEVAB1
   20 ESTIF(IEVAB,JEVAB)=ESTIF(JEVAB,IEVAB)
C
C***CALCULATE THE NODE FREEDOMS CORRESPONDING
C   TO THE NODES AT THE ENDS OF THE ELEMENT AND
C   STORE IT IN THE ARRAY MEMDIS(NDOFN).
C
      DO 30 IDOFN=1,NDOFN
      MEMDIS(IDOFN)=NODFRE(IEND,IDOFN)
   30 MEMDIS(IDOFN+NDOFN)=NODFRE(JEND,IDOFN)
C
C***ASSEMBLE THE STRUCTURAL STIFFNESS MATRIX IN
C   THE BANDED FORM WITH THE DIAGONAL ELEMENTS
C   IN THE FIRST COLUMN
C
      DO 40 IEVAB=1,NEVAB
      IF(MEMDIS(IEVAB).EQ.0) GOTO 40
      DO 40 JEVAB=1,NEVAB
      IF(MEMDIS(JEVAB).EQ.0) GOTO 40
C
C***CALCULATE THE COLUMN POSITION IN THE BANDED
C   FORMAT.
C
      NEWCOL=MEMDIS(JEVAB)-MEMDIS(IEVAB)+1
      IF(NEWCOL.LE.0) GOTO 40
      GSTIF(MEMDIS(IEVAB),NEWCOL)=GSTIF(MEMDIS(IEVAB),
     @NEWCOL)+ESTIF(IEVAB,JEVAB)
   40 CONTINUE
   10 CONTINUE
      RETURN
      END
```

```
C
C*********************************************
C
      SUBROUTINE STIFFRJ(NODFRE,LNODS,MATNO)
C
C***THIS SUBROUTINE COMPUTES THE ELEMENT
C   STIFFNESS MATRIX AND ASSEMBLES THE
C   STRUCTURAL STIFFNESS MATRIX IN BANDED FORMAT
C   WITH THE DIAGONAL TERM IN THE FIRST COLUMN
C   FOR A 2-D RIGID-JOINTED PLANE FRAME
C
      COMMON/CONTRO/NPOIN,NELEM,NNODE,NDOFN,
     @NDIME,NPROP,NMATS,NRESND,NVFIX,NEVAB,
     @NFAIL,NLODEL,NEQNS,NBAND,ITEMP
      COMMON/LGDATA/COORD(100,2),PROPS(20,11),
     @PRESC(80,3),TEMP(100,2),NDFFIX(80),NDVFIX(80),
     @IFPRE(80,3),FORSEL(80,4),LODEL(80),
     @ASDIS(800),GSTIF(100,30),GLOAD(100)
```

```fortran
      DIMENSION ESTIF(6,6),MEMDIS(6),
     @NODFRE(100,3),LNODS(100,2),MATNO(100,2)
C
C*** CYCLE ON ALL ELEMENTS  AND CALCULATE ELEMENT
C    STIFFNESS MATRIX AND ADD TO THE STRUCTURAL
C    STIFFNESS MATRIX.
C
      DO 10 IELEM=1,NELEM
      LPROP=MATNO(IELEM,1)
      NTYPE=MATNO(IELEM,2)
C
C*** CALCULATE THE PROPERTIES OF ELEMENTS.
C
      YOUNG=PROPS(LPROP,1)
      SHEARM=PROPS(LPROP,2)
      YINERA=PROPS(LPROP,3)
      AREA=PROPS(LPROP,4)
      ZASF=PROPS(LPROP,5)
C
C***CALCULATE THE NODES CORRESPONDING TO THE TWO
C    ENDS OF THE ELEMENT, THE ELEMENT LENGTH AND
C    THE DIRECTION COSINES.
C
      IEND=LNODS(IELEM,1)
      JEND=LNODS(IELEM,2)
      XPROJ=COORD(JEND,1)-COORD(IEND,1)
      YPROJ=COORD(JEND,2)-COORD(IEND,2)
      SPAN=SQRT(XPROJ*XPROJ + YPROJ*YPROJ)
      XL=XPROJ/SPAN
      YM=YPROJ/SPAN
      EI=YOUNG*YINERA
      GA=SHEARM*AREA
      EIBYL=EI/SPAN
      EIBYL2=EIBYL/SPAN
      EIBYL3=EIBYL2/SPAN
      BETA=12.0*EIBYL2/GA*ZASF
      AEBYL=AREA*YOUNG/SPAN
      IF(NTYPE.EQ.2) GOTO 11
      IF(NTYPE.EQ.3) GOTO 12
      IF(NTYPE.EQ.4) GOTO 13
C
C***BENDING STIFFNESS COEFFICIENTS  FOR AN
C    ORTHODOX BEAM ELEMENT
C
      S11=(4.0+BETA)/(1.0+BETA)
      S22=S11
      S12=(2.0-BETA)/(1.0+BETA)
      Q11=S11+S12
      Q22=Q11
      Q12=Q11
      Q21=Q11
      T11=2.0*Q11
      T22=T11
      T12=T11
      GOTO 14
C
C***BENDING STIFFNESS COEFFICIENTS OF A BEAM ON
C    ELASTIC FOUNDATION
C
   11 SUBGRM=PROPS(LPROP,NPROP+1)
```

```
      AMDAL=SPAN*SQRT(SQRT(SUBGRM/(4.0*EI)))
      SH=SINH(AMDAL)
      CH=COSH(AMDAL)
      S=SIN(AMDAL)
      C=COS(AMDAL)
      FACTO1=AMDAL/(SH*SH-S*S)
      FACTO2=FACTO1*AMDAL
      FACTO3=FACTO2*AMDAL
      S11=2.0*(SH*CH-S*C)*FACTO1
      S22=S11
      S12=2.0*(CH*S-SH*C)*FACTO1
      T11=4.0*(SH*CH+S*C)*FACTO3
      T22=T11
      T12=4.0*(CH*S+SH*C)*FACTO3
      Q11=2.0*(CH*CH-C*C)*FACTO2
      Q22=Q11
      Q12=4.0*(SH*S)*FACTO2
      Q21=Q12
      GOTO 14
C
C***BENDING STIFFNESS COEFFICIENTS  FOR A
C    SEMI-RIGID BEAM
C
   12 SPRNG1=PROPS(LPROP,NPROP+2)
      SPRNG2=PROPS(LPROP,NPROP+3)
      ALPHA1=EIBYL/SPRNG1
      ALPHA2=EIBYL/SPRNG2
      DENOM=1.0+BETA+(4.0+BETA)*(ALPHA1+ALPHA2)
     @+12.0*ALPHA1*ALPHA2
      S11=(4.0+BETA+12.0*ALPHA2)/DENOM
      S12=(2.0-BETA)/DENOM
      S22=(4.0+BETA+12.0*ALPHA1)/DENOM
      Q11=S11+S12
      Q12=Q11
      Q22=S12+S22
      Q21=Q22
      T11=Q11+Q22
      T22=T11
      T12=T11
      GOTO 14
C
C***BENDING STIFFNESS COEFFICIENTS FOR A
C    NONPRISMATIC ELEMENT FOR WHICH THE BENDING
C    STIFFNESS COEFFICIENTS ARE READ AS DATA
C
   13 WRITE(6,910)
  910 FORMAT(//,'BENDING STIFFNESS COEFFICIENTS
     @FOR A NONPRISMATIC BEAM ELEMENT')
      READ(5,900)S11,S12,S22
      READ(5,900)T11,T12,T22
      READ(5,900)Q11,Q12,Q22,Q21
      WRITE(6,900)S11,S12,S22
      WRITE(6,900)T11,T12,T22
      WRITE(6,900)Q11,Q12,Q22,Q21
   14 T11=T11*EIBYL3
      T22=T22*EIBYL3
      T12=T12*EIBYL3
      S11=S11*EIBYL
      S22=S22*EIBYL
      S12=S12*EIBYL
```

```fortran
      Q11=Q11*EIBYL2
      Q22=Q22*EIBYL2
      Q12=Q12*EIBYL2
      Q21=Q21*EIBYL2
C
C***CALCULATE THE ELEMENT STIFFNESS MATRIX.
C   ONLY THE UPPER TRIANGLE IS CALCULATED.
C
      ESTIF(1,1)= T11*YM*YM + AEBYL*XL*XL
      ESTIF(1,2)=-T11*XL*YM + AEBYL*XL*YM
      ESTIF(1.3)=Q11*YM
      ESTIF(1,4)=-T12*YM*YM - AEBYL*XL*XL
      ESTIF(1,5)= T12*XL*YM - AEBYL*XL*YM
      ESTIF(1,6)=Q21*YM
      ESTIF(2,2)= T11*XL*XL + AEBYL*YM*YM
      ESTIF(2,3)=-Q11*XL
      ESTIF(2,4)= T12*XL*YM - AEBYL*XL*YM
      ESTIF(2,5)=-T12*XL*XL - AEBYL*YM*YM
      ESTIF(2,6)=-Q21*XL
      ESTIF(3,3)=S11
      ESTIF(3,4)=-Q12*YM
      ESTIF(3,5)=Q12*XL
      ESTIF(3,6)=S12
      ESTIF(4,4)= T22*YM*YM + AEBYL*XL*XL
      ESTIF(4,5)=-T22*XL*YM + AEBYL*XL*YM
      ESTIF(4,6)=-Q22*YM
      ESTIF(5,5)= T22*XL*XL + AEBYL*YM*YM
      ESTIF(5,6)=Q22*XL
      ESTIF(6,6)=S22
C
C***FILL UP THE SYMMETRICAL LOWER HALF OF THE
C   ESTIF MATRIX
C
      DO 20 IEVAB=2,NEVAB
      IEVAB1=IEVAB-1
      DO 20 JEVAB=1,IEVAB1
   20 ESTIF(IEVAB,JEVAB)=ESTIF(JEVAB,IEVAB)
C
C***CALCULATE THE NODE FREEDOMS CORRESPONDING
C   TO THE NODES AT THE ENDS OF THE ELEMENT.
C   THIS IS STORED IN ARRAY MEMDIS(NDOFN).
C
      DO 30 IDOFN=1,NDOFN
      MEMDIS(IDOFN)=NODFRE(IEND,IDOFN)
   30 MEMDIS(IDOFN+NDOFN)=NODFRE(JEND,IDOFN)
C
C***ASSEMBLE THE STRUCTURAL STIFFNESS MATRIX IN
C   THE BANDED FORM WITH THE DIAGONAL ELEMENTS
C   IN THE FIRST COLUMN
C
      DO 40 IEVAB=1,NEVAB
      IF(MEMDIS(IEVAB).EQ.0) GOTO 40
      DO 40 JEVAB=1,NEVAB
      IF(MEMDIS(JEVAB).EQ.0) GOTO 40
C
C***CALCULATE THE COLUMN POSITION IN THE BANDED
C   FORMAT.IN BANDED FORMAT THE LOWER TRIANGULAR
C   ELEMENTS ARE NOT STORED AND THE
C   UPPER TRIANGULAR ELEMENTS ARE SHIFTED TO THE
C   LEFT BY (ROW -1) POSITIONS.
```

```fortran
C
      NEWCOL=MEMDIS(JEVAB)-MEMDIS(IEVAB)+1
      IF(NEWCOL.LE.0) GOTO 40
      GSTIF(MEMDIS(IEVAB),NEWCOL)=GSTIF(MEMDIS
     @(IEVAB),NEWCOL)+ESTIF(IEVAB,JEVAB)
   40 CONTINUE
   10 CONTINUE
  900 FORMAT(4E14.6)
      RETURN
      END
```

```fortran
C
C***********************************************
C
      SUBROUTINE STIFFGRID(NODFRE,LNODS,MATNO)
C
C***THIS SUBROUTINE COMPUTES THE ELEMENT
C    STIFFNESS MATRIX AND ASSEMBLES THE STRUCTURAL
C    STIFFNESS MATRIX IN BANDED FORM WITH THE
C    DIAGONAL TERM IN THE FIRST COLUMN FOR A
C    PLANE GRID STRUCTURE WITH OR WITHOUT WARPING
C    RESTRAINT
C
      COMMON/CONTRO/NPOIN,NELEM,NNODE,NDOFN,
     @NDIME,NPROP,NMATS,NRESND,NVFIX,NEVAB,
     @NFAIL,NLODEL,NEQNS,NBAND,ITEMP,IWARP
      COMMON/LGDATA/COORD(100,2),PROPS(10,12),
     @PRESC(80,4),TEMP(100,2),NDFFIX(80),
     @NDVFIX(80),IFPRE(80,4),FORSEL(80,2),
     @LODEL(80),ASDIS(800),GSTIF(100,30),
     @GLOAD(100)
      DIMENSION ESTIF(8,8),MEMDIS(8),
     @NODFRE(100,4),LNODS(100,2),MATNO(100,2)
C
C***CYCLE ON ALL THE ELEMENTS, CALCULATE THE
C    ELEMENT STIFFNESS MATRIX AND ADD AS
C    APPROPRIATE TO THE STRUCTURAL STIFFNESS
C    MATRIX.
C
      DO 10 IELEM=1,NELEM
      LPROP=MATNO(IELEM,1)
      NTYPE=MATNO(IELEM,2)
C
C***CALCULATE THE PROPERTIES OF THE ELEMENT.
C
      YOUNG=PROPS(LPROP,1)
      SHEARM=PROPS(LPROP,2)
      YINERA=PROPS(LPROP,3)
      TINERA=PROPS(LPROP,4)
      AREA=PROPS(LPROP,5)
      ZASF=PROPS(LPROP,6)
C
C***CALCULATE THE NODES CORRESPONDING TO THE TWO
C    ENDS OF THE ELEMENT, THE ELEMENT LENGTH,
C    DIRECTION COSINES ETC.
C
      IEND=LNODS(IELEM,1)
      JEND=LNODS(IELEM,2)
      XPROJ=COORD(JEND,1)-COORD(IEND,1)
      YPROJ=COORD(JEND,2)-COORD(IEND,2)
```

```
      SPAN=SQRT(XPROJ*XPROJ + YPROJ*YPROJ)
      XL=XPROJ/SPAN
      YM=YPROJ/SPAN
      EI=YOUNG*YINERA
      GA=SHEARM*AREA
      EIBYL=EI/SPAN
      EIBYL2=EIBYL/SPAN
      EIBYL3=EIBYL2/SPAN
      BETA=12.0*EIBYL2/GA*ZASF
      GJBYL=TINERA*SHEARM/SPAN
      IF(NTYPE.EQ.2) GOTO 11
      IF(NTYPE.EQ.3) GOTO 12
      IF(NTYPE.EQ.4) GOTO 13
C
C***BENDING STIFFNESS COEFFICIENTS FOR AN
C    ORTHODOX BEAM ELEMENT
C
      S11=(4.0+BETA)/(1.0+BETA)
      S22=S11
      S12=(2.0-BETA)/(1.0+BETA)
      Q11=S11+S12
      Q22=Q11
      Q12=Q11
      Q21=Q11
      T11=2.0*Q11
      T22=T11
      T12=T11
      GOTO 14
C
C***BENDING STIFFNESS COEFFICIENTS OF A BEAM ON
C    ELASTIC FOUNDATION
C
   11 SUBGRM=PROPS(LPROP,NPROP+1)
      AMDAL=SPAN*SQRT(SQRT(SUBGRM/(4.0*EI)))
      SH=SINH(AMDAL)
      CH=COSH(AMDAL)
      S=SIN(AMDAL)
      C=COS(AMDAL)
      FACTO1=AMDAL/(SH*SH-S*S)
      FACTO2=FACTO1*AMDAL
      FACTO3=FACTO2*AMDAL
      S11=2.0*(SH*CH-S*C)*FACTO1
      S22=S11
      S12=2.0*(CH*S-SH*C)*FACTO1
      T11=4.0*(SH*CH+S*C)*FACTO3
      T22=T11
      T12=4.0*(CH*S+SH*C)*FACTO3
      Q11=2.0*(CH*CH-C*C)*FACTO2
      Q22=Q11
      Q12=4.0*(SH*S)*FACTO2
      Q21=Q12
      GOTO 14
C
C***BENDING STIFFNESS COEFFICIENTS FOR A
C    SEMI-RIGID BEAM
C
   12 SPRNG1=PROPS(LPROP,NPROP+2)
      SPRNG2=PROPS(LPROP,NPROP+3)
      ALPHA1=EIBYL/SPRNG1
      ALPHA2=EIBYL/SPRNG2
```

```
      DENOM=1.0+BETA+(4.0+BETA)*(ALPHA1+
     @ALPHA2)+12.0*ALPHA1*ALPHA2
      S11=(4.0+BETA+12.0*ALPHA2)/DENOM
      S12=(2.0-BETA)/DENOM
      S22=(4.0+BETA+12.0*ALPHA1)/DENOM
      Q11=S11+S12
      Q12=Q11
      Q22=S12+S22
      Q21=Q22
      T11=Q11+Q22
      T22=T11
      T12=T11
      GOTO 14
C
C***BENDING STIFFNESS COEFFICIENTS FOR A
C   NONPRISMATIC ELEMENT FOR WHICH THE BENDING
C   STIFFNESS COEFFICIENTS ARE READ AS DATA
C
   13 WRITE(6,910)
  910 FORMAT(/,'ROTATIONAL, TRANSLATIONAL AND
     @CROSS STIFFNESS COEFFICIENTS FOR A NON
     @PRISMATIC BEAM ELEMENT')
      READ(5,900)S11,S12,S22
      READ(5,900)T11,T12,T22
      READ(5,900)Q11,Q12,Q22,Q21
      WRITE(6,900)S11,S12,S22
      WRITE(6,900)T11,T12,T22
      WRITE(6,900)Q11,Q12,Q22,Q21
   14 CONTINUE
      T11=T11*EIBYL3
      T22=T22*EIBYL3
      T12=T12*EIBYL3
      S11=S11*EIBYL
      S22=S22*EIBYL
      S12=S12*EIBYL
      Q11=Q11*EIBYL2
      Q22=Q22*EIBYL2
      Q12=Q12*EIBYL2
      Q21=Q21*EIBYL2
C
C***THE STIFFNESS MATRIX FOR THE ELEMENT DEPENDS
C   ON WHETHER WARPING IS CONSIDERED OR NOT.
C
      IF(IWARP.EQ.1) GOTO 15
C
C***CALCULATE THE ELEMENT STIFFNESS MATRIX IN
C   THE CASE WHEN WARPING EFFECT IS IGNORED.
C   ONLY UPPER TRIANGLE IS CALCULATED.
C
      ESTIF(1,1)=T11
      ESTIF(1,2)=Q11*YM
      ESTIF(1,3)=-Q11*XL
      ESTIF(1,4)=-T12
      ESTIF(1,5)=Q21*YM
      ESTIF(1,6)=-Q21*XL
      ESTIF(2,2)= S11*YM*YM+GJBYL*XL*XL
      ESTIF(2,3)=-S11*XL*YM+GJBYL*XL*YM
      ESTIF(2,4)=-Q12*YM
      ESTIF(2,5)= S12*YM*YM-GJBYL*XL*XL
      ESTIF(2,6)=-S12*XL*YM-GJBYL*XL*YM
```

```
      ESTIF(3,3)= S11*XL*XL+GJBYL*YM*YM
      ESTIF(3,4)=Q12*XL
      ESTIF(3,5)=-S12*XL*YM-GJBYL*XL*YM
      ESTIF(3,6)= S12*XL*XL-GJBYL*YM*YM
      ESTIF(4,4)=T22
      ESTIF(4,5)=-Q22*YM
      ESTIF(4,6)=Q22*XL
      ESTIF(5,5)= S22*YM*YM+GJBYL*XL*XL
      ESTIF(5,6)=-S22*XL*YM+GJBYL*XL*YM
      ESTIF(6,6)= S22*XL*XL+GJBYL*YM*YM
      GOTO 16
C
C***CALCULATE THE ELEMENT STIFFNESS MATRIX WHEN
C   THE WARPING EFFECTS ARE CONSIDERED.
C
   15 WINERA=PROPS(LPROP,NPROP)
      GAMAL=SPAN*SQRT(SHEARM*TINERA/(YOUNG*
     @WINERA))
      SH=SINH(GAMAL)
      CH=COSH(GAMAL)
      D=SH*GAMAL-2.0*(CH - 1.0)
      GJBYLD=GJBYL/D
      TW11=GJBYLD*SH*GAMAL
      TW12=GJBYLD*(CH-1.0)*SPAN
      TW22=GJBYLD*(CH-SH/GAMAL)*SPAN*SPAN
      TW24=GJBYLD*(SH/GAMAL-1.0)*SPAN*SPAN
      ESTIF(1,1)=T11
      ESTIF(1,2)=Q11*YM
      ESTIF(1,3)=-Q11*XL
      ESTIF(1,4)=0.0
      ESTIF(1,5)=-T12
      ESTIF(1,6)=Q21*YM
      ESTIF(1,7)=-Q21*XL
      ESTIF(1,8)=0.0
      ESTIF(2,2)= S11*YM*YM+TW11*XL*XL
      ESTIF(2,3)=-S11*XL*YM+TW11*XL*YM
      ESTIF(2,4)=TW12*XL
      ESTIF(2,5)=-Q12*YM
      ESTIF(2,6)= S12*YM*YM-TW11*XL*XL
      ESTIF(2,7)=-S12*XL*YM-TW11*XL*YM
      ESTIF(2,8)=TW12*XL
      ESTIF(3,3)= S11*XL*XL+TW11*YM*YM
      ESTIF(3,4)=TW12*YM
      ESTIF(3,5)=Q12*XL
      ESTIF(3,6)=-S12*XL*YM-TW11*XL*YM
      ESTIF(3,7)= S12*XL*XL-TW11*YM*YM
      ESTIF(3,8)=TW12*YM
      ESTIF(4,4)=TW22
      ESTIF(4,5)=0.0
      ESTIF(4,6)=-TW12*XL
      ESTIF(4,7)=-TW12*YM
      ESTIF(4,8)=TW24
      ESTIF(5,5)=T22
      ESTIF(5,6)=-Q22*YM
      ESTIF(5,7)=Q22*XL
      ESTIF(5,8)=0.0
      ESTIF(6,6)= S22*YM*YM+TW11*XL*XL
      ESTIF(6,7)=-S22*XL*YM+TW11*XL*YM
      ESTIF(6,8)=-TW12*XL
      ESTIF(7,7)= S22*XL*XL+TW11*YM*YM
```

```
          ESTIF(7,8)=-TW12*YM
          ESTIF(8,8)=TW22
C
C***FILL UP THE SYMMETRICAL LOWER HALF OF THE
C    ESTIF MATRIX
C
    16 DO 20 IEVAB=2,NEVAB
          IEVAB1=IEVAB-1
          DO 20 JEVAB=1,IEVAB1
    20 ESTIF(IEVAB,JEVAB)=ESTIF(JEVAB,IEVAB)
C
C***CALCULATE THE NODE FREEDOMS CORRESPONDING TO
C    THE NODES AT THE ENDS OF THE ELEMENT. THIS
C    IS STORED IN THE ARRAY MEMDIS(NDOFN)
C
          DO 30 IDOFN=1,NDOFN
          MEMDIS(IDOFN)=NODFRE(IEND,IDOFN)
    30 MEMDIS(IDOFN+NDOFN)=NODFRE(JEND,IDOFN)
C
C***ASSEMBLE THE STRUCTURAL STIFFNESS MATRIX IN
C    THE BANDED FORM WITH THE DIAGONAL ELEMENTS
C    IN THE FIRST COLUMN
C
          DO 40 IEVAB=1,NEVAB
          IF(MEMDIS(IEVAB).EQ.0) GOTO 40
          DO 40 JEVAB=1,NEVAB
          IF(MEMDIS(JEVAB).EQ.0) GOTO 40
C
C***CALCULATE THE COLUMN POSITION IN THE BANDED
C    FORMAT. IN BANDED FORMAT THE LOWER TRIANGLE
C    IS NOT STORED.  THE UPPER TRIANGLE IS
C    STORED WITH THE DIAGONAL IN THE FIRST COLUMN.
C    THEREFORE ALL ELEMENTS IN A ROW ARE SHIFTED
C    TO THE LEFT BY (ROW - 1) POSITIONS.
C
          NEWCOL=MEMDIS(JEVAB)-MEMDIS(IEVAB)+1
          IF(NEWCOL.LE.0) GOTO 40
          GSTIF(MEMDIS(IEVAB),NEWCOL)=GSTIF(MEMDIS
         @(IEVAB),NEWCOL)+ESTIF(IEVAB,JEVAB)
    40 CONTINUE
    10 CONTINUE
   900 FORMAT(6E14.6)
          RETURN
          END
```

CHAPTER 4 Calculations of the Load Vector

After the structural stiffness matrix GSTIF is assembled, the next step in the stiffness method is the calculation of the load vector GLOAD. In general the static external loads acting on a structure can be due to the following causes.

(1) Loads applied at the nodes of the structure.
(2) Loads applied on the elements.
(3) Loads caused by partial restraint to free thermal expansion.

The load vector GLOAD, which stands for Global LOAD, is assembled from the effect of the above three types of loads.

4.1 LOADS APPLIED AT THE NODES

These loads are the easiest to handle. As the loads are applied directly at the nodes, all that is necessary is to associate the load at a node with the corresponding node freedom number in which direction it acts. The steps involved in assembling the vector GLOAD are as follows.

(1) Input NLODPT which stands for Number of LOaDed PoinTs. This shows how many nodes are loaded by concentrated loads. This will considerably reduce the amount of data input as data about only those nodes where loads act need to be input. If NLODPT is equal to zero, then skip this section.
(2) Input for all the NLODPT nodes, the node number LODPT which stands for LOaDed PoinT and the loads corresponding to the NDOFN degrees of freedom at the LODPT. The loads are stored in a temporary array FORSPT(NDOFN). FORSPT stands for FORce at a PoinT.
(3) The freedom numbers corresponding to LODPT are given by NODFRE(LODPT,NDOFN). The elements of the vector FORSPT(NDOFN) are added to the elements of the vector GLOAD(NODFRE(LODFT,NDOFN)) taking care to exclude those freedoms which are restrained, i.e. have zero freedom numbers.
(4) The above steps (2) and (3) are repeated for all the NLODPT nodes.

This completes the calculation of the load vector due to the loads applied directly at the joints.

4.2 LATERAL LOADS ACTING ON A BEAM ELEMENT

Consider the inclined beam element shown in Fig. 4.1. There are four types of lateral load that are considered in this book. The loads are positive in the positive direction of the coordinates. They are as follows.

(1) A uniformly distributed load of intensity q_x in the x direction. The total distributed load in the x direction is UDLXT = q_x|projection of the element on the y axis|.

(2) A uniformly distributed load of intensity q_y in the y direction. The total distributed load in the y direction is UDLYT = q_y|projection of the element on the x axis|.

(3) The concentrated load C_x at mid-span in the x direction.

(4) The concentrated load C_y at mid-span in the y direction.

The applied loads can be resolved into normal and tangenrial loads as shown in Fig. 4.2 as follows

(a) The total Uniformly Distributed Normal Load is

$$UDLN = (-UDLXT \cdot m + UDLYT \cdot l).$$

(b) The mid-span normal load is CN $= -C_x m + C_y l$

(c) Total TANgential COMponent of load TANCOM = (UDLXT + C_x)l + (UDLYT + C_y)m.

where l and m are the direction cosines of the line 1–2.

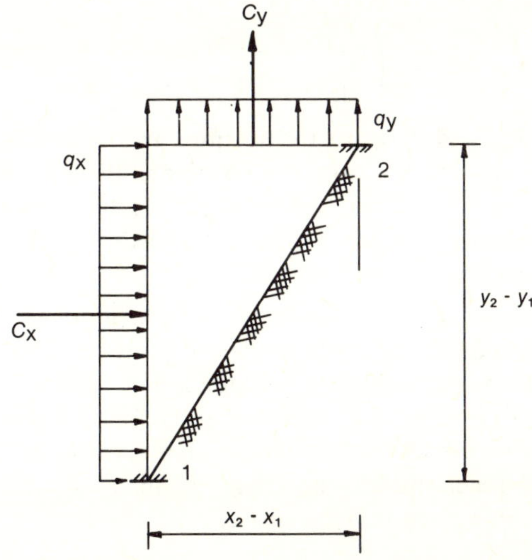

Fig. 4.1   Lateral forces on an element in a 2-D rigid-jointed structure.

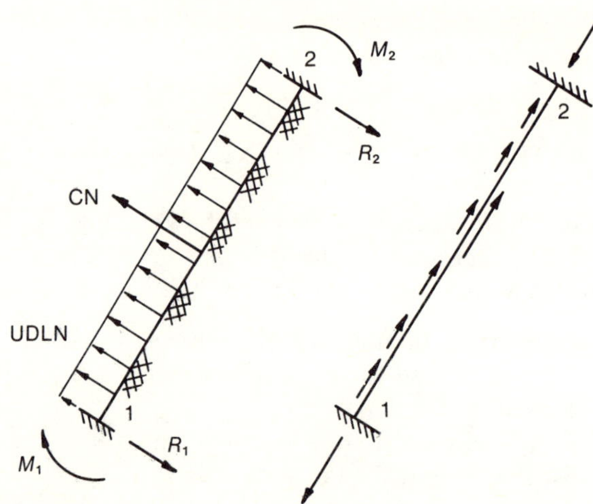

Fig. 4.2   Fixed end reactions due to normal and tangential applied load components.

4.3   FIXED END REACTIONS DUE TO LATERAL LOADS ON A BEAM ELEMENT

The fixed end reactions due to the loads on the beam depend upon the type of beam element that one is dealing with, i.e. whether it is an orthodox beam element, a beam on an elastic foundation, etc. The fixed end reactions as shown in Fig. 4.2 for the various types of element are calculated as follows.

(i) *An orthodox prismatic beam element.*   The reactions normal to the beam are $R_1 = R_2 = 0.5(\text{UDLN} + \text{CN})$. The reactions in the x and y directions and fixed end moments are given as shown in Fig. 4.3 by

$$R_{x1} = R_{x2} = -0.5(\text{UDLXT} + C_x), \quad R_{y1} = R_{y2} = -0.5(\text{UDLYT} + C_y)$$

$$M_1 = -M_2 = L\left(\frac{\text{UDLN}}{12.0} + \frac{\text{CN}}{8.0}\right)$$

where R_{x1} and R_{x2} are the rections in the x direction at ends 1 and 2 respectively. The reactions are positive in the positive x direction. Similarly, R_{y1} and R_{y2} are the reactions in the y direction and M_1 and M_2 are the fixed end moments at ends 1 and 2 respectively. The moments are positive clockwise.

(ii) *A prismatic beam element on an elastic foundation.*   The beam on an elastic foundation differs from the orthodox beam element because part of the lateral load is absorbed by the foundation 'springs'. Therefore the reactions at the ends tend to be smaller than those for an orthodox beam depending on the relative stiffness of the foundation and the beam. The following formulae give the reactions for the beam on an elastic foundation. The detailed derivation is given in Appendix 2.

$$M_1 = -M_2 = L\left(f_1\frac{\text{UDLN}}{12.0} + f_3\frac{\text{CN}}{8.0}\right)$$

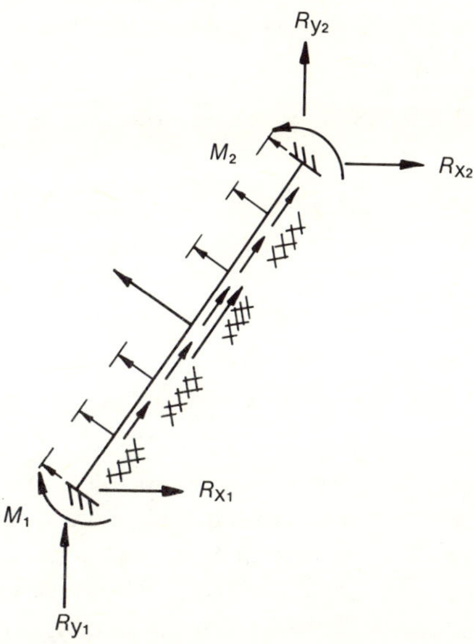

Fig. 4.3   Fixed end reactions in a 2-D rigid-joined structure element.

The normal reactions are

$$R_1 = R_2 = 0.5(f_2 \text{ UDLN} + f_4 \text{ CN})$$

where $f_1 = 6(\text{S-s})D$, $f_2 = 2\lambda L(\text{C} - \text{c})D$, $f_3 = 8\lambda L(\text{SH sh})D$, $f_4 = 2(\lambda L)^2(\text{SH ch} + \text{CH sh})D$, $D = 1.0/(\lambda L)^2(\text{S-s})$, $\text{S} = \sinh \lambda L$, $\text{C} = \cosh \lambda L$, $\text{s} = \sin \lambda L$, $\text{c} = \cos \lambda L$, $\text{SH} = \sinh 0.5\lambda L$, $\text{CH} = \cosh 0.5\lambda L$, $\text{sh} = \sin 0.5\lambda L$, $\text{ch} = \cos 0.5\lambda L$. The parameter λL which is a measure of the relative foundation stiffness and beam stiffness was defined in section 3.3 and is also explained in Appendix 2.

The reactions in the x and y directions are given by

$$R_{x1} = R_{x2} = -0.5 \cdot \text{TANCOM} \cdot l + R_1 m$$
$$R_{y1} = R_{y2} = -(0.5 \cdot \text{TANCOM}) + R_1 l \cdot m$$

(iii) *A prismatic beam with semi-rigid ends.* In this case the fixed end moments are always smaller than those of an orthodox beam element. The reactions are given by

$$M_1 = L\left(f_1\frac{\text{UDLN}}{12.0} + 2f_1\frac{\text{CN}}{8.0}\right), \quad M_2 = L\left(f_2\frac{\text{UDLN}}{12.0} + 2f_2\frac{\text{CN}}{8.0}\right)$$

where $f_1 = 3(1 + 6\alpha_2)D$, $f_2 = 3(1 + 6\alpha_1)D$, $D = 1.0/[(2 + 6\alpha_1)(2 + 6\alpha_2) - 1]$, and α_1 and α_2 are a measure of the relative beam stiffness to the stiffness of the 'springs' at the ends of the beam. The above formulae are derived in Appendix 3. From the expressions for the end moments, the normal reactions at the ends can be derived as

$$R_1 = 0.5(\text{UDLN} + \text{CN}) + \frac{M_1 + M_2}{L},$$

$$R_2 = 0.5(\text{UDLN} + \text{CN}) - \frac{M_1 + M_2}{l}$$

From the above normal reactions, the reactions in the x and y directions can be expressed as

$$R_{x1} = (-0.5 \cdot \text{TANCOM} \cdot l + R_1 m), \quad R_{x2} = (-0.5 \cdot \text{TANCOM} \cdot l + R_2 m)$$
$$R_{y1} = -(0.5 \cdot \text{TANCOM} \cdot m + R_1 l), \quad R_{y2} = -(0.5 \cdot \text{TANCOM} \cdot m + R_2 l)$$

(iv) *Nonprismatic beams*   In this case the reactions have to be input as data using the references quoted in section 3.8

4.4   LATERAL LOADS ACTING ON A PLANE GRID ELEMENT

For a plane grid element, the lateral loads consist of uniformly distributed load of intensity q_z and concentrated load C_z at the mid-span of the beam element. The total distributed normal load UDLN is $q_z L$ and the total concentrated load CN is C_z. Using the above values of UDLN and CN, the normal reactions R_1 and R_2 and the fixed end moments M_1 and M_2 can be calculated using the formulae given in section 4.3. The reactions in the x, y and z directions can then be expressed as follows:

$$F_{z1} = -R_1, \quad M_{x1} = -M_1 m, \quad M_{y1} = M_1 l,$$
$$F_{z2} = -R_2, \quad M_{x2} = -M_2 m, \quad M_{y2} = M_2 l$$

4.5   FIXED END REACTIONS DUE TO THERMAL LOADS

The fixed end reactions due to restraining of the free thermal expansion was derived in Chapter 1. Figures 4.4, 4.5 and 4.6 show the reactions developed in the various elements which are given as follows.

(i) *A 2-D or 3-D pin-jointed element.*   As shown in Fig. 4.4, if the fixed end axial force is

$$F = -AE\alpha_t T$$

where AE is the axial rigidity, α_t is the coefficient of linear thermal expansion and T is the rise in temperature, then the fixed end reactions are

$$R_{x1} = -Fl, \quad R_{y1} = -Fm, \quad R_{z1} = -Fn,$$
$$R_{x2} = Fl, \quad R_{y2} = Fm, \quad R_{z2} = Fn$$

(ii) *A 2-D rigid-jointed structure element.*   In this case it is necessary to take into consideration the change in temperature at the bottom and top faces which are given by T_b and T_u respectively. The fixed end axial force F and moment M are given by

$$F = \frac{-AE\alpha_t(T_u Y_b + T_b Y_u)}{d}, \quad M = \frac{EI\alpha_t(T_u - T_b)}{d}$$

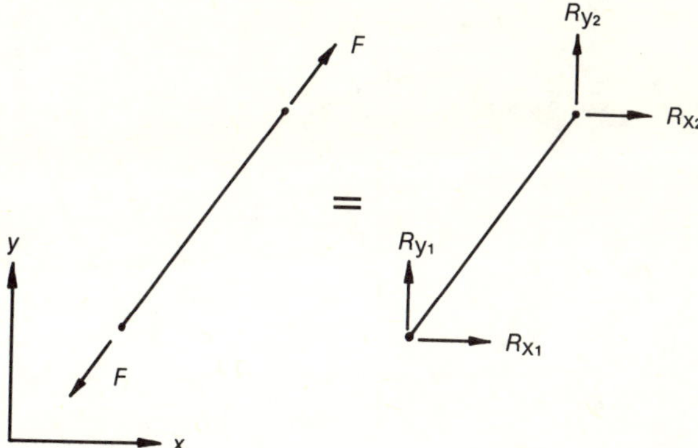

Fig. 4.4   Forces developed as a result of restraining the thermal expansion in a 2-D pin-jointed element.

where Y_u and Y_b are the distances from the neutral axis to the top and bottom faces respectively of the beam and $d = Y_u + Y_b$ as shown in Fig. 1.9. The other symbols have their usual meanings. The reactions are, as shown in Fig. 4.5, given by

$$R_{x1} = -Fl, \quad R_{y1} = -Fm, \quad M_1 = M,$$
$$R_{x2} = Fl, \quad R_{y2} = Fm, \quad M_2 = -M$$

(iii) *A plane grid element.*   In this case, one needs to consider only the fixed end moment M due to thermal restraint given above. The reactions are then, as shown in Fig. 4.6, given by

$$F_{z1} = 0, \quad M_{x1} = -Mm, \quad M_{y1} = Ml,$$
$$F_{z2} = 0, \quad M_{x2} = Mm, \quad M_{y2} = -Ml$$

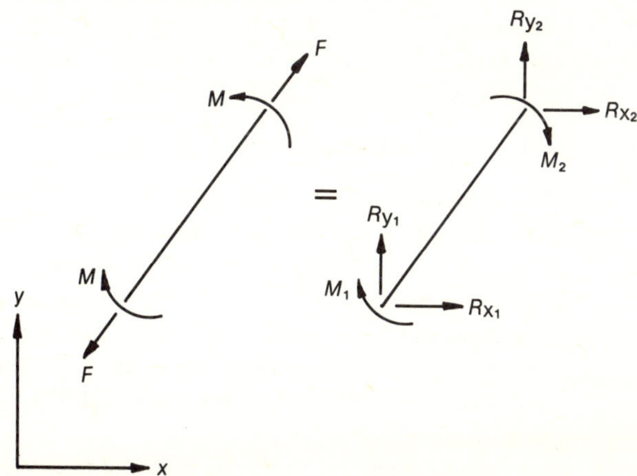

Fig. 4.5   Forces developed as a result of restraining the thermal expansion in a 2-D rigid-jointed structure element.

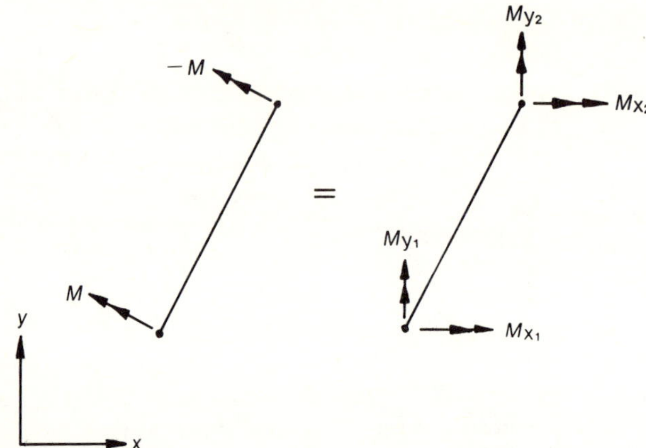

Fig. 4.6   Forces developed as a result of restraining the thermal expansion in a plane grid element.

4.6 CALCULATION OF THE LOAD VECTOR DUE TO THE LATERAL LOADS

The steps involved in the calculation of the load vector GLOAD due to the lateral loads on the elements can be summarized as follows.

(1)   Input NLODEL which shows how many elements carry lateral loads. NLODEL stands for Number of LOaDed ELements.

(2)   For each loaded element input the following.

    (a)   The element number and store it in the array LODEL(NLODEL). LODEL stands for LOaDed ELement.

    (b)   The intensity of the distributed load q_x and q_y and the mid-span concentrated loads C_x and C_y in the x and y directions respectively. Store the four loads on the element in an array FORSEL (NLODEL,4). The name FORSEL stands for FORces on ELement.

(3)   For each of the NLODEL elements calculate the following.

    (a)   The element number from the array LODEL(NLODEL). The node numbers defining the element from the array LNODS (NLODEL). From this information, calculate the coordinates of the nodes corresponding to the element from the array COORD (NPOIN,NDIME). Also calculate the group number and the type number of the element from the array MATNO(NELEM,2) and from the array PROPS(NMATS,NPROP + ?) the material and geometric properties of the element. Finally, calculate the freedom numbers corresponding to the nodes of the element from the array NODFRE(NPOIN,NDOFN) and store them in a temporary array MEMDIS(NEVAB) as explained in section 3.5.

(b)   The loads acting on the element from the array FORSEL (NLODEL,4).

(c)   From the data in (a) and (b) and using the formulae given in section 4.3, calculate the fixed end reactions. Store these fixed end reactions in a temporary array ELOAD(NEVAB). The name ELOAD stands for Element LOAD.

(d)   Add the negative of elements of the array ELOAD(NEVAB) to the elements of the array GLOAD(MEMDIS(NEVAB)) ignoring the addition if any elements of MEMDIS(NEVAB) are zero, indicating that the corresponding degree of freedom is fully restrained. It should be appreciated that it is the negative of ELOAD(NEVAB) that is added because it is the removal of restraint which causes the joints to displace. For more explanation, see section 1.20.

(e)   Repeat the calculation for all the NLODEL loaded elements.

4.7 CALCULATION OF THE LOAD VECTOR DUE TO THERMAL RESTRAINT

The calculation of the load vector due to thermal changes is very similar to the calculation presented in the previous section. The main change is that the fixed end forces calculated are due to thermal restraint and stored in a temporary array FORSTM(NEVAB) instead of ELOAD(NEVAB). The array name FORSTM stands for FORces due to TeMperature. The steps involved in calculating the array FORSTM(NEVAB) are as follows.

(1)   Input the coefficient of thermal expansion α_t, the depth d, the distances Y_u and Y_b to the top and the bottom fibre respectively from the neutral axis for all the NMATS groups of elements and store it in the array PROPS(NMATS,NPROP + ?).

(2)   Input the temperature changes in all the elements and store them in the array TEMP(NELEM,2). The data input is organized so as to minimize the data input. At the start of the calculations, the temperature change in all the elements is set to zero. The data are input only for those elements which have a nonzero change in temperature.

(3)   Using the available data, the reactions due to the restraining of thermal expansion can be calculated using the formulae given in section 4.5 and stored in a temporary array FORSTM(NEVAB).

(4)   The elements of FORSTM(NEVAB) are added to the elements of GLOAD(MEMDIS(NEVAB)) as explained in the previous section.

4.8 SUBROUTINES LOADPJ, LOADRJ, LOADGRID, THERMPJ, THERMRJ AND THERMGRID

There are two sets of subroutines used for the calculation of the load vector. The first set is used to handle the load vector due to loads at the joints and on the elements. The second set is used for the calculation of the load vector due to thermal loads. Under the first set we have LOADPJ, LOADRJ and LOADGRID which, as the name indicates, are used respectively for pin-jointed, 2-D rigid-jointed and grid structures. The corresponding subroutines in the second category are THERMPJ, THERMRJ and THERMGRID.

It should be noted in connection with LOADGRID that it is very similar to LOADRJ except for the fact that the GRID program handles both free and restrained warping analyses using the descriminator IWARP. However, the loads acting on the element, i.e. the distributed and concentrated loads do not cause any fixed end torque at the ends and therefore no bimoment either. Therefore, apart from the fact that if $IWARP = 0$ (i.e. ignore the effects of warping restraint) then $NEVAB = 6$ and if $IWARP = 1$ (i.e. include the effects of warping restraint) then $NEVAB = 8$, there are no real differences in terms of calculating the fixed end forces.

```
C
C**********************************************
C
      SUBROUTINE LOADPJ(NODFRE)
C
C***THIS SUBROUTINE READS THE LOADS AT THE NODES
C   AND CALCULATES THE GLOBAL LOAD VECTOR.
C
      COMMON/CONTRO/NPOIN,NELEM,NNODE,NDOFN,
     @NDIME,NPROP,NMATS,NRESND,NVFIX,NEVAB,
     @NFAIL,NOPTN,NEQNS,NBAND,ITEMP
      COMMON/LGDATA/COORD(100,3),PROPS(10,3),
     @PRESC(80,3),TEMP(100),NDFFIX(80),
     @NDVFIX(80),IFPRE(80,3),ASDIS(400),
     @GSTIF(100,30),GLOAD(100)
      DIMENSION FORSPT(3),NODFRE(100,3)
C
C***ZERO LOAD VECTOR.
C
      DO 10 IEQNS=1,NEQNS
   10 GLOAD(IEQNS)=0.0
C
C***READ HOW MANY POINTS ARE LOADED.
C
      READ(5,900)NLODPT
      WRITE(6,910)NLODPT
  910 FORMAT(/, ' NUMBER OF LOADED NODES = ',I5)
      WRITE (6,'("-------------------------
     @----")')
      IF(NLODPT.EQ.0) GOTO 40
      WRITE(6,920)
  920 FORMAT(//,' NODE',6X,'X-LOAD',6X,'Y-LOAD'
     @,6X,'Z-LOAD')
      WRITE (6,'("----------------------------
     @--------------------")')
C
C***READ AND WRITE THE APPLIED LOADS AT THE NODES.
C
      DO 20 ILODPT=1,NLODPT
      READ(5,930)LODPT,(FORSPT(IDOFN),IDOFN=1,
     @NDOFN)
      WRITE(6,931)LODPT,(FORSPT(IDOFN),IDOFN=1,
     @NDOFN)
C
C***ADD THE NODAL LOADS TO THE LOAD VECTOR AS
C   APPROPRIATE.
C
      DO 30 IDOFN=1,NDOFN
      IF(NODFRE(LODPT,IDOFN).EQ.0) GOTO 30
      GLOAD(NODFRE(LODPT,IDOFN))=FORSPT(IDOFN)
   30 CONTINUE
   20 CONTINUE
  900 FORMAT(5I5)
  930 FORMAT(I3,3F10.4)
  931 FORMAT(I3,3(3X,F10.4),/)
   40 RETURN
      END
```

```
C
C**********************************************
```

```fortran
C
      SUBROUTINE LOADRJ(NODFRE,LNODS,MATNO)
C
C***THIS SUBROUTINE READS THE APPLIED LOADS AT
C   THE NODES AND ON THE MEMBERS AND CALCULATES
C   THE GLOBAL NODAL LOAD VECTOR.
C
      COMMON/CONTRO/NPOIN,NELEM,NNODE,NDOFN,
     @NDIME,NPROP,NMATS,NRESND,NVFIX,NEVAB,NFAIL,
     @NLODEL,NEQNS,NBAND,ITEMP
      COMMON/LGDATA/COORD(100,2),PROPS(20,11),
     @PRESC(80,3),TEMP(100,2),NDFFIX(80),
     @NDVFIX(80),IFPRE(80,3),FORSEL(80,4),
     @LODEL(80),ASDIS(800),GSTIF(100,30),
     @GLOAD(100)
      DIMENSION FORSPT(3),ELOAD(6),MEMDIS(6),
     @NODFRE(100,3),LNODS(100,2),MATNO(100,2)
C
C***ZERO LOAD VECTOR
C
      DO 10 IEQNS=1,NEQNS
   10 GLOAD(IEQNS)=0.0
C
C***READ HOW MANY POINTS ARE LOADED.
C
      READ(5,900)NLODPT
      IF(NLODPT.EQ.0) GOTO 25
      WRITE(6,910)NLODPT
  910 FORMAT(/, ' NUMBER OF LOADED NODES = ',I5)
      WRITE(6,'("--------------------------")')
      WRITE(6,920)
  920 FORMAT(//,' NODE',6X,'X-LOAD',10X,'Y-LOAD'
     @,6X,'THETAZ-LOAD',/,'------------------
     @----------------------------------------
     @',//)
C
C***READ AND WRITE LOADS ACTING AT THE JOINTS
C   ONLY.
C
      DO 20 ILODPT=1,NLODPT
      READ(5,930)LODPT,(FORSPT(IDOFN),IDOFN=1,
     @NDOFN)
      WRITE(6,932)LODPT,(FORSPT(IDOFN),IDOFN=1,
     @NDOFN)
  932 FORMAT(I4,3(4X,F10.4))
C
C***ADD THE LOADS AT JOINTS TO THE APPROPRIATE
C   POSITIONS IN THE GLOBAL LOAD VECTOR.
C
      DO 30 IDOFN=1,NDOFN
C
C***IF A DEGREE OF FREEDOM IS RESTRAINED DO NOT
C   CALCULATE THE LOAD VECTOR.
C
      IF(NODFRE(LODPT,IDOFN).EQ.0) GOTO 30
      GLOAD(NODFRE(LODPT,IDOFN))=FORSPT(IDOFN)
   30 CONTINUE
   20 CONTINUE
C
C***READ HOW MANY ELEMENTS ARE LOADED BY LATERAL
```

```
C     LOADS.
C
   25 READ(5,900)NLODEL
      IF(NLODEL.EQ.0) GOTO 80
      WRITE(6,940)NLODEL
  940 FORMAT(//,' NO. OF ELEMENTS CARRYING
     @DISTRIBUTED OR CONCENTRATED LATERAL LOAD =
     @',I5,/,'--------------------------------
     @--------------------------------
     @--------',//)
      WRITE(6,950)
  950 FORMAT(//,' ELEMENT',6X,'UDL-X',6X,'UDL-Y',
     @6X,'CONLOAD-X',6X,'CONLOAD-Y',/,'------
     @------------------------------------
     @--------------',//)
C
C***READ AND WRITE THE LATERAL DISTRIBUTED AND
C   CONCENTRATED LOADS ON THE ELEMENTS
C
      DO 40 ILODEL=1,NLODEL
      READ(5,930)LODEL(ILODEL),(FORSEL(ILODEL,J)
     @,J=1,4)
   40 WRITE(6,931)LODEL(ILODEL),(FORSEL(ILODEL,J)
     @,J=1,4)
  931 FORMAT(I3,5X,1F10.4,3(F10.4,5X))
C
C***CALCULATE THE FIXED END FORCES ALONG X,Y AND
C   THETAZ DIRECTIONS DUE TO THE  LATERAL LOADS
C   ON ELEMENTS.
C
      DO 50 ILODEL=1,NLODEL
      IELEM=LODEL(ILODEL)
      NTYPE=MATNO(IELEM,2)
      IEND=LNODS(IELEM,1)
      JEND=LNODS(IELEM,2)
      XPROJ=COORD(JEND,1)-COORD(IEND,1)
      YPROJ=COORD(JEND,2)-COORD(IEND,2)
      SPAN=SQRT(XPROJ*XPROJ+YPROJ*YPROJ)
      XL=XPROJ/SPAN
      YM=YPROJ/SPAN
C
C***CALCULATE THE TOTAL DISTRIBUTED LOAD AND
C   CONCENTRATED LOAD ON THE ELEMENT.
C
      UDLXT=FORSEL(ILODEL,1)*ABS(YPROJ)
      UDLYT=FORSEL(ILODEL,2)*ABS(XPROJ)
      CONLX=FORSEL(ILODEL,3)
      CONLY=FORSEL(ILODEL,4)
      IF(NTYPE.EQ.2) GOTO 52
      IF(NTYPE.EQ.3) GOTO 54
      IF(NTYPE.EQ.4) GOTO 56
C
C***CALCULATE THE FIXED END FORCES ALONG X,Y AND
C   THETAZ DIRECTIONS FOR AN ORDINARY FIXED BEAM.
C
      ELOAD(1)=-0.5*(UDLXT+CONLX)
      ELOAD(4)=ELOAD(1)
      ELOAD(2)=-0.5*(UDLYT+CONLY)
      ELOAD(5)=ELOAD(2)
      ELOAD(3)=SPAN/12.0*(UDLYT*XL-UDLXT*YM)+
```

```
     @SPAN/8.0*(CONLY*XL-CONLX*YM)
      ELOAD(6)=-ELOAD(3)
      GOTO 58
C
C***CALCULATE THE FIXED END REACTIONS ALONG X,Y
C   AND THETAZ DIRECTIONS FOR A BEAM ON ELASTIC
C   SUBGRADE. NOTE THAT PART OF THE NORMAL LOAD
C   ON THE BEAM IS ABSORBED BY THE 'FOUNDATION'.
C
   52 LPROP=MATNO(IELEM,1)
      SUBGRM=PROPS(LPROP,NPROP+1)
      YOUNG=PROPS(LPROP,1)
      YINERA=PROPS(LPROP,3)
      EI=YOUNG*YINERA
      AMDAL=SPAN*(SQRT(SQRT(SUBGRM/(4.0*EI))))
      SH=SINH(AMDAL)
      CH=COSH(AMDAL)
      S=SIN(AMDAL)
      C=COS(AMDAL)
      AMDAL2=0.5*AMDAL
      SH2=SINH(AMDAL2)
      CH2=COSH(AMDAL2)
      S2=SIN(AMDAL2)
      C2=COS(AMDAL2)
C
C***CALCULATE THE MODIFICATION FACTORS FOR THE
C   FIXED END MOMENT AND NORMAL REACTION FOR
C   UNIFORMLY DISTRIBUTED NORMAL LOAD.
C
      FACTO1=6.0/AMDAL/AMDAL*(SH-S)/(SH+S)
      FACTO2=2.0/AMDAL*(CH-C)/(SH+S)
C
C***CALCULATE THE MODIFICATION FACTORS FOR THE
C   FIXED END MOMENT AND NORMAL REACTION FOR A
C   CONCENTRATED NORMAL LOAD AT MID-SPAN.
C
      FACTO3=8.0/AMDAL*(SH2*S2)/(SH+S)
      FACTO4=2.0*(SH2*C2+CH2*S2)/(SH+S)
C
C***CALCULATE THE NORMAL COMPONENTS OF THE TOTAL
C   DISTRIBUTED AND CONCENTRATED LOAD AND THE
C   TOTAL TANGENTIAL COMPONENT OF ALL
C   THE APPLIED LOADS.
C
      UDLN=-UDLXT*YM+UDLYT*XL
      CN=-CONLX*YM+CONLY*XL
      TANCOM=(UDLXT+CONLX)*XL+(UDLYT+CONLY)*YM
      ELOAD(1)=-0.5*TANCOM*XL+0.5*(UDLN*FACTO2+
     @CN*FACTO4)*YM
      ELOAD(4)=ELOAD(1)
      ELOAD(2)=-0.5*TANCOM*YM-0.5*(UDLN*FACTO2+
     @CN*FACTO4)*XL
      ELOAD(5)=ELOAD(2)
      ELOAD(3)=SPAN/12.0*UDLN*FACTO1+SPAN/8.0
     @*CN*FACTO3
      ELOAD(6)=-ELOAD(3)
      GOTO 58
C
C***FIXED END REACTIONS ALONG X,Y AND THETAZ
C   DIRECTIONS FOR A SEMI-RIGIDLY CONNECTED BEAM.
```

```
C
   54 LPROP=MATNO(IELEM,1)
      SPRNG1=PROPS(LPROP,NPROP+2)
      SPRNG2=PROPS(LPROP,NPROP+3)
      YOUNG=PROPS(LPROP,1)
      YINERA=PROPS(LPROP,3)
      EIBYL=YOUNG*YINERA/SPAN
      ALPHA1=EIBYL/SPRNG1
      ALPHA2=EIBYL/SPRNG2
C
C***CALCULATE THE MODIFICATION FACTORS FOR FIXED
C   END MOMENT AND NORMAL REACTIONS FOR A
C   SEMI-RIGID BEAM.
C
      D=(2.0+6.0*ALPHA1)*(2.0+6.0*ALPHA2)-1
      FACTO1=3.0*(1.0+6.0*ALPHA2)/D
      FACTO2=3.0*(1.0+6.0*ALPHA1)/D
C
C***CALCULATE THE NORMAL COMPONENTS OF THE
C   DISTRIBUTED LOAD AND THE CONCENTRATED LOAD
C   AND THE TOTAL TANGENTIAL COMPONENT OF
C   ALL THE APPLIED LOAD.
C
      UDLN=-UDLXT*YM+UDLYT*XL
      CN=-CONLX*YM+ CONLY*XL
      TANCOM=(UDLXT+CONLX)*XL+(UDLYT+CONLY)*YM
      ELOAD(3)=SPAN/12.0*UDLN*FACTO1+SPAN/8.0*
     @CN*(2.0*FACTO1)
      ELOAD(6)=-SPAN/12.0*UDLN*FACTO2-SPAN/8.0*
     @CN*(2.0*FACTO2)
      ELOAD(1)=-0.5*TANCOM*XL+0.5*(UDLN+CN)*YM
     @+(ELOAD(3)+ELOAD(6))/SPAN*YM
      ELOAD(4)=-0.5*TANCOM*XL+0.5*(UDLN+CN)*YM-
     @(ELOAD(3)+ELOAD(6))/SPAN*YM
      ELOAD(2)=-0.5*TANCOM*YM-0.5*(UDLN+CN)*XL-
     @(ELOAD(3)+ELOAD(6))/SPAN*XL
      ELOAD(5)=-0.5*TANCOM*YM-0.5*(UDLN+CN)*XL+
     @(ELOAD(3)+ELOAD(6))/SPAN*XL
      GOTO 58
C
C***READ THE FIXED END FORCES FOR THE
C   NONPRISMATIC BEAM ELEMENT.
C
   56 WRITE(6,960)
  960 FORMAT(/,'THE FIXED END REACTIONS FOR A
     @NONPRISMATIC BEAM')
      READ(5,970)(ELOAD(IEVAB),IEVAB=1,NEVAB)
      WRITE(6,970)(ELOAD(IEVAB),IEVAB=1,NEVAB)
  970 FORMAT(6F10.4)
   58 CONTINUE
C
C***CALCULATE THE NODE FREEDOMS AT THE TWO NODES
C   AT THE ENDS OF THE BEAM AND STORE IT IN THE
C   ARRAY MEMDIS(NDOFN).
C
      DO 60 IDOFN=1,NDOFN
      MEMDIS(IDOFN)=NODFRE(IEND,IDOFN)
      MEMDIS(IDOFN+NDOFN)=NODFRE(JEND,IDOFN)
   60 CONTINUE
C
```

```fortran
C***ADD THE FIXED END FORCES TO THE APPROPRIATE
C   LOAD VECTOR
C
      DO 70 IEVAB=1,NEVAB
      IF(MEMDIS(IEVAB).EQ.0) GOTO 70
      GLOAD(MEMDIS(IEVAB))=GLOAD(MEMDIS(IEVAB))
     @-ELOAD(IEVAB)
   70 CONTINUE
   50 CONTINUE
  900 FORMAT(5I5)
  930 FORMAT(I5,4(F10.4))
   80 RETURN
      END
C
C***********************************************
C
      SUBROUTINE FIXDISP(NODFRE)
C
C***THIS SUBROUTINE MODIFIES THE LOAD VECTOR
C   CORRESPONDING TO THOSE FREEDOMS WHICH HAVE A
C   PRESCRIBED VALUE.
C
      COMMON/CONTRO/NPOIN,NELEM,NNODE,NDOFN,
     @NDIME,NPROP,NMATS,NRESND,NVFIX,NEVAB,NFAIL
     @,NLODEL,NEQNS,NBAND,ITEMP
      COMMON/LGDATA/COORD(100,2),PROPS(20,11),
     @PRESC(80,3),TEMP(100,2),NDFFIX(80),
     @NDVFIX(80),IFPRE(80,3),FORSEL(80,4),
     @LODEL(80),ASDIS(800),GSTIF(100,30),
     @GLOAD(100)
      DIMENSION  NODFRE(100,3)
C
C***THERE ARE NVFIX NODES WHERE THERE ARE ONE OR
C   MORE FREEDOMS HAVE A FIXED VALUE.  THE NODE
C   NUMBER IS IN THE ARRAY NDVFIX(NVFIX) AND THE
C   VALUE OF THE PRESCRIBED DISPLACEMENT IN THE
C   ARRAY PRESC(NVFIX, NDOFN).  IF PRESC(IVFIX,
C   IDOFN) = 0 IT MEANS THAT IT IS A FREE DEGREE
C   OF FREEDOM.
C
      DO 10 IVFIX=1,NVFIX
C
C***DETERMINE THE NODE NUMBER OF THE NODE
C
      NODFIX=NDVFIX(IVFIX)
C
C***CHECK THROUGH THE NDOFN FREEDOMS AT NODFIX
C   TO SEE WHICH HAVE A PRESCRIBED VALUE.  IF
C   THE PRESCRIBED VALUE OF DISPLACEMENT IS ZERO
C   IT MEANS IT IS NOT- REPEAT NOT- FIXED.
C
      DO 20 IDOFN=1,NDOFN
      IF(PRESC(IVFIX,IDOFN).EQ.0) GOTO 20
C
C***CALCULATE THE NODE FREEDOM NUMBER OF THE
C   PRESCRIBED DISPLACEMENT.
C
      IFRDOM=NODFRE(NODFIX,IDOFN)
C
C***THE LOAD CORRESPONDING TO THE FIXED DEGREE
```

```
C    OF DISPLACEMENT IS EQUAL TO DIAGONAL TERM
C    TIMES  FIXED DISPLACEMENT VALUE
C
        GLOAD(IFRDOM)=GSTIF(IFRDOM,1)*PRESC(
       @IVFIX,IDOFN)
     20 CONTINUE
     10 CONTINUE
        RETURN
```

```
C
C*********************************************
C
        SUBROUTINE LOADGRID(NODFRE,LNODS,MATNO)
C
C***THIS SUBROUTINE READS THE LOADS AT THE NODES
C    AND ON THE ELEMENTS AND ASSEMBLES THE GLOBAL
C    LOAD VECTOR.
C
        COMMON/CONTRO/NPOIN,NELEM,NNODE,NDOFN,
       @NDIME,NPROP,NMATS,NRESND,NVFIX,NEVAB,
       @NFAIL,NLODEL,NEQNS,NBAND,ITEMP,IWARP
        COMMON/LGDATA/COORD(100,2),PROPS(10,12),
       @PRESC(80,4),TEMP(100,2),NDFFIX(80),
       @NDVFIX(80),IFPRE(80,4),FORSEL(80,2),
       @LODEL(80),ASDIS(800),GSTIF(100,30),
       @GLOAD(100)
        DIMENSION FORSPT(4),ELOAD(8),MEMDIS(8),
       @NODFRE(100,4),LNODS(100,2),MATNO(100,2)
C
C***ZERO LOAD VECTOR
C
        DO 10 IEQNS=1,NEQNS
     10 GLOAD(IEQNS)=0.0
C
C***READ HOW MANY POINTS ARE LOADED.
C
        READ(5,900)NLODPT
        IF(NLODPT.EQ.0) GOTO 25
        WRITE(6,910)NLODPT
    910 FORMAT(/, ' NUMBER OF LOADED NODES = ',I5)
        WRITE(6,'("------------------------")')
        WRITE(6,920)
    920 FORMAT(//,' NODE',8X,'Z-LOAD',10X,'THETA
       @-LOAD',6X,'THETAY-LOAD',6X,'THETAZ-LOAD',
       @/,'-----------------------------------
       @--------------------',//)
C
C***READ AND WRITE THE LOADS ACTING AT NODES
C    ONLY.
C
        DO 20 ILODPT=1,NLODPT
        READ(5,930)LODPT,(FORSPT(IDOFN),IDOFN=1,
       @NDOFN)
        WRITE(6,931)LODPT,(FORSPT(IDOFN),IDOFN=1,
       @NDOFN)
    931 FORMAT(I4,4(6X,F10.4))
C
C***ADD THE LOADS AT THE JOINTS TO THE
C    APPROPRIATE POSITION IN THE GLOBAL LOAD
C    VECTOR.
```

```
C
      DO 30 IDOFN=1,NDOFN
      IF(NODFRE(LODPT,IDOFN).EQ.0) GOTO 30
      GLOAD(NODFRE(LODPT,IDOFN))=FORSPT(IDOFN)
   30 CONTINUE
   20 CONTINUE
C
C***READ HOW MANY ELEMENTS CARRY LATERAL
C   DISTRIBUTED AND CONCENTRATED LOADS IN THE
C   Z-DIRECTION.
C
   25 READ(5,900)NLODEL
      IF(NLODEL.EQ.0) GOTO 80
      WRITE(6,940)NLODEL
  940 FORMAT(//,' NO. OF ELEMENTS CARRYING
     @DISTRIBUTED OR CONCENTRATED LATERAL LOAD =
     @',I3,/,'--------------------------------
     @----------------------------------',//)
      WRITE(6,950)
  950 FORMAT(//,' ELEMENT',6X,'UDL-Z',6X,'CON-Z'
     @,/,'------------------------------------
     @-------------',//)
C
C***READ AND WRITE  THE LATERAL DISTRIBUTED AND
C   CONCENTRATED LOADS IN THE Z-DIRECTION ON THE
C   ELEMENTS
C
      DO 40 ILODEL=1,NLODEL
      READ(5,930)LODEL(ILODEL),(FORSEL(ILODEL,J)
     @,J=1,2)
   40 WRITE(6,933)LODEL(ILODEL),(FORSEL(ILODEL,J)
     @,J=1,2)
  933 FORMAT(I5,7X,2(F10.4))
C
C***CALCULATE THE FIXED END FORCES ALONG W,
C   THETAX AND THETAY DIRECTIONS FROM THE
C   LATERAL LOADS ON THE ELEMENTS.
C
      DO 50 ILODEL=1,NLODEL
      IELEM=LODEL(ILODEL)
      NTYPE=MATNO(IELEM,2)
      LPROP=MATNO(IELEM,1)
      IEND=LNODS(IELEM,1)
      JEND=LNODS(IELEM,2)
      XPROJ=COORD(JEND,1)-COORD(IEND,1)
      YPROJ=COORD(JEND,2)-COORD(IEND,2)
      SPAN=SQRT(XPROJ*XPROJ+YPROJ*YPROJ)
      XL=XPROJ/SPAN
      YM=YPROJ/SPAN
C
C***CALCULATE THE TOTAL DISTRIBUTED AND
C   CONCENTRATED LOAD ON THE ELEMENT.
C
      UDLZT=FORSEL(ILODEL,1)*SPAN
      CONLZ=FORSEL(ILODEL,2)
      IF(NTYPE.EQ.2) GOTO 52
      IF(NTYPE.EQ.3) GOTO 54
      IF(NTYPE.EQ.4) GOTO 56
C
C***CALCULATE THE FIXED END REACTIONS ALONG W,
```

```
C    THETAX AND THETAY DIRECTIONS FOR AN ORDINARY
C    FIXED END BEAM.
C
      ELOAD(1)=-0.5*(UDLZT+CONLZ)
      ELOAD(NDOFN+1)=ELOAD(1)
      FEM=SPAN*(UDLZT/12.0+CONLZ/8.0)
      ELOAD(2)=-FEM*YM
      ELOAD(NDOFN+2)=-ELOAD(2)
      ELOAD(3)=FEM*XL
      ELOAD(NDOFN+3)=-ELOAD(3)
      IF(IWARP.EQ.1)ELOAD(NDOFN)=0.0
      IF(IWARP.EQ.1)ELOAD(NDOFN+NDOFN)=0.0
      GOTO 58
C
C***CALCULATE THE FIXED END FORCES FOR A BEAM ON
C    ELASTIC SUBGRADE. NOTE THAT PART OF THE
C    NORMAL LOAD ON THE BEAM IS ABSORBED BY THE
C    'FOUNDATION'.
C
   52 SUBGRM=PROPS(LPROP,NPROP+1)
      YOUNG=PROPS(LPROP,1)
      YINERA=PROPS(LPROP,3)
      EI=YOUNG*YINERA
      AMDAL=SPAN*(SQRT(SQRT(SUBGRM/(4.0*EI))))
      SH=SINH(AMDAL)
      CH=COSH(AMDAL)
      S=SIN(AMDAL)
      C=COS(AMDAL)
      AMDAL2=0.5*AMDAL
      SH2=SINH(AMDAL2)
      CH2=COSH(AMDAL2)
      S2=SIN(AMDAL2)
      C2=COS(AMDAL2)
C
C***CALCULATE THE MODIFICATION FACTORS FOR THE
C    FIXED END REACTIONS DUE TO DISTRIBUTED LOAD
C
      FACTO1=6.0/AMDAL/AMDAL*(SH-S)/(SH+S)
      FACTO2=2.0/AMDAL*(CH-C)/(SH+S)
C
C***CALCULATE THE MODIFICATION FACTORS FOR THE
C    FIXED END REACTIONS DUE TO CONCENTRATED LOAD
C    AT MID-SPAN.
C
      FACTO3=8.0/AMDAL*(SH2*S2)/(SH+S)
      FACTO4=2.0*(SH2*C2+CH2*S2)/(SH+S)
      FEM=SPAN*(UDLZT*FACTO1/12.0+CONLZ*FACTO3
     @/8.0)
      ELOAD(1)=-0.5*(UDLZT*FACTO2+CONLZ*FACTO4)
      ELOAD(NDOFN+1)=ELOAD(1)
      ELOAD(2)=-FEM*YM
      ELOAD(NDOFN+2)=-ELOAD(2)
      ELOAD(3)=FEM*XL
      ELOAD(NDOFN+3)=-ELOAD(3)
      IF(IWARP.EQ.1)ELOAD(NDOFN)=0.0
      IF(IWARP.EQ.1)ELOAD(NDOFN+NDOFN)=0.0
      GOTO 58
C
C***FIXED END REACTIONS  FOR A SEMI-RIGIDLY
C    CONNECTED BEAM
```

```fortran
C
   54 SPRNG1=PROPS(LPROP,NPROP+2)
      SPRNG2=PROPS(LPROP,NPROP+3)
      YOUNG=PROPS(LPROP,1)
      YINERA=PROPS(LPROP,3)
      EIBYL=YOUNG*YINERA/SPAN
      ALPHA1=EIBYL/SPRNG1
      ALPHA2=EIBYL/SPRNG2
      D=(2.0+6.0*ALPHA1)*(2.0+6.0*ALPHA2)-1
C
C***CALCULATE THE MODIFICATION FACTORS FOR THE
C   FIXED END MOMENTS AT THE TWO ENDS WHICH ARE
C   NOT NECESSARILY EQUAL.
C
      FACTO1=3.0*(1.0+6.0*ALPHA2)/D
      FACTO2=3.0*(1.0+6.0*ALPHA1)/D
      FEM1=SPAN*(UDLZT*FACTO1/12.0+CONLZ*2.0*
     @FACTO1/8.0)
      FEM2=SPAN*(UDLZT*FACTO2/12.0+CONLZ*2.0*
     @FACTO2/8.0)
      ELOAD(2)=-FEM1*YM
      ELOAD(3)=FEM1*XL
      ELOAD(NDOFN+2)=FEM2*YM
      ELOAD(NDOFN+3)=-FEM2*XL
      ELOAD(1)=-0.5*(UDLZT+CONLZ)-(FEM1-FEM2)
     @/SPAN
      ELOAD(NDOFN+1)=-0.5*(UDLZT+CONLZ)+(FEM1
     @-FEM2)/SPAN
      IF(IWARP.EQ.1)ELOAD(NDOFN)=0.0
      IF(IWARP.EQ.1)ELOAD(NDOFN+NDOFN)=0.0
      GOTO 58
C
C***READ THE FIXED END FORCES FOR THE
C   NONPRISMATIC ELEMENT.
C
   56 WRITE(6,960)
  960 FORMAT(/,'FIXED END REACTIONS FOR A
     @NONPRISMATIC ELEMENT')
      READ(5,970)(ELOAD(IEVAB),IEVAB=1,NEVAB)
      WRITE(6,970)(ELOAD(IEVAB),IEVAB=1,NEVAB)
  970 FORMAT(8F10.4)
   58 CONTINUE
C
C***CALCULATE THE NODE FREEDOMS AT THE ENDS OF
C   THE ELEMENT AND
C   STORE IT IN THE MEMDIS(NDOFN) ARRAY.
C
      DO 60 IDOFN=1,NDOFN
      MEMDIS(IDOFN)=NODFRE(IEND,IDOFN)
      MEMDIS(IDOFN+NDOFN)=NODFRE(JEND,IDOFN)
   60 CONTINUE
C
C***ADD THE FIXED END FORCES AS APPROPRIATE TO
C   THE GLOBAL LOAD VECTOR
C
      DO 70 IEVAB=1,NEVAB
      IF(MEMDIS(IEVAB).EQ.0) GOTO 70
      GLOAD(MEMDIS(IEVAB))=GLOAD(MEMDIS(IEVAB))
     @-ELOAD(IEVAB)
   70 CONTINUE
```

```
    50 CONTINUE
   900 FORMAT(5I5)
   930 FORMAT(I5,4(F10.4))
    80 RETURN
       END
```

```
C
C**********************************************
C
       SUBROUTINE THERMPJ(NODFRE,LNODS,MATNO)
C
C***THIS SUBROUTINE READS THE TEMPERATURE IN
C   SOME ELEMENTS AND CALCULATES THE RESTRAINT
C   FORCE AND ADDS IT TO THE GLOBAL LOAD VECTOR
C   AS APPROPRIATE.
C
       COMMON/CONTRO/NPOIN,NELEM,NNODE,NDOFN,
      @NDIME,NPROP,NMATS,NRESND,NVFIX,NEVAB,NFAIL
      @,NOPTN,NEQNS,NBAND,ITEMP
       COMMON/LGDATA/COORD(100,3),PROPS(10,3),
      @PRESC(80,3),TEMP(100),NDFFIX(80),NDVFIX(80),
      @IFPRE(80,3),ASDIS(400),GSTIF(100,30),
      @GLOAD(100)
       DIMENSION NODFRE(100,3),LNODS(100,2),
      @MATNO(100,2),FORSTM(6),MEMDIS(6)
C
C***READ AND WRITE COEFFICIENT OF THERMAL
C   EXPANSION FOR ALL MATERIALS
C
       WRITE(6,900)
   900 FORMAT(//,' MATERIAL NO.',6X,'COEFF. OF
      @THERMAL EXPANSION')
       DO 10 IMATS=1,NMATS
       READ(5,905)PROPS(IMATS,NPROP+1)
    10 WRITE(6,910)IMATS,PROPS(IMATS,NPROP+1)
   905 FORMAT(E14.6)
   910 FORMAT(I5,6X,E14.6)
       WRITE(6,920)
   920 FORMAT(//,' TEMPERATURE CHANGES IN THE
      @ELEMENTS',/,'---------------------------
      @---------------------------------',//,
      @' ELEMENT',6X,'TEMPERATURE')
C
C***ZERO THE TEMPERATURE CHANGE IN ALL THE
C   ELEMENTS.
C
       DO 20 IELEM=1,NELEM
    20 TEMP(IELEM)=0.0
C
C***READ THE NONZERO CHANGE IN TEMPERATURES
C   ONLY.
C
    30 READ(5,930) IELEM,TEMP(IELEM)
   930 FORMAT(I5,F10.4)
       IF(IELEM.NE.NELEM) GOTO 30
       DO 40 IELEM=1,NELEM
    40 WRITE(6,940)IELEM,TEMP(IELEM)
   940 FORMAT(I5,7X,F10.4)
C
C***CALCULATE THE THERMAL RESTRAINT FORCES ALONG
```

```
C    X, Y AND Z DIRECTIONS AND ADD IT AS
C    APPROPRIATE TO THE GLOBAL LOAD VECTOR.
C
     DO 50 IELEM=1,NELEM
     IF(TEMP(IELEM).EQ.0.0) GO TO 50
C
C***EXTRACT FROM THE VARIOUS ARRAYS THE
C    INFORMATION NEEDED TO CALCULATE THE THERMAL
C    RESTRAINT REACTIONS.
C
     LPROP=MATNO(IELEM,1)
     YOUNG=PROPS(LPROP,1)
     AREA=PROPS(LPROP,2)
     ALPHAT=PROPS(LPROP,NPROP+1)
     RAXIAL=-AREA*YOUNG*ALPHAT*TEMP(IELEM)
C
C***CALCULATE THE NODES CORRESPONDING TO THE TWO
C    ENDS OF THE ELEMENT
C
     IEND=LNODS(IELEM,1)
     JEND=LNODS(IELEM,2)
     IF(NDIME.EQ.3)GOTO 65
C
C***CALCULATE THE COORDINATES OF THE NODES AT THE
C    ENDS OF THE ELEMENT, LENGTH, DIRECTION
C    COSINES ETC FOR A 2-D ELEMENT.
C
     XPROJ=COORD(JEND,1)-COORD(IEND,1)
     YPROJ=COORD(JEND,2)-COORD(IEND,2)
     SPAN=SQRT(XPROJ*XPROJ+YPROJ*YPROJ)
     XL=XPROJ/SPAN
     YM=YPROJ/SPAN
C
C***THERMAL RESTRAINT REACTIONS ARE STORED IN A
C    TEMPORARY ARRAY CALLED FORSTM(NEVAB).
C
     FORSTM(1)=-RAXIAL*XL
     FORSTM(2)=-RAXIAL*YM
     FORSTM(3)=-FORSTM(1)
     FORSTM(4)=-FORSTM(2)
     GOTO 75
C
C***CALCULATE THE COORDINATES OF THE NODES AT
C    THE ENDS OF THE ELEMENT, LENGTH, DIRECTION
C    COSINES ETC FOR A 3-D ELEMENT.
C
  65 XPROJ=COORD(JEND,1)-COORD(IEND,1)
     YPROJ=COORD(JEND,2)-COORD(IEND,2)
     ZPROJ=COORD(JEND,3)-COORD(IEND,3)
     SPAN=SQRT(XPROJ*XPROJ+YPROJ*YPROJ+
    @ZPROJ*ZPROJ)
     XL=XPROJ/SPAN
     YM=YPROJ/SPAN
     ZN=ZPROJ/SPAN
C
C***THE THERMAL RESTRAINT REACTIONS ARE STORED
C    IN A TEMPORARY ARRAY CALLED FORSTM(NEVAB).
C
     FORSTM(1)=-RAXIAL*XL
     FORSTM(2)=-RAXIAL*YM
```

```
      FORSTM(3)=-RAXIAL*ZN
      FORSTM(4)=-FORSTM(1)
      FORSTM(5)=-FORSTM(2)
      FORSTM(6)=-FORSTM(3)
C
C***DETERMINE THE DEGREES OF FREEDOMS AT THE
C   ENDS OF THE ELEMENT AND STORE IT IN THE
C   ARRAY MEMDIS(NDOFN).
C
   75 DO 70 IDOFN=1,NDOFN
      MEMDIS(IDOFN)=NODFRE(IEND,IDOFN)
      MEMDIS(IDOFN+NDOFN)=NODFRE(JEND,IDOFN)
   70 CONTINUE
C
C***ADD THE NEGATIVE OF THE THERMAL RESTRAINT
C   FORCES TO THE GLOBAL LOAD VECTOR.
C
      DO 80 IEVAB=1,NEVAB
      IF(MEMDIS(IEVAB).EQ.0) GOTO 80
      GLOAD(MEMDIS(IEVAB))=GLOAD(MEMDIS(IEVAB))-
     @FORSTM(IEVAB)
   80 CONTINUE
   50 CONTINUE
      RETURN
      END
```

```
C
C********************************************
C
      SUBROUTINE THERMRJ(NODFRE,LNODS,MATNO)
C
C***THIS SUBROUTINE CALCULATES THE NODAL FORCES
C   DUE TO THE PREVENTION OF THERMAL EXPANSION
C
      COMMON/CONTRO/NPOIN,NELEM,NNODE,NDOFN,
     @NDIME,NPROP,NMATS,NRESND,NVFIX,NEVAB,
     @NFAIL,NLODEL,NEQNS,NBAND,ITEMP
      COMMON/LGDATA/COORD(100,2),PROPS(20,11),
     @PRESC(80,3),TEMP(100,2),NDFFIX(80),
     @NDVFIX(80),IFPRE(80,3),FORSEL(80,4),
     @LODEL(80),ASDIS(800),GSTIF(100,30),
     @GLOAD(100)
      DIMENSION NODFRE(100,3),LNODS(100,2),
     @MATNO(100,2),FORSTM(6),MEMDIS(6)
C
C***READ AND WRITE THE INFORMATION ABOUT COEFF.
C   OF THERMAL EXPANSION, DEPTH AND THE DISTANCE
C   TO TOP FACE FROM THE NEUTRAL AXIS.
C
      DO 10 IMATS=1,NMATS
   10 READ(5,900)(PROPS(IMATS,NPROP+JPROP),
     @JPROP=4,6)
  900 FORMAT(E14.6,2F10.4)
      WRITE(6,910)
  910 FORMAT(//,' MATERIAL NO.',8X,'ALPHA T',
     @8X,'DEPTH',6X,'DIST. TO TOP FIBRE')
      DO 20 IMATS=1,NMATS
   20 WRITE(6,920)IMATS,(PROPS(IMATS,NPROP+
     @JPROP),JPROP=4,6)
  920 FORMAT(I5,12X,E14.6,2F10.4)
```

```
      C
      C***ZERO TEMPERATURE IN ALL THE ELEMENTS AND
      C    READ ONLY NONZERO TEMPERATURE CHANGES IN
      C    THE TWO FACES OF THE ELEMENTS.
      C
            DO 30 IELEM=1,NELEM
            DO 30 JFACE=1,2
         30 TEMP(IELEM,JFACE)=0.0
         40 READ(5,925)IELEM,(TEMP(IELEM,JFACE),
           @JFACE=1,2)
        925 FORMAT(I5,2F10.4)
            IF(IELEM.NE.NELEM) GOTO 40
        930 FORMAT(I5,6X,2F10.4)
            WRITE(6,940)
        940 FORMAT(//,'TEMPERATURE VARIATION IN
           @ELEMENTS')
            DO 50 IELEM=1,NELEM
         50 WRITE(6,930)IELEM,(TEMP(IELEM,JFACE),
           @JFACE=1,2)
      C
      C***CALCULATE THE FIXED END FORCES DUE TO
      C    RESTRAINT TO THERMAL EXPANSION AND ADD AS
      C    APPROPRIATE TO NODAL FORCES.
      C
            DO 60 IELEM=1,NELEM
      C
      C***EXTRACT INFORMATION FROM THE DIFFERENT
      C    ARRAYS ABOUT THE PROPERTIES NEEDED TO
      C    CALCULATE THE THERMAL RESTRAINT REACTIONS.
      C
            TOPTEM=TEMP(IELEM,1)
            BOTTEM=TEMP(IELEM,2)
            IF(TOPTEM.EQ. 0.0 .AND.BOTTEM.EQ. 0.0)
           @ GOTO 60
            LPROP=MATNO(IELEM,1)
            YOUNG=PROPS(LPROP,1)
            YINERA=PROPS(LPROP,3)
            AREA=PROPS(LPROP,4)
            ALPHAT=PROPS(LPROP,NPROP+4)
            DEPTH=PROPS(LPROP,NPROP+5)
            YUP=PROPS(LPROP,NPROP+6)
            YDOWN=DEPTH-YUP
      C
      C***DETERMINE THE NODES CORRESPONDING TO THE TWO
      C    ENDS OF THE ELEMENT
      C
            IEND=LNODS(IELEM,1)
            JEND=LNODS(IELEM,2)
      C
      C***CALCULATE THE COORDINATES OF THE NODES AT THE
      C    ENDS OF THE ELEMENT, THE LENGTH, DIRECTION
      C    COSINES ETC.
      C
            XPROJ=COORD(JEND,1)-COORD(IEND,1)
            YPROJ=COORD(JEND,2)-COORD(IEND,2)
            SPAN=SQRT(XPROJ*XPROJ+YPROJ*YPROJ)
            XL=XPROJ/SPAN
            YM=YPROJ/SPAN
            AE=AREA*YOUNG
            EI=YINERA*YOUNG
```

```
C
C***CALCULATE THE AXIAL AND MOMENT RESTRAINT
C    FORCES.
C
      RAXIAL=-ALPHAT*AE/DEPTH*(TOPTEM*YDOWN+
     @BOTTEM*YUP)
      RMOMENT=ALPHAT*EI/DEPTH*(TOPTEM-BOTTEM)
C
C***DETERMINE THE COMPONENTS OF THE RESTRAINT
C    FORCES ALONG X,Y AND THETAZ DIRECTIONS.STORE
C    IT IN A TEMPORARY ARRAY FORSTM(NEVAB).
C
      FORSTM(1)=-RAXIAL*XL
      FORSTM(2)=-RAXIAL*YM
      FORSTM(3)=RMOMENT
      FORSTM(4)=-FORSTM(1)
      FORSTM(5)=-FORSTM(2)
      FORSTM(6)=-FORSTM(3)
C
C***DETERMINE THE DEGREES OF FREEDOM AT THE ENDS
C    OF THE ELEMENT AND STORE IT IN THE ARRAY
C    MEMDIS(NDOFN)
C
      DO 70 IDOFN=1,NDOFN
      MEMDIS(IDOFN)=NODFRE(IEND,IDOFN)
   70 MEMDIS(IDOFN+NDOFN)=NODFRE(JEND,IDOFN)
C
C***ADD THE NEGATIVE OF RESTRAINT FORCES TO THE
C    GLOBAL LOAD ARRAY.
C
      DO 80 IEVAB=1,NEVAB
      IF(MEMDIS(IEVAB).EQ.0) GOTO 80
      GLOAD(MEMDIS(IEVAB))=GLOAD(MEMDIS(IEVAB))
     @-FORSTM(IEVAB)
   80 CONTINUE
   60 CONTINUE
      RETURN
      END
```

```
C
C*********************************************
C
      SUBROUTINE THERMGRID(NODFRE,LNODS,MATNO)
C
C***THIS SUBROUTINE CALCULATES THE NODAL FORCES
C    DUE TO THE PREVENTION OF THERMAL EXPANSION
C    AND ADDS IT AS APPROPRIATE TO THE GLOBAL
C    LOAD VECTOR.
C
      COMMON/CONTRO/NPOIN,NELEM,NNODE,NDOFN,
     @NDIME,NPROP,NMATS,NRESND,NVFIX,NEVAB,
     @NFAIL,NLODEL,NEQNS,NBAND,ITEMP,IWARP
      COMMON/LGDATA/COORD(100,2),PROPS(10,12),
     @PRESC(80,4),TEMP(100,2),NDFFIX(80),
     @NDVFIX(80),IFPRE(80,4),FORSEL(80,2),
     @LODEL(80),ASDIS(800),GSTIF(100,30),
     @GLOAD(100)
      DIMENSION NODFRE(100,4),LNODS(100,2),
     @MATNO(100,2),FORSTM(8),MEMDIS(8)
C
```

```
C***READ THE INFORMATION ABOUT COEFF. OF THERMAL
C   EXPANSION,DEPTH AND THE DISTANCE TO TOP FACE
C   FROM THE NEUTRAL AXIS.
C
      DO 10 IMATS=1,NMATS
   10 READ(5,900)(PROPS(IMATS,NPROP+JPROP),
     @JPROP=4,5)
  900 FORMAT(E14.6,2F10.4)
      WRITE(6,910)
  910 FORMAT(//,' MATERIAL NO.',6X,'ALPHA T',
     @10X,'DEPTH')
      DO 20 IMATS=1,NMATS
   20 WRITE(6,920)IMATS,(PROPS(IMATS,NPROP+
     @JPROP),JPROP=4,5)
  920 FORMAT(I5,11X,E14.6,2F10.4)
C
C***ZERO TEMPERATURE IN ALL THE ELEMENTS AND
C   READ ONLY NONZERO TEMPERATURE IN THE
C   ELEMENTS.
C
      DO 30 IELEM=1,NELEM
      DO 30 JFACE=1,2
   30 TEMP(IELEM,JFACE)=0.0
   40 READ(5,925)IELEM,(TEMP(IELEM,JFACE),
     @JFACE=1,2)
  925 FORMAT(I5,2F10.4)
      IF(IELEM.NE.NELEM) GOTO 40
  930 FORMAT(I5,6X,2F10.4)
      WRITE(6,940)
  940 FORMAT(//,'TEMPERATURE VARIATIONS IN
     @ELEMENTS',//,'ELEMENT',6X,'TOP TEMP',6X,'
     @BOTTOM TEMP')
      DO 50 IELEM=1,NELEM
   50 WRITE(6,930)IELEM,(TEMP(IELEM,JFACE),
     @JFACE=1,2)
C
C***CALCULATE THE FIXED END FORCES DUE TO
C   RESTRAINT TO THERMAL EXPANSION AND ADD AS
C   APPROPRIATE TO NODAL FORCES.
C
      DO 60 IELEM=1,NELEM
C
C***EXTRACT FROM THE VARIOUS ARRAYS THE
C   INFORMATION NEEDED FOR CALCULATING THE
C   THERMAL RESTRAINT FORCES.
C
      TOPTEM=TEMP(IELEM,1)
      BOTTEM=TEMP(IELEM,2)
      IF(TOPTEM.EQ. 0.0 .AND.BOTTEM.EQ. 0.0)
     @GOTO 60
      LPROP=MATNO(IELEM,1)
      YOUNG=PROPS(LPROP,1)
      YINERA=PROPS(LPROP,3)
      ALPHAT=PROPS(LPROP,NPROP+4)
      DEPTH=PROPS(LPROP,NPROP+5)
C
C***DETERMINE THE NODES CORRESPONDING TO THE TWO
C   ENDS OF THE ELEMENT
C
      IEND=LNODS(IELEM,1)
```

```
      JEND=LNODS(IELEM,2)
C
C***CALCULATE   COORDINATES ETC. OF THE NODES AT
C    THE ENDS OF THE ELEMENT, LENGTH, DIRECTION
C    COSINES ETC.
C
      XPROJ=COORD(JEND,1)-COORD(IEND,1)
      YPROJ=COORD(JEND,2)-COORD(IEND,2)
      SPAN=SQRT(XPROJ*XPROJ+YPROJ*YPROJ)
      XL=XPROJ/SPAN
      YM=YPROJ/SPAN
      EI=YINERA*YOUNG
C
C***CALCULATE THE MOMENT RESTRAINING THE THERMAL
C    DEFORMATIONS.
C
      RMOMENT=ALPHAT*EI/DEPTH*(TOPTEM-BOTTEM)
C
C***DETERMINE THE COMPONENTS OF THE RESTRAINING
C    MOMENT ALONG THE THETAX AND THETAY
C    DIRECTIONS. STORE IT IN A TEMPORARY ARRAY
C    FORSTM
C
      FORSTM(2)=-RMOMENT*YM
      FORSTM(3)=RMOMENT*XL
      FORSTM(NDOFN+2)=-FORSTM(2)
      FORSTM(NDOFN+3)=-FORSTM(3)
C
C***DETERMINE THE DEGREES OF FREEDOM AT THE ENDS
C    OF THE MOMENT AND STORE IT IN THE ARRAY
C    MEMDIS(NDOFN).
C
      DO 70 IDOFN=1,NDOFN
      MEMDIS(IDOFN)=NODFRE(IEND,IDOFN)
   70 MEMDIS(IDOFN+NDOFN)=NODFRE(JEND,IDOFN)
C
C***ADD THE NEGATIVE OF THE RESTRAINT OF THE
C    FORCES TO THE GLOBAL LOAD VECTOR.
C
      DO 80 IEVAB=1,NEVAB
      IF(MEMDIS(IEVAB).EQ.0) GOTO 80
      GLOAD(MEMDIS(IEVAB))=GLOAD(MEMDIS(IEVAB))
     @-FORSTM(IEVAB)
   80 CONTINUE
   60 CONTINUE
      RETURN
      END
```

CHAPTER 5 Solution of Simultaneous Equations

After the structural stiffness matrix GSTIF and the load vector GLOAD have been calculated, the next step is to solve for the displacements of the joints of the structure. The displacements will be stored in a vector called ASDIS which stands for ASsembled DISplacements. The three matrices are related by the stiffness relationship

$$GSTIF(NEQNS,NEQNS).ASDIS(NEQNS) = GLOAD(NEQNS)$$

where NEQNS is the Number of EQuatioNS or the number of unknown displacements to be solved. In the following sections, in order to simplify writing, the above four unknowns will be represented as shown in Table 5.1.

Table 5.1

Representation of the four unknowns.

Actual name	Substitute name
GSTIF	A
GLOAD	F
ASDIS	X
NEQNS	N

The displacements are calculated using the Gaussian elimination method for the solution of simultaneous equations to be described in the next section.

5.1 THE GAUSSIAN ELIMINATION METHOD

In the solution of structural analysis problems, the Gaussian elimination method is divided into three distinct stages as follows.

(1) The structural stiffness matrix GSTIF is modified by a series of systematic operations such that the last row contains only the last

unknown and in general the ith row contains only the unknowns from the ith to the last unknown. This step is known as the Gaussian reduction of the structural stiffness matrix.

(2) The load vector GLOAD is also modified such that the new load vector is compatible with the modified GSTIF. The reason for not 'reducing' both GSTIF and GLOAD in one operation is that a structure frequently needs to be analysed under different load cases and it is therefore convenient and economical to keep the reduction of GSTIF and GLOAD as two separate operations. The reduction of GSTIF needs to be done only once, while the reduction of GLOAD has to be done as many times as there are load cases to be analysed. It is worth mentioning that, in many nonlinear structural analysis problems, the 'current' load vector is a function of the displacements caused by the 'previous' load vector. In such cases, in order to save the effort involved in assembling and reducing the GSTIF, it is better to keep the reduction of GSTIF and GLOAD as totally independent operations.

(3) After the reduction of GSTIF and GLOAD, the displacements are calculated by working backwards from the last equation which contains only the last unknown. This step is known as back substitution.

The above three steps will be described in detail in the following sections.

5.2 DEVELOPMENT OF THE SOLUTION ALGORITHM FOR FULL MATRIX REPRESENTATION

Consider the set of equations shown in Fig. 5.1. If it is necessary to eliminate the unknown $x(k)$ from the ith equation, where $i > k$, then all that is needed is to multiply the kth equation by $a(i, k)/(ak, k)$ and to subtract the modified kth

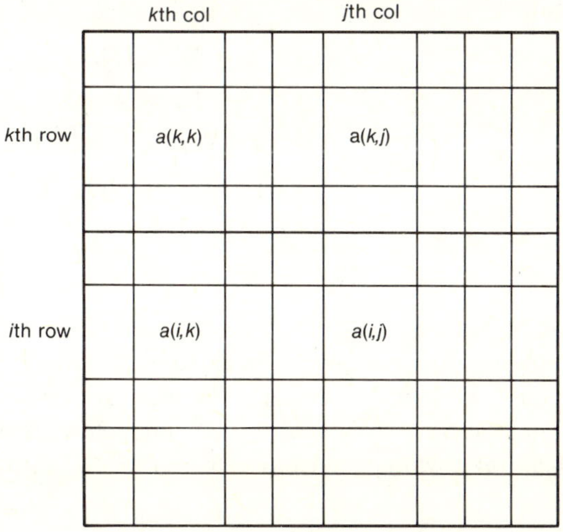

Fig. 5.1   Gaussian elimination scheme.

equation from the ith equation. This procedure modifies an element $a(i, j)$ to the new value

$$a(i, j) - \frac{a(i, k)}{a(k, k)} a(k, j)$$

Since the matrix A is symmetric and it can be shown that, if the unknown $x(k)$ say, is being eliminated from the equations $k + 1$ to N, then the equations $k + 1$ to N remain symmetric after modification, it is convenient to replace $a(i, k)$ in the multiplication factor for the ith equation by $a(k, i)$.

The above calculations can be cast into the following program segment.

```
C       CHOOSE THE KTH UNKNOWN TO BE
        ELIMINATED FROM ALL THE SUCCEEDING
C       EQUATIONS
        DO 10 K=1,(N-1)
C       ELIMINATE THE KTH UNKNOWN FROM ALL THE
        (K+1) TO N EQUATIONS
        DO 20 I=(K+1),N
C       CALCULATING THE MULTIPLYING FACTOR
        QUOT=A(K,I)/A(K,K)
C       MODIFY THE ITH EQUATION ELEMENTS
        DO 30 J=(K+1),N
        A(I,J)=A(I,J)-QUOT*A(K,J)
     30 CONTINUE
     20 CONTINUE
     10 CONTINUE
```

The load vector is also reduced in a similar way by the following program segment.

```
        DO 10 K=1,(N-1)
        DO 20 I=(K+1),N
        QUOT=A(K,I)/A(K,K)
        F(I)=F(I)-QUOT*F(K)
     20 CONTINUE
     10 CONTINUE
```

Having reduced the stiffness matrix and the load vector, the next step is to calculate the displacements by back substitution. Using the reduced stiffness matrix and the load vector, the Nth unknown is determined from the Nth equation as $x(N) = f(N)/a(N, N)$. Having determined the Nth unknown, the $(N-1)th$ unknown is determined from the $(N-1)$th equation as

$$x(N-1) = \frac{f(N-1) - a(N-1, N) * x(N)}{a(N-1, N-1)}$$

In a similar way the Ith unknown is determined from the Ith equation by

$$x(I) = \frac{f(I) - a(I, N) * x(N) - a(I, N-1) * x(N-1) - \cdots - a(I, I+1) * x(I+1)}{a(I, I)}$$

The above equation can be compactly expressed as

$$x(I) = \frac{f(I) - \sum a(I, J) * x(J)}{a(I, I)}$$

The summation $\sum$ is over $J = I+1$ to N.

The above equations can be cast into the following program segment.

```
      X(N) = F(N)/A(N,N)
      DO 40 I=(N-1),1,-1
      SUM = 0,0
      DO 50 J=(I+1),N
      SUM = SUM + A(I,J)*X(J)
   50 CONTINUE
      X(I) = (F(I) - SUM)/A(I,I)
   40 CONTINUE
```

The three program segments complete the three steps of the Gaussian elimination method for the solution of simultaneous equations.

5.3 MODIFICATION OF THE ALGORITHM FOR BANDED FORMAT STORING OF THE STRUCTURAL STIFFNESS MATRIX

As shown in Fig. 3.5, the structural stiffness matrix remains highly banded and, in order to save space in the high speed core of the machine, only a symmetrical half of the stiffness matrix is stored in rectangular form with the diagonal terms in the first column. If the structural stiffness matrix is stored in this form, then the program segments given in the previous section have to be modified to take account of the altered position of the elements of the matrix. In order to assist in the modifications, Fig. 5.2 shows the positions occupied by the elements in the new rectangular representation. The main points to be kept in mind about the rectangular representation are as follows.

(1) The diagonal terms are in the first column.
(2) Only the upper triangular part of the stiffness matrix is stored.
(3) In any row the elements occupy the position which correspond to the shifting of the original elements to the left by (row number $-$ 1) positions from their original positions in the full matrix representation.

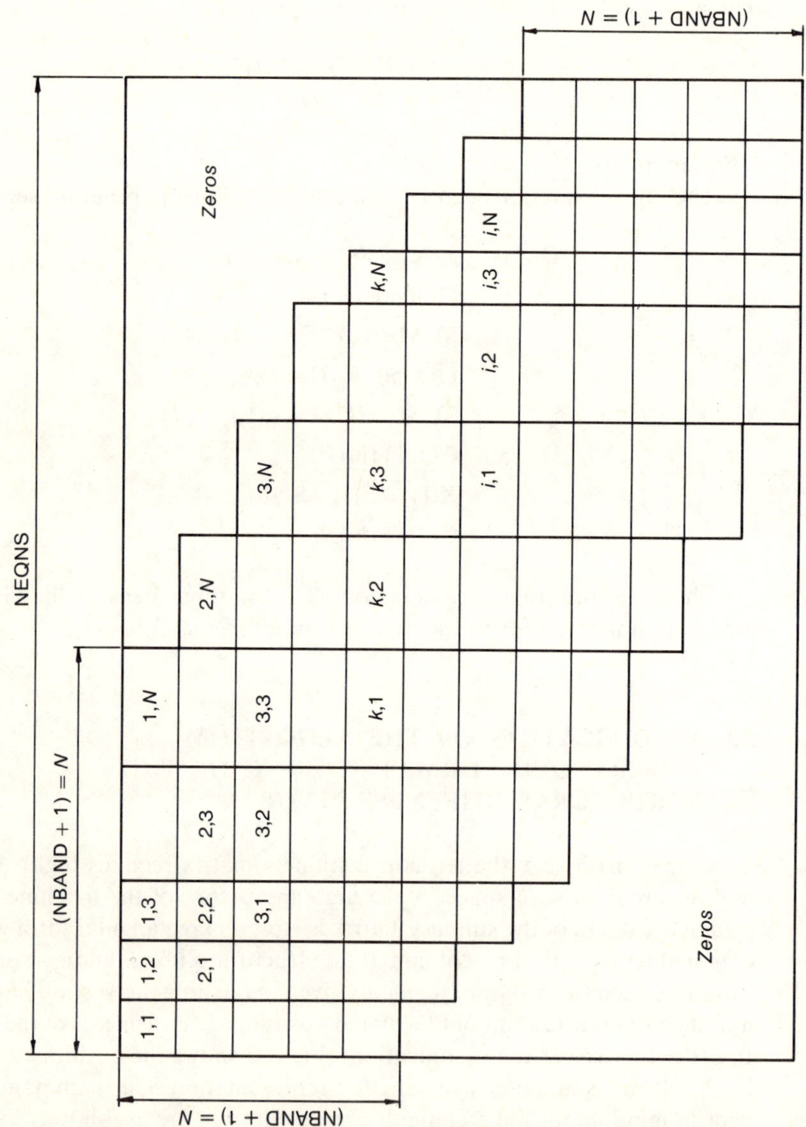

Fig. 5.2 Position occupied by the elements in the rectangular representation in the original full matrix location.

(4) The maximum number of nonzero elements in any row is equal to NBAND + 1 where NBAND is the maximum half-bandwidth as shown in Fig. 3.5 and Fig. 5.2.
(5) Because of the banded nature of the matrix, if the unknown $x(k)$ is being eliminated from all the equations from $k+1$ onwards, then it is necessary to consider only equations $k+1$ to $k+$NBAND. This is because the unknown $x(k)$ does not appear in equations beyond $k+$NBAND as can be seen from Fig. 5.2.
(6) Note that bandedness implies that, in the square representation, in the kth equation $a(k,j) = 0.0$ if $j > k+$NBAND.

The above points can be used to modify the previously given program segments as follows.

(a) The segment for the reduction of the stiffness matrix is as follows.

```
        DO 10 K=1,(N-1)
        DO 20 I=(K+1),(K+NBAND)
        IF (I>N) GOTO 10
        QUOT=A(K,I-K+1)/A(K,1)
        DO 30 J=1,(K+NBAND-(I-1))
        IF (J>(NBAND+1)) GOTO 20
        A(I,J)=A(I,J)-QUOT*A(K,(J-K+1))
   30   CONTINUE
   20   CONTINUE
   10   CONTINUE
```

Note that the element $a(k,j)$ in the full representation occupies the position $a[k,j-(k-1)]$ in the rectangular representation. The elements affected by eliminating $a(k)$ from the equations $k+1$ to $k+$NBAND are shown shaded in Fig. 5.2. Thus in the $(k+1)$th row the farthest element affected is in the column NBAND. Similarly, in the ith equation, the farthest element affected is in column $k+$ NBAND $-(i-1)$ provided that $i \not> k+$ NBAND.

(b) The modified program segment for the reduction of load vector is as follows.

```
        DO 10 K=1,(N-1)
        DO 20 I=(K+1),(K+NBAND)
        IF (I>N) GOTO 10
        QUOT=A(K,I-K+1)/A(K,1)
        F(I)=F(I)-QUOT*F(K)
   20   CONTINUE
   10   CONTINUE
```

(c)   The program segment for back substitution is as follows.

```
          X(N) = F(N)/A(N,1)
          DO  50  I = (N − 1),1,− 1
          SUM = 0.0
          DO  60  J = 2,(NBAND + 1)
   C  (J + I − 1) IS THE TRUE COLUMN POSITION IN
          RECTANGULAR REPRESENTATION
          IF  (J + I − 1) > N)  GOTO  55
          SUM = SUM + A(I,J)*X(J + I − 1)
   60  CONTINUE
   55  CONTINUE
   50  CONTINUE
```

5.4 MINIMIZATION OF THE BANDWIDTH

It is obvious that the storage requirements can be minimized by keeping the half-bandwidth NBAND as small as possible. This can be achieved by remembering the following two points.

(1)   The half-bandwidth is equal to the maximum difference between the nonzero freedom numbers of the nodal displacements at the ends of all the elements of a structure.

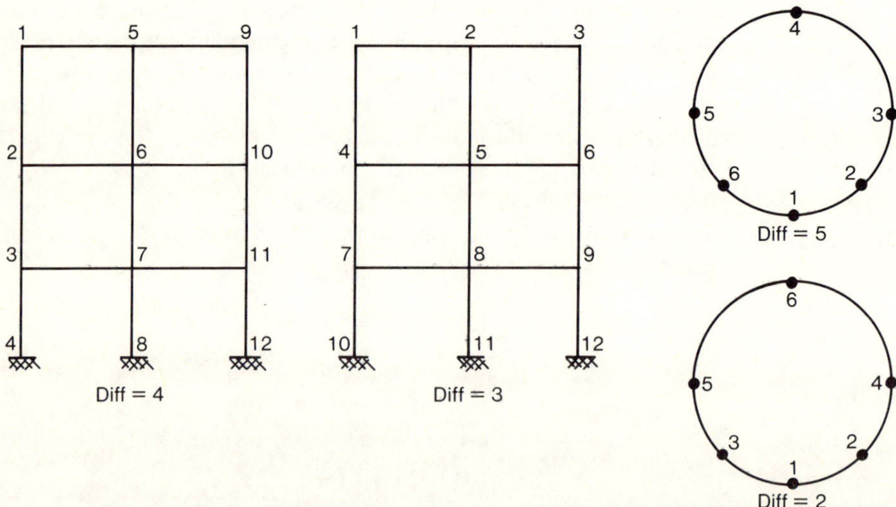

Fig. 5.3   Different node numbering schemes to minimize the difference between the node numbers at the ends of elements.

(2) In the majority of structures, the node freedom numbers at a higher
     numbered node are generally higher than those at a lowered numbered
     node.

In order to minimize the half-bandwidth it is necessary to ensure that the
node numbers at the ends of an element are as close as possible. This can be
achieved by numbering the nodes of a structure in the 'shortest direction' as
shown in Fig. 5.3.

5.5 SUBROUTINES BANDWIDTH, MAXEQNS, GAUSRED AND BACKSUB

There are four important subroutines used in connection with the solution of
simultaneous equations. They are as follows.

(1) BANDWIDTH. This subroutine calculates the maximum value of the
     half-bandwidth NBAND by calculating the maximum difference be-
     tween the nonzero node freedom numbers for the nodes defining the
     element. The steps in the calculation are as follows.

   (a) Choose an element IELEM.
   (b) Determine the node numbers of the nodes at the ends of this
        element from the array LNODS(NELEM,NNODF).
   (c) Calculate the node freedom numbers of the nodes defining the
        element from the array NODFRE(NPOIN,NDOFN).
   (d) Store the node freedom numbers calculated in (3) in a temporary
        array MEMDIS(NEVAB) as explained in section 3.5.
   (e) Calculate the maximum difference between the nonzero elements
        of the array MEMDIS(NEVAB). Call this LBAND.
   (f) Repeat steps (1) to (5) for all NELEM elements. The maximum
        value of LBAND is equal to NBAND.

(2) MAXEQNS. This subroutine calculates the number of equations (i.e.
     the number of nonzero displacements) to be solved. The value NEQNS
     is equal to the maximum freedom number in the array NODFRE-
     (NPOIN,NDOFN). All that is involved is the scanning of the array
     NODFRE(NPOIN,NDOFN) to determine the largest number in the
     array.
(3) GAUSRED. This subroutine carries out the Gaussian reduction of the
     structural stiffness matrix GSTIF. The steps involved have already been
     explained in section 5.3. Table 5.2 shows the names as used in short
     program segments and the subroutine GAUSRED.
(4) BACKSUB. This subroutine carries out the reduction of the load vector
     GLOAD and also the back substitution to determine the displacements.
     The steps involved have been explained in section 5.3.

Table 5.2

Relationship between the names of variables.

Names as used in section 5.3	Names used in subroutines
A	GSTIF
F	GLOAD
X	ASDIS
N	NEQNS
K	IEQNS
(I,J)	(IROWS,JCOLMS) in GAUSRED
(I,J)	(IVARBL,JBAND) in BACKSUB

5.6 FIXED VALUE OF DISPLACEMENT AT CERTAIN NODES

It often happens that, when analysing a structure under given loads, it is necessary to ensure that the value of the displacement corresponding to a particular node freedom at a point is equal to a given value. A common situation is where it is required that the displacement at a support is equal to the estimated foundation settlement. There are many ways of ensuring this when solving for the displacements, but the simplest approach is the so-called 'stiff spring' approach. The main advantage of this approach is that it requires minimum modification to the general procedure described in the previous sections. The basic concept behind the method can be explained with respect to a simple example. Consider the following row of structural stiffness matrix corresponding to the equilibrium in the direction of displacement d_i:

$$F_i = k_{1i}d_1 + k_{2i}d_2 + \cdots + k_{ii}d_i + \cdots + k_{Ni}d_N$$

As opposed to the normal case where F_i is known but not any of the joint displacements, in this case the load F_i is not known but the displacement d_i is given a fixed value. It is necessary that, during the solution of simultaneous equations, d_i must be held at a given value. In other words, while in the normal case all the components of the load vector are known and none of the displacement values is known, in this case we have a 'mixed' situation in the sense that certain displacements are known to have a given value but to compensate for this the corresponding components of the load vector are unknown. Theoretically the problem can be solved but the standard procedure requires a great deal of rearranging of the structural stiffness matrix. This can be avoided with a little thought. As given, the value of F_i depends on the values of all the displacements d_j, $j = 1$, N. However, if F_i is made to depend essentially on d_i only, then since d_i is known then F_i can be determined and the standard procedure can be used for the solution of simultaneous equations, pretending that d_i is unknown but F_i is known. This can be done by making k_{ii} much larger than the other coefficients in the equation. If this is done, then $F_i = k_{ii}d_i$ for all practical purposes. The steps involved can be summarized as follows.

(1) Form the structural stiffness matrix.
(2) Multiply the diagonal terms of the stiffness matrix corresponding to the node freedom numbers whose values are fixed by a 'large' number such as 10^6. In the program this is accomplished by the subroutine STFSPRG which stands for <u>ST</u>i<u>Ff</u> <u>SPRinG</u>.
(3) The load vector term corresponding to the fixed displacement is made equal to the modified diagonal term multiplied by the fixed value of displacement. In the program this operation is accomplished by the subroutine FIXDISP which stands for <u>FIX</u>ed <u>DISP</u>lacement.
(4) Solve for the unknown displacement by using the subroutines GAUSRED and BACKSUB.

It should be appreciated that the procedure adopted does modify the original structural stiffness matrix. If a structure has to be analysed not only for given external loads but also for fixed displacements, it is necessary to treat the case of fixed displacement as a separate structure and not as a part of the case when the structure needs to be analysed for given loads only.

5.7 REFERENCES

[1] P. Bhatt, *Problems in Matrix Analysis of Structures*, Construction Press, 1981, pp. 182–185.

```
C
C**********************************************
C
      SUBROUTINE MAXEQNS(NODFRE)
C
C***THIS SUBROUTINE CALCULATES THE NUMBER OF
C   SIMULTANEOUS EQUATIONS TO BE SOLVED BY
C   SIMPLY DETERMINING THE HIGHEST NODE FREEDOM
C   NUMBER EXISTING ANYWHERE IN THE STRUCTURE.
C
C
**************************************************
*                                                *
*   INCLUDE APPROPRIATE COMMON BLOCKS HERE !!!!   *
*                                                *
**************************************************
      DIMENSION NODFRE(100,3)
      NEQNS=0
      DO 10 IPOIN=1,NPOIN
      DO 10 JDOFN=1,NDOFN
      IF(NODFRE(IPOIN,JDOFN).GT.NEQNS)NEQNS=
     @NODFRE(IPOIN,JDOFN)
   10 CONTINUE
      RETURN
      END
```

```
C
C***********************************************
C
      SUBROUTINE BANDWIDTH(NODFRE,LNODS)
C
C***THIS SUBROUTINE CALCULATES THE HALF-BANDWIDTH
C   FROM THE MAXIMUM DIFFERENCE BETWEEN THE
C   NODE FREEDOMS OF THE NODES DEFINING THE
C   ENDS OF ELEMENTS
CC
C
**************************************************
*                                                *
*   INCLUDE APPROPRIATE COMMON BLOCKS HERE !!!!   *
*                                                *
**************************************************
C
      DIMENSION NODFRE(100,3),LNODS(100,2),
     @MEMDIS(6)
      NBAND=0
C
C***CYCLE ON ALL THE ELEMENTS.
C
      DO 1 IELEM=1,NELEM
C
C***DETERMINE THE NODES AT THE ENDS OF THE
C   ELEMENT.
C
      IEND=LNODS(IELEM,1)
      JEND=LNODS(IELEM,2)
C
C***DETERMINE THE NODE FREEDOMS CORRESPONDING
C   TO THE NODES AT THE ENDS OF THE ELEMENT AND
C   STORE IT IN ARRAY MEMDIS(NDOFN).
```

```fortran
C
        DO 2 IDOFN=1,NDOFN
        MEMDIS(IDOFN)=NODFRE(IEND,IDOFN)
      2 MEMDIS(IDOFN+NDOFN)=NODFRE(JEND,IDOFN)
C
C***CALCULATE THE HALF BAND-WIDTH LBAND FOR THE
C   ELEMENT BY CALCULATING THE MAXIMUM
C   DIFFERENCE BETWEEN ANY TWO ELEMENTS OF THE
C   ARRAY MEMDIS(NEVAB) EXCLUDING ZEROS.THE
C   MAXIMUM OF ALL THE MAXIMUM NODE FREEDOMS SO
C   CALCULATED GIVES THE VALUE OF NBAND.
        DO 3 IEVAB=1,NEVAB
        IF(MEMDIS(IEVAB).EQ.0) GOTO 3
        DO 4 JEVAB=1,NEVAB
        IF(MEMDIS(JEVAB).EQ.0) GOTO 4
        LBAND = ABS(MEMDIS(IEVAB)-MEMDIS(JEVAB))
        IF(LBAND.GT.NBAND) NBAND=LBAND
      4 CONTINUE
      3 CONTINUE
      1 CONTINUE
        RETURN
        END
```

```fortran
C
C**********************************************
C
        SUBROUTINE GAUSRED
C
C***THIS SUBROUTINE REDUCES THE BANDED STIFFNESS
C   MATRIX TO THE 'UPPER TRIANGULAR' FORM.THE
C   STIFFNESS MATRIX IS STORED IN THE RECTANGULAR
C   FORM WITH THE DIAGONAL IN THE FIRST COLUMN.
C
C
******************************************************
*                                                    *
*   INCLUDE APPROPRIATE COMMON BLOCKS HERE !!!!       *
*                                                    *
******************************************************
C
C***START THE DO LOOP WITH THE PIVOT EQUATION
C   K CALLED IEQNS VARYING FROM 1 TO (NEQNS-1).
C
        DO 1 IEQNS=1,NEQNS-1
C
C***PIVOT IS THE DIAGONAL TERM OF THE PIVOT
C   EQUATION. IF THE PIVOT IS NEAR TO ZERO THEN
C   THE MATRIX IS SINGULAR.  THEREFORE STOP ALL
C   THE CALCULATIONS. NFAIL = 1 IS A FLAG TO
C   STOP THE COMPUTATIONS.
C
        PIVOT=GSTIF(IEQNS,1)
        IF(ABS(PIVOT).LT.1.0E-10) GOTO 5
C
C***GAUSSIAN ELIMINATION NEEDS TO BE DONE ONLY
C   FOR EQUATIONS (PIVOT EQUATION + IBAND) WHERE
C   IBAND = 1 TO NBAND WITH THE PROVISO THAT
C   (PIVOT EQUATION + IBAND) IS = OR < NEQNS.
C
        DO 2 IBAND=1,NBAND
```

```
      IROWS=IEQNS+IBAND
      IF(IROWS.GT.NEQNS) GOTO 1
C
C***THE COEFFICIENT OF X(PIVOT EQUATION) OF THE
C   EQUATION (PIVOT EQUATION + IBAND) IS,
C   BECAUSE OF SYMMETRY, GSTIF(IEQNS,IBAND+1).
C   IF THIS IS ZERO THEN THERE IS NO NEED
C   TO MODIFY THE EQUATION (PIVOT EQUATION
C   + IBAND).
C
      IF(GSTIF(IEQNS,IBAND+1).EQ.0) GOTO 2
      QUOT=GSTIF(IEQNS,IBAND+1)/PIVOT
C
C***FOR THE EQUATION BEING MODIFIED VIZ. (PIVOT
C   EQUATION + IBAND) ONLY ELEMENTS UP TO
C   (NBAND-IBAND +1) NEED TO BE CONSIDERED.
C
      NCOLMS=NBAND-IBAND+1
      DO 3 JCOLMS=1,NCOLMS
      IF(JCOLMS.GT.NBAND+1) GOTO 3
      GSTIF(IROWS,JCOLMS)=GSTIF(IROWS,JCOLMS)
     @-QUOT*
     @GSTIF(IEQNS,JCOLMS+IBAND)
    3 CONTINUE
    2 CONTINUE
    1 CONTINUE
      GOTO 6
    5 NFAIL=1
      WRITE(6,900)IEQNS
  900 FORMAT(//,' MATRIX IS SINGULAR',//,'
     @ERROR IN DIAGONAL TERM OF EQUATION = ',I5)
    6 RETURN
      END
```

```
C
C*********************************************
C
      SUBROUTINE STFSPRG(NODFRE)
C
C***THIS SUBROUTINE USES THE STIFF SPRING
C   APPROACH TO INCORPORATE THE FIXED
C   DISPLACEMENT VALUES. THE DIAGONAL TERMS
C   CORRESPONDING TO THE FREEDOMS WHICH HAVE A
C   PRESCRIBED VALUE ARE MULTIPLIED BY 10**6.
C   THE SUBROUTINE FIXDISP MODIFIES THE GLOAD
C   VECTOR.
C
C
**************************************************
*                                                *
*   INCLUDE APPROPRIATE COMMON BLOCKS HERE !!!!   *
*                                                *
**************************************************
C
      DIMENSION NODFRE(100,3)
C
C***THERE ARE NVFIX NODES WHERE ONE OR MORE
C   FREEDOMS HAVE A FIXED VALUE.  THE NODE
C   NUMBER IS IN THE ARRAY NDVFIX(NVFIX)
C   AND THE VALUE OF THE PRESCRIBED DISPLACEMENT
```

```
C     IN THE ARRAY PRESC(NVFIX, NDOFN).   IF
C     PRESC(IVFIX,IDOFN) = 0 IT MEANS THAT IT
C     IS A FREE DEGREE OF FREEDOM.
C
      DO 10 IVFIX=1,NVFIX
C
C***DETERMINE THE NODE NUMBER OF THE NODE
C
      NODFIX=NDVFIX(IVFIX)
C
C***CHECK THROUGH THE NDOFN FREEDOMS AT NODFIX
C     TO SEE WHICH HAVE A PRESCRIBED VALUE.   IF
C     THE PRESCRIBED VALUE OF DISPLACEMENT IS ZERO
C     IT MEANS IT IS NOT- REPEAT NOT- FIXED.
C
      DO 20 IDOFN=1,NDOFN
      IF(PRESC(IVFIX,IDOFN).EQ.0.0) GOTO 20
C
C***CALCULATE THE NODE FREEDOM NUMBER OF THE
C     PRESCRIBED DISPLACEMENT.
C
      IFRDOM=NODFRE(NODFIX,IDOFN)
C
C***MODIFY THE DIAGONAL ELEMENT OF THE STIFFNESS
C     MATRIX.
C
      GSTIF(IFRDOM,1)=GSTIF(IFRDOM,1)*(1.0E06)
   20 CONTINUE
   10 CONTINUE
      RETURN
      END
```

```
C
C********************************************
C
      SUBROUTINE FIXDISP(NODFRE)
C
C***THIS SUBROUTINE MODIFIES THE LOAD VECTOR
C     CORRESPONDING TO THOSE FREEDOMS WHICH HAVE
C     A PRESCRIBED VALUE.
C
**********************************************
*                                            *
*   INCLUDE APPROPRIATE COMMON BLOCKS HERE !!!!   *
*                                            *
**********************************************
C
      DIMENSION  NODFRE(100,3)
C
C***THERE ARE NVFIX NODES WHERE ONE OR MORE
C     FREEDOMS HAVE A FIXED VALUE.   THE NODE
C     NUMBER IS IN THE ARRAY NDVFIX(NVFIX) AND THE
C     VALUE OF THE PRESCRIBED DISPLACEMENT IN THE
C     ARRAY PRESC(NVFIX, NDOFN).   IF PRESC(IVFIX,
C     IDOFN) = 0 IT MEANS THAT IT IS A FREE DEGREE
C     OF FREEDOM.
C
      DO 10 IVFIX=1,NVFIX
C
C***DETERMINE THE NODE NUMBER OF THE NODE
```

```
C
      NODFIX=NDVFIX(IVFIX)
C
C***CHECK THROUGH THE NDOFN FREEDOMS AT NODFIX
C   TO SEE WHICH HAVE A PRESCRIBED VALUE.   IF
C   THE PRESCRIBED VALUE OF DISPLACEMENT IS ZERO
C   IT MEANS IT IS NOT- REPEAT NOT- FIXED.
C
      DO 20 IDOFN=1,NDOFN
      IF(PRESC(IVFIX,IDOFN).EQ.0) GOTO 20
C
C***CALCULATE THE NODE FREEDOM NUMBER OF THE
C   PRESCRIBED DISPLACEMENT.
C
      IFRDOM=NODFRE(NODFIX,IDOFN)
C
C***THE LOAD CORRESPONDING TO THE FIXED DEGREE
C   OF DISPLACEMENT IS EQUAL TO DIAGONAL TERM X
C   FIXED DISPLACEMENT VALUE
C
      GLOAD(IFRDOM)=GSTIF(IFRDOM,1)*
     @PRESC(IVFIX,IDOFN)
   20 CONTINUE
   10 CONTINUE
      RETURN
      END
```

```
C
C*********************************************
C
      SUBROUTINE BACKSUB
C
C***THIS SUBROUTINE REDUCES THE LOAD VECTOR BY
C   GAUSSIAN ELIMINATION AND CALCULATES THE
C   DISPLACEMENTS BY BACK SUBSTITUTION.
C
**********************************************
*                                            *
*   INCLUDE APPROPRIATE COMMON BLOCKS HERE !!!!   *
*                                            *
**********************************************
C
C***CHOOSE PIVOT EQUATION VARYING FROM 1 TO
C   (NEQNS-1).  GAUSSIAN ELIMINATION ON THE LOAD
C   VECTOR.
C
      DO 10 IEQNS=1,NEQNS-1
      PIVOT=GSTIF(IEQNS,1)
C
C***GAUSSIAN ELIMINATION ONLY FOR LOAD VECTOR
C   COMPONENTS UP TO NBAND FROM THE PIVOT
C   EQUATION.
C
      DO 20 IBAND=1,NBAND
      IROWS=IEQNS+IBAND
      IF(IROWS.GT.NEQNS) GOTO 10
      QUOT=GSTIF(IEQNS,IBAND+1)/PIVOT
      GLOAD(IROWS)=GLOAD(IROWS)-QUOT*
     @GLOAD(IEQNS)
   20 CONTINUE
```

```
      10 CONTINUE
C
C***THE LAST UNKNOWN IS DIRECTLY CALCULATED.
C
      ASDIS(NEQNS)=GLOAD(NEQNS)/GSTIF(NEQNS,1)
C
C***WORK BACKWARDS TO CALCULATE THE VALUE OF
C   THE OTHER UNKNOWNS.
C
      DO 30 IVARBL=(NEQNS-1),1,-1
      SUM=0.0
      DO 40 JBAND=2,(NBAND+1)
C
C***(IVARBL+JBAND-1) IS THE COLUMN POSITION IN
C   THE FULL MATRIX REPRESENTATION.
C
      JCOLMN=IVARBL+JBAND-1
      IF(JCOLMN.GT.NEQNS) GOTO 45
      SUM=SUM+GSTIF(IVARBL,JBAND)*ASDIS(JCOLMN)
   40 CONTINUE
   45 ASDIS(IVARBL)=(GLOAD(IVARBL)-SUM)/
     @GSTIF(IVARBL,1)
   30 CONTINUE
      RETURN
```

CHAPTER 6   Calculation of the Stresses in the Elements

After the displacements of the joints of the structure have been calculated, the next step is to calculate 'stresses' in the elements of the structure. The stresses in the context of the skeletal structures are the axial force, the shear force, and the bending and twisting moments at the ends of the element. The final stresses are a summation of the stresses due to the following causes.

(1) The fixed end state with the lateral loads on the element.
(2) The fixed end state with all 'expansion' due to temperature changes fully restrained.
(3) The displacements of the joints of the structure.

The state of stress due to each of the above three causes will be discussed in the following sections.

6.1 THE FIXED END STATE WITH LATERAL LOADS ON THE ELEMENT

The stresses from this source are applicable only to elements of 2-D rigid-jointed structures and plane grids as discussed below.

(i) *A 2-D rigid-jointed structure element.* Figure 4.1 shows the general loading on an element. The applied loading can be resolved, as discussed in section 4.3 into normal and tangential loads as shown in Fig. 4.2. The normal loads induce shear forces and moments while the tangential load induces only axial forces. Using the convention that axial forces are positive if tensile, that the shear forces are positive if they produce 'clockwise rotation' and that moments are positive if clockwise, the stresses at the ends of an element due to lateral loads are as follows. At end 1, the axial tension is $0.5 \cdot$ total tangential load, the shear force is $-R_1$ and the moment is M_1. Similarly, at end 2, the axial tension is $-0.5 \cdot$ total tangential load, the shear force is R_2 and the moment is M_2.

(ii) *A plane grid element.* In a manner similar to the 2-D rigid-jointed structure element, the applied load on the element induces only bending moment and shear force in the element. It is assumed that the external load is concentric w.r.t. the element cross section so that no torsional moments are induced. Using the convention described in (i), the stresses at the ends of the element are as follows. At end 1, the shear force is $-R_1$, the moment M_1 and the torque is zero. Similarly, at end 2, the shear force is R_2, the moment M_2 and the torque zero.

In the program the above six quantities are stored in a temporary array called ELOAD(NEVAB).

6.2   THE FIXED END STATE WITH THERMAL EXPANSION FULLY RESTRAINED

The expressions for fixed end forces due to thermal expansion were derived in Chapter 1. They are repeated here for convenience. Referring to Fig. 4.3 and section 4.5 where the notation used is defined, the fixed end forces in the various types of elements as shown in Figs 4.4(a)–(c) are as follows.

(i) *A 2-D or 3-D pin-jointed element.*   The axial tension is given by $F = -AE\alpha_t T$.

(ii) *A 2-D rigid-jointed structure element.*   The axial tension and moments at the ends are follows. At end 1, the axial tension is $-AE\alpha_t(T_u Y_b + T_b Y_u)/d$, the shear force is zero and the moment is $EI\alpha_t(T_u - T_b)/d$. Similarly, at end 2, the axial tension is the same as the axial tension at 1, the shear force is zero and the moment is minus the moment at end 1.

(iii) *A plane grid element.*   Only moments are induced but the shear force and the twisting moments are zero at both ends. Therefore the moments are given as follows. At end 1, the moment is $EI\alpha_t(T_u - T_b)/d$. Similarly, at end 2, the moment is minus the moment at end 1.

In the program the above six values are stored in a temporary array FORSTM(NEVAB) where FORSTM stands for <u>FOR</u>ces due to <u>TeM</u>perature changes.

6.3   STRESSES DUE TO DISPLACEMENTS AT THE ENDS OF THE ELEMENT

The stresses due to displacements were derived in Chapter 1 and are repeated here for convenience. They are as follows.

(i) *A 2-D pin-jointed element.*   Referring to Fig. 1.10, the axial tensile force is given by

$$F = \frac{AE}{L}[(u_2 - u_1)l + (v_2 - v_1)m]$$

Similarly for a 3-D pin-jointed element the axial force is given by

$$F = \frac{AE}{L}\left[(u_2 - u_1)l + (v_2 - v_1)m + (w_2 - w_1)n\right]$$

(ii) *A 2-D rigid-jointed structure element.*   Referring to Figs. 1.11 and 3.2, the stresses at the ends of the element are given as follows. The axial tension is given by

$$F = \frac{AE}{L}\left[(u_2 - u_1)l + (v_2 - v_1)m\right]$$

At end 1 the moment M_1 and the shear force F_{n1} are given by

$$M_1 = \frac{EI}{L}\left(S_{11}\theta_1 + S_{12}\theta_2 - \frac{Q_{11}}{L}V_{n1} + \frac{Q_{12}}{L}V_{n2}\right)$$

$$F_{n1} = \frac{EI}{L^2}\left(\frac{T_{11}}{L}V_{n1} - Q_{11}\theta_1 - \frac{T_{12}}{L}V_{n2} - Q_{21}\theta_2\right)$$

At end 2, the moment M_2 and the shear force F_{n2} are given by

$$M_2 = \frac{EI}{L}\left(S_{12}\theta_1 + S_{22}\theta_2 - \frac{Q_{21}}{L}V_{n1} + \frac{Q_{22}}{L}V_{n2}\right)$$

$$F_{n2} = \frac{EI}{L^2}\left(-\frac{T_{12}}{L}V_{n1} + Q_{12}\theta_1 + \frac{T_{22}}{L}V_{n2} + Q_{22}\theta_2\right)$$

where the normal translations V_{n1} and V_{n2} at ends 1 and 2 respectively are given in terms of the displacements along the coordinate axes by $V_{n1} = v_1 l - u_1 m$ and $V_{n2} = v_2 l - u_2 m$.

(iii) *A plane grid element.*   Referring to Fig. 1.12, the moments and shear forces at the ends of the element are given by the equations in (ii). The bending rotations θ_1 and θ_2 and normal translations V_{n1} and V_{n2} are given in terms of the rotations about the coordinate axes by

$$\theta_1 = \theta_{y1}l - \theta_{x1}m, \quad \theta_2 = \theta_{y2}l - \theta_{x2}m, \quad V_{n1} = w_1, \quad V_{n2} = w_2$$

The expressions for the torques and bimoments depend on whether warping restraint is ignored (i.e. IWARP = 0) or considered (i.e. IWARP = 1). The two cases are handled as follows.

(a)   IWARP = 0. In this case the torque is given by

$$T = \frac{GJ}{L}(\psi_2 - \psi_1)$$

where the twists ψ_1 and ψ_2 are given in terms of rotations about the coordinate axes by

$$\psi_1 = \theta_{x1}l + \theta_{y1}m, \quad \psi_2 = \theta_{x2}l + \theta_{y2}m$$

(b)   IWARP $= 1$. In this case the torques and bimoments are calculated using the formulae derived in Appendix 4 and are given by

$$T = \frac{GJ}{L}[TW_{11}(\psi_2 - \psi_1) + TW_{12}(\psi'_1 + \psi'_2)]$$

The bimoments at the ends are given by

$$(\text{bimoment})_2 = \frac{GJ}{L}[TW_{12}(\psi_2 - \psi_1) + TW_{24}\psi'_1 + TW_{22}\psi'_2]$$

$$(\text{bimoment})_2 = \frac{GJ}{L}[TW_{12}(\psi_2 - \psi_1) + TW_{24}\psi'_1 + TW_{22}\psi'_2]$$

In the above equations, ψ' is the first derivative of ψ w.r.t. x, i.e. $d\psi/dx$. The coefficients TW_{11}, TW_{22}, TW_{12} and TW_{24} are defined in Appendix 4.

The above forces are stored in a temporary array DISFOR(NEVAB) which stands for <u>DIS</u>placement <u>FOR</u>ces.

6.4   THE PROCEDURE FOR CALCULATION OF THE FINAL STRESSES IN AN ELEMENT

The steps involved in the calculation of the stresses in an element can be summarized as follows.

(1)   Choose an element IELEM.

(2)   Determine the node numbers at the ends of the element which are stored in the array LNODS(NELEM,NNODE).

(3)   Determine the coordinates of nodes at the ends of the element from the array COORD(NPOIN,NDIME). Similarly, determine the node freedom numbers of the nodes defining the element from the array NODFRE(NPOIN,NDOFN).

(4)   Determine the group number and, where relevant, the type number of the element from the array MATNO(NELEM,2).

(5)   Determine the material and cross-sectional properties of the element from the array PROPS(NMATS,NPROP + ?).

(6)   From the information in (2) to (5), calculate the axial rigidity AE, the flexural rigidity EI, the direction cosines l and m, etc., of the element.

(7)   Where relevant, from the array LODEL(NLODEL) check whether the element has any lateral loads on it. If it has, determine the lateral load values from the array FORSEL(NLODEL,4). Calculate the fixed end stresses at the ends of the element as described in section 6.1, and store them in a temporary array ELOAD(NEVAB).

(8)   From the array ASDIS(NEQNS), determine the nonzero joint displacements of the nodes at the ends of the element. Calculate the stresses at the ends due to displacements as described in section 6.3 and store them in a temporary array DISFOR(NEVAB).

(9)   From the array TEMP(NPOIN,2), determine the temperature change in the element and from the array PROPS(NMATS,NPROP+?), determine the coefficient of linear thermal expansion α_t, the depth d of the element, etc. From the data, calculate the restraint forces due to the restraining of thermal expansion as explained in section 6.2. Store these stresses in a temporary array FORSTM(NEVAB).

(10)   The sum of the three forces described in (7) to (9) are stored in a temporary array TOTFOR(NEVAB). The name TOTFOR stands for <u>TOT</u>al <u>FOR</u>ce which is the final force at the ends of the element. Print the value of these forces.

(11)   Choose the next element and repeat the calculations (2) to (10).

Note that for a pin-jointed member it is not necessary to use the arrays ELOAD, FORSTM, DISFOR and TOTFOR because there is only one tensile force which is valid for the whole element.

6.5 STRESS CALCULATION FOR A BEAM ON AN ELASTIC FOUNDATION

For a beam element without lateral elastic support, once the forces at the ends of the element are known, then the forces at any section along its length can be detemined from statics. For a beam on an elastic foundation, because of the complex distribution of the foundation pressure, the determination of the stresses along the element from simple statics alone is not possible. The procedure followed in the program is as follows.

(1)   Choose an element IELEM. From the array MATNO(NELEM,2), determine whether the element is a beam on an elastic foundation. If it is not, then skip all the following steps. If it is, then determine from the array PROPS(NMATS,NPROP+?) the elastic and geometrical properties including the modulus of subgrade reaction.

(2)   From the arrays LNODS(NELEM,NNODE), COORD(NPOIN,-NDIME), NODFRE(NPOIN,NDOFN), etc., determine the node numbers at the ends of the element, their coordinates and the node freedom numbers corresponding to them. Calculate the length of the element, direction cosines, etc.

(3)   By scanning the array LODEL(NLODEL), determine whether the element carries any lateral loads and, if it does, determine their values from the array FORSEL(NLODEL,4).

(4)   From the array ASDIS(NEQNS), determine the displacements at the ends of the element.

(5)   Choose, say, ten sections along the element.

(6)   At each of these sections in turn, using the formulae derived in Appendix 2, do the following.

  (a)   Using the shape functions and the values of the displacements at the ends of the element determined in (4), calculate the lateral displacement of the element due to end displacements.

   (b) Determine the lateral displacement due to the lateral loads on the element when the ends are fully fixed.

   (c) The bending moment M and shear force Q at a section are obtained from the derivatives of displacement in (a) and (b) as

$$M = EI\,\frac{d^2v}{dx^2}, \quad Q = -EI\,\frac{d^3v}{dx^3}$$

Note that the formulae given in the Appendix 2 for the displacements due to a concentrated load at mid-span apply only to one half of the element. However, because of symmetry (the displacement and bending moment are symmetrical while the shear force is antisymmetrical w.r.t. the line of symmetry of the element); the displacement and forces throughout the element can be calculated.

6.6 SUBROUTINES STRESSPJ, STRESSRJ, STRESSGRID, PRESURRJ AND PRESURGRID

There are five subroutines used for the calculation the stresses in the elements. The following is a brief description of the subroutines.

(1) STRESSPJ. This calculates the axial tensile force in 2-D and 3-D pin-jointed elements.

(2) STRESSRJ. This calculates the axial force, shear force and bending moment in the elements of 2-D rigid-jointed structures.

(3) STRESSGRID. This calculates the bending and twisting moments and shear force in the elements of plane grid structures.

(4) PRESSURRJ. This calculates the foundation pressure, displacement, bending moment and shear force in a beam on an elastic foundation which is a part of a 2-D rigid-jointed structure.

(5) PRESSURGRID. This is similar to PRESURRJ except that it is applicable to a beam on an elastic foundation forming part of a plane grid structure, e.g. a foundation grillage.

The reader should be able to follow the subroutines without any problems.

```
C
C***************************************************
C
      SUBROUTINE STRESSPJ(NODFRE,LNODS,MATNO)
C
C***THIS SUBROUTINE CALCULATES THE AXIAL FORCE
C   IN THE ELEMENTS
C
      COMMON/CONTRO/NPOIN,NELEM,NNODE,NDOFN,
     @NDIME,NPROP,NMATS,NRESND,NVFIX,NEVAB,
     @NFAIL,NOPTN,NEQNS,NBAND,ITEMP
      COMMON/LGDATA/COORD(100,3),PROPS(10,3),
     @PRESC(80,3),TEMP(100),NDFFIX(80),
     @NDVFIX(80),IFPRE(80,3),ASDIS(400),
     @GSTIF(100,30),GLOAD(100)
      DIMENSION NODFRE(100,3),LNODS(100,2),
     @MATNO(100,2)
C
C***STRESS CALCULATION IN THE MEMBERS
C
      DO 10 IELEM=1,NELEM
      LPROP=MATNO(IELEM,1)
      YOUNG=PROPS(LPROP,1)
      AREA=PROPS(LPROP,2)
C
C***CALCULATE THE NODES CORRESPONDING TO THE TWO
C   ENDS OF THE ELEMENT
C
      IEND=LNODS(IELEM,1)
      JEND=LNODS(IELEM,2)
      IF(NDIME.EQ.3) GOTO 15
C
C***CALCULATE THE COORDINATES OF THE NODES AT
C   THE ENDS OF THE ELEMENT, THE LENGTH,
C   DIRECTION COSINES ETC. FOR A 2-D ELEMENT.
C
      XPROJ=COORD(JEND,1)-COORD(IEND,1)
      YPROJ=COORD(JEND,2)-COORD(IEND,2)
      SPAN=SQRT(XPROJ*XPROJ+YPROJ*YPROJ)
      XL=XPROJ/SPAN
      YM=YPROJ/SPAN
      AEBYL=AREA*YOUNG/SPAN
C
C***SET THE DISPLACEMENTS AT THE ENDS OF THE
C   ELEMENT AS ZERO.
C
      U1=0.0
      V1=0.0
      U2=0.0
      V2=0.0
C
C***CALCULATE THE NONZERO VALUES OF
C   DISPLACEMENTS.
C
      IF(NODFRE(IEND,1).NE.0)U1=
     @ASDIS(NODFRE(IEND,1))
      IF(NODFRE(IEND,2).NE.0)V1=
     @ASDIS(NODFRE(IEND,2))
      IF(NODFRE(JEND,1).NE.0)U2=
     @ASDIS(NODFRE(JEND,1))
```

```
      IF(NODFRE(JEND,2).NE.0)V2=
      @ASDIS(NODFRE(JEND,2))
         GOTO 25
C
C***CALCULATE THE COORDINATES AT THE ELEMENT
C   ENDS, LENGTH ETC.
C   FOR A 3-D ELEMENT.
C
   15 XPROJ=COORD(JEND,1)-COORD(IEND,1)
      YPROJ=COORD(JEND,2)-COORD(IEND,2)
      ZPROJ=COORD(JEND,3)-COORD(IEND,3)
      SPAN=SQRT(XPROJ*XPROJ+YPROJ*YPROJ +
      @ZPROJ*ZPROJ)
      XL=XPROJ/SPAN
      YM=YPROJ/SPAN
      ZN=ZPROJ/SPAN
      AEBYL=AREA*YOUNG/SPAN
C
C***SET THE DISPLACEMENTS AT THE ENDS OF THE
C   ELEMENT AS ZERO.
C
      U1=0.0
      V1=0.0
      W1=0.0
      U2=0.0
      V2=0.0
      W2=0.0
C
C***CALCULATE THE NONZERO VALUES OF
C   DISPLACEMENTS.
C
      IF(NODFRE(IEND,1).NE.0)U1=
      @ASDIS(NODFRE(IEND,1))
      IF(NODFRE(IEND,2).NE.0)V1=
      @ASDIS(NODFRE(IEND,2))
      IF(NODFRE(IEND,3).NE.0)W1=
      @ASDIS(NODFRE(IEND,3))
      IF(NODFRE(JEND,1).NE.0)U2=
      @ASDIS(NODFRE(JEND,1))
      IF(NODFRE(JEND,2).NE.0)V2=
      @ASDIS(NODFRE(JEND,2))
      IF(NODFRE(JEND,3).NE.0)W2=
      @ASDIS(NODFRE(JEND,3))
C
C***CALCULATE THE VALUE OF THE THERMAL RESTRAINT
C   FORCE.
C
   25 RAXIAL=0.0
      IF(ITEMP.EQ.0) GOTO 20
      IF(TEMP(IELEM).EQ. 0.0) GOTO 20
      ALPHAT=PROPS(LPROP,3)
      RAXIAL=-AREA*YOUNG*ALPHAT*TEMP(IELEM)
C
C***TOTAL FORCE = FORCE DUE TO DISPLACEMENTS +
C   THERMAL RESTRAINT
C   FORCE.
C
   20 IF(NDIME.EQ.2)AXFORS=AEBYL*((U2-U1)*XL+
      @(V2-V1)*YM)+RAXIAL
      IF(NDIME.EQ.3)AXFORS=AEBYL*((U2-U1)*XL+
```

```
      @(V2-V1)*YM+(W2-W1)*ZN)+RAXIAL
       WRITE(6,900)IELEM,IEND,JEND,AXFORS
   10 CONTINUE
  900 FORMAT(I5,7X,I5,5X,I5,8X,E14.6)
       RETURN
       END
```

```
C
C**********************************************
C
       SUBROUTINE STRESSRJ(NODFRE,LNODS,MATNO)
C
C***THIS SUBROUTINE CALCULATES THE FINAL AXIAL,
C   BENDING AND SHEAR FORCES AT THE ENDS OF
C   ELEMENTS.
C
       COMMON/CONTRO/NPOIN,NELEM,NNODE,NDOFN,
      @NDIME,NPROP,NMATS,NRESND,NVFIX,NEVAB,
      @NFAIL,NLODEL,NEQNS,NBAND,ITEMP
       COMMON/LGDATA/COORD(100,2),PROPS(20,11),
      @PRESC(80,3),TEMP(100,2),NDFFIX(80),NDVFIX(80),
      @IFPRE(80,3),FORSEL(80,4),LODEL(80),
      @ASDIS(800),GSTIF(100,30),GLOAD(100)
       DIMENSION NODFRE(100,3),LNODS(100,2),
      @MATNO(100,2),DISFOR(6),TOTFOR(6),ELOAD(6),
      @FORSTM(6)
C
C***STRESS CALCULATION IN THE MEMBERS
C
       DO 10 IELEM=1,NELEM
       LPROP=MATNO(IELEM,1)
       NTYPE=MATNO(IELEM,2)
C
C***CALCULATE THE PROPERTIES OF THE ELEMENT.
C
       YOUNG=PROPS(LPROP,1)
       SHEARM=PROPS(LPROP,2)
       YINERA=PROPS(LPROP,3)
       AREA=PROPS(LPROP,4)
       ZASF=PROPS(LPROP,5)
C
C***CALCULATE THE NODES CORRESPONDING TO THE TWO
C   ENDS OF THE ELEMENT
C
       IEND=LNODS(IELEM,1)
       JEND=LNODS(IELEM,2)
C
C***CALCULATE THE COORDINATES OF THE NODES AT THE
C   ENDS OF THE ELEMENT, THE LENGTH AND THE
C   DIRECTION COSINES.
C
       XPROJ=COORD(JEND,1)-COORD(IEND,1)
       YPROJ=COORD(JEND,2)-COORD(IEND,2)
       SPAN=SQRT(XPROJ*XPROJ+YPROJ*YPROJ)
       XL=XPROJ/SPAN
       YM=YPROJ/SPAN
       EI=YOUNG*YINERA
       GA=SHEARM*AREA
       EIBYL=EI/SPAN
       EIBYL2=EIBYL/SPAN
```

```
      EIBYL3=EIBYL2/SPAN
      AEBYL=AREA*YOUNG/SPAN
      BETA=12.0*EIBYL2/GA*ZASF
C
C***ZERO THE VECTOR OF FIXED END FORCES
C
      DO 20 IEVAB=1,NEVAB
   20 ELOAD(IEVAB)=0.0
C
C***CALCULATE THE LATERAL LOADS ON THE ELEMENT
C   IF ANY.  AT FIRST ASSUME THAT THE ELEMENT
C   IS AN UNLOADED ELEMENT IE. JLODEL=0
C
      JLODEL=0
C
C***START CYCLE ON CHECKING IF IELEM IS A LOADED
C   ELEMENT.  IF IT IS CALCULATE THE LATERAL
C   LOADS ON IT.
C
      DO 50 ILODEL=1,NLODEL
      IF(LODEL(ILODEL).EQ.IELEM) JLODEL=ILODEL
      IF(JLODEL.NE.0) GOTO 55
   50 CONTINUE
      IF(JLODEL.EQ.0) GOTO 60
   55 UDLXT=FORSEL(JLODEL,1)*ABS(YPROJ)
      UDLYT=FORSEL(JLODEL,2)*ABS(XPROJ)
      CONLX=FORSEL(JLODEL,3)
      CONLY=FORSEL(JLODEL,4)
C
C***CALCULATE THE NORMAL COMPONENTS OF UDL AND
C   CONCENTRATED LOADS AND THE TOTAL TANGENTIAL
C   COMPONENT OF ALL APPLIED LOADS.
C
      UDLN=-UDLXT*YM+UDLYT*XL
      CN= -CONLX*YM+CONLY*XL
      TANCOM=(UDLXT+CONLX)*XL+(UDLYT+CONLY)*YM
   60 IF(NTYPE.EQ.2) GOTO 11
      IF(NTYPE.EQ.3) GOTO 12
      IF(NTYPE.EQ.4) GOTO 13
C
C***BENDING STIFFNESS COEFFICIENTS  FOR AN
C   ORTHODOX BEAM ELEMENT
C
      S11=(4.0+BETA)/(1.0+BETA)
      S22=S11
      S12=(2.0-BETA)/(1.0+BETA)
      Q11=S11+S12
      Q22=Q11
      Q12=Q11
      Q21=Q11
      T11=2.0*Q11
      T22=T11
      T12=T11
C
C***IF THE ELEMENT IS NOT CARRYING ANY LATERAL
C   LOAD NO NEED TO CALCULATE THE FIXED END
C   FORCES AT THE ENDS.
C
      IF(JLODEL.EQ.0) GOTO 14
C
```

```
C***CALCULATE THE FIXED END AXIAL, BENDING AND
C    SHEAR FORCES FOR AN ORDINARY BEAM.
C
        ELOAD(1)=0.5*TANCOM
        ELOAD(4)=-ELOAD(1)
        ELOAD(2)=-0.5*(UDLN+CN)
        ELOAD(5)=ELOAD(2)
        ELOAD(3)=SPAN/12.0*UDLN+SPAN/8.0*CN
        ELOAD(6)=-ELOAD(3)
        GOTO 14
C
C***BENDING STIFFNESS COEFFICIENTS OF A BEAM ON
C    ELASTIC FOUNDATION
C
    11  SUBGRM=PROPS(LPROP,NPROP+1)
        AMDAL=SPAN*(SQRT(SQRT(SUBGRM/(4.0*EI))))
        SH=SINH(AMDAL)
        CH=COSH(AMDAL)
        S=SIN(AMDAL)
        C=COS(AMDAL)
        FACTO1=AMDAL/(SH*SH-S*S)
        FACTO2=FACTO1*AMDAL
        FACTO3=FACTO2*AMDAL
        S11=2.0*(SH*CH-S*C)*FACTO1
        S22=S11
        S12=2.0*(CH*S-SH*C)*FACTO1
        T11=4.0*(SH*CH+S*C)*FACTO3
        T22=T11
        T12=4.0*(CH*S+SH*C)*FACTO3
        Q11=2.0*(CH*CH-C*C)*FACTO2
        Q22=Q11
        Q12=4.0*(SH*S)*FACTO2
        Q21=Q12
C
C***IF THE ELEMENT CARRIES NO LATERAL LOAD THERE
C    IS NO NEED TO CALCULATE THE FIXED END FORCES
C    AT THE ENDS.
C
        IF(JLODEL.EQ.0) GOTO 14
C
C***CALCULATE THE FIXED END AXIAL, SHEAR AND
C    BENDING FORCES FOR A BEAM ON ELASTIC SUBGRADE.
C    NOTE THAT PART OF THE NORMAL LOAD ON THE
C    BEAM IS ABSORBED BY THE 'FOUNDATION'.
C
        AMDAL2=0.5*AMDAL
        SH2=SINH(AMDAL2)
        CH2=COSH(AMDAL2)
        S2=SIN(AMDAL2)
        C2=COS(AMDAL2)
C
C***CALCULATE THE MODIFICATION FACTORS FOR FIXED
C    END MOMENT AND SHEAR FOR NORMAL DISTRIBUTED
C    LOAD.
C
        FACTO1=6.0/AMDAL/AMDAL*(SH-S)/(SH+S)
        FACTO2=2.0/AMDAL*(CH-C)/(SH+S)
C
C***CALCULATE THE MODIFICATION FACTORS FOR FIXED
C    END REACTIONS DUE TO CENTRAL NORMAL
```

```
C    CONCENTRATED LOAD.
C
        FACTO3=8.0/AMDAL*(SH2*S2)/(SH+S)
        FACTO4=2.0*(SH2*C2+CH2*S2)/(SH+S)
        ELOAD(1)=-0.5*TANCOM
        ELOAD(2)=-0.5*(UDLN*FACTO2+CN*FACTO4)
        ELOAD(4)=-ELOAD(1)
        ELOAD(5)=ELOAD(2)
        ELOAD(3)=SPAN/12.0*UDLN*FACTO1+
       @SPAN/8.0*CN*FACTO3
        ELOAD(6)=-ELOAD(3)
        GOTO 14
C
C***BENDING STIFFNESS COEFFICIENTS FOR A
C    SEMI-RIGID BEAM
C
     12 SPRNG1=PROPS(LPROP,NPROP+2)
        SPRNG2=PROPS(LPROP,NPROP+3)
        ALPHA1=EIBYL/SPRNG1
        ALPHA2=EIBYL/SPRNG2
        DENOM=1.0+BETA+(4.0+BETA)*(ALPHA1+ALPHA2)
       @+12.0*ALPHA1*ALPHA2
        S11=(4.0+BETA+12.0*ALPHA2)/DENOM
        S12=(2.0-BETA)/DENOM
        S22=(4.0+BETA+12.0*ALPHA1)/DENOM
        Q11=S11+S12
        Q12=Q11
        Q22=S12+S22
        Q21=Q22
        T11=Q11+Q22
        T22=T11
        T12=T11
C
C***IF THE ELEMENT DOES NOT CARRY ANY LATERAL
C    LOADS SKIP THE CALCULATION OF FIXED END
C    FORCES.
C
        IF(JLODEL.EQ.0) GOTO 14
C
C***CALCULATE THE FIXED END AXIAL, SHEAR AND
C    BENDING FORCES FOR A SEMI-RIGID BEAM.
C
        D=(2.0+6.0*ALPHA1)*(2.0+6.0*ALPHA2)-1
C
C***CALCULATE THE MODIFICATION FACTORS FOR FIXED
C    END FORCES DUE TO SEMI-RIGID EFFECT.
C
        FACTO1=3.0*(1.0+6.0*ALPHA2)/D
        FACTO2=3.0*(1.0+6.0*ALPHA1)/D
        ELOAD(3)=SPAN/12.0*UDLN*FACTO1+
       @SPAN/8.0*CN*(2.0*FACTO1)
        ELOAD(6)=-SPAN/12.0*UDLN*FACTO2
       @+SPAN/8.0*CN*(2.0*FACTO2)
        ELOAD(2)=-0.5*(UDLN+CN)+
       @(ELOAD(3)+ELOAD(6))/SPAN
        ELOAD(5)=-0.5*(UDLN+CN)-
       @(ELOAD(3)+ELOAD(6))/SPAN
        ELOAD(1)=-0.5*TANCOM
        ELOAD(4)=-ELOAD(1)
        GOTO 14
```

```
C
C***STIFFNESS MATRIX FOR A NONPRISMATIC ELEMENT
C   FOR WHICH THE BENDING STIFFNESS COEFFICIENTS
C   ARE READ AS DATA
C
   13 WRITE(6,955)
  955 FORMAT(//,'BENDING STIFFNESS COEFFICIENTS
     @FOR A NONPRISMATIC BEAM ELEMENT')
      READ(5,960)S11,S12,S22
      READ(5,960)T11,T12,T22
      READ(5,960)Q11,Q12,Q22,Q21
      WRITE(6,960)S11,S12,S22
      WRITE(6,960)T11,T12,T22
      WRITE(6,960)Q11,Q12,Q22,Q21
C
C***IF THE ELEMENT DOES NOT CARRY ANY LATERAL
C   LOAD SKIP CALCULATING THE FIXED END FORCES.
C
      IF(JLODEL.EQ.0) GOTO 14
C
C***READ THE FIXED END FORCES FOR THE UNUSUAL
C   ELEMENT
C
      WRITE(6,965)
  965 FORMAT(//,'FIXED END AXIAL FORCE, BM AND
     @SF FOR A NONPRISMATIC BEAM ELEMENT')
      READ(5,960)(ELOAD(IEVAB),IEVAB=1,NEVAB)
      WRITE(6,960)(ELOAD(IEVAB),IEVAB=1,NEVAB)
  960 FORMAT(6F10.4)
   14 CONTINUE
      T11=T11*EIBYL3
      T22=T22*EIBYL3
      T12=T12*EIBYL3
      S11=S11*EIBYL
      S22=S22*EIBYL
      S12=S12*EIBYL
      Q11=Q11*EIBYL2
      Q22=Q22*EIBYL2
      Q12=Q12*EIBYL2
      Q21=Q21*EIBYL2
C
C***SET THE VALUES OF THE END DISPLACEMENTS OF
C   THE ELEMENT AS ZERO.
C
      U1=0.0
      V1=0.0
      THETA1=0.0
      U2=0.0
      V2=0.0
      THETA2=0.0
C
C***CALCULATE THE NONZERO END DISPLACEMENTS
C   OF THE ELEMENT.
C
      IF(NODFRE(IEND,1).NE.0)U1=
     @ASDIS(NODFRE(IEND,1))
      IF(NODFRE(IEND,2).NE.0)V1=
     @ASDIS(NODFRE(IEND,2))
      IF(NODFRE(JEND,1).NE.0)U2=
     @ASDIS(NODFRE(JEND,1))
```

```
      IF(NODFRE(JEND,2).NE.0)V2=
     @ASDIS(NODFRE(JEND,2))
       IF(NODFRE(IEND,3).NE.0)THETA1=
     @ASDIS(NODFRE(IEND,3))
       IF(NODFRE(JEND,3).NE.0)THETA2=
     @ASDIS(NODFRE(JEND,3))
C
C***CALCULATE THE AXIAL DISPLACEMENTS.
C
      DISPAX1=U1*XL+V1*YM
      DISPAX2=U2*XL+V2*YM
C
C***CALCULATE THE NORMAL DISPLACEMENTS.
C
      DISPN1=-U1*YM+V1*XL
      DISPN2=-U2*YM+V2*XL
C
C***CALCULATE THE AXIAL FORCE DUE TO
C   DISPLACEMENTS.
C
      DISFOR(1)=AEBYL*(DISPAX2-DISPAX1)
      DISFOR(4)=DISFOR(1)
C
C***CALCULATE THE BENDING MOMENTS AT THE ENDS
C   DUE TO DISPLACEMENTS.
C
      DISFOR(3)=S11*THETA1+S12*THETA2-Q11*DISPN1
     @+Q12*DISPN2
      DISFOR(6)=S12*THETA1+S22*THETA2-Q21*DISPN1
     @+Q22*DISPN2
C
C***CALCULATE THE SHEAR FORCES DUE TO
C   DISPLACEMENTS.
C
      DISFOR(2)=-Q11*THETA1-Q21*THETA2+T11*
     @DISPN1-T12*DISPN2
      DISFOR(5)=+Q12*THETA1+Q22*THETA2-T12*
     @DISPN1+T22*DISPN2
C
C***ZERO THE VECTOR OF THERMAL RESTRAINT FORCES.
C
      DO 80 IEVAB=1,NEVAB
   80 FORSTM(IEVAB)=0.0
C
C***IF THERE ARE NO THERMAL LOADS OR IF THE
C   THERMAL LOAD IS ZERO FOR A PARTICULAR
C   ELEMENT SKIP THERMAL LOAD CALCULATIONS.
C
      IF(ITEMP.EQ.0) GOTO 90
      IF(TEMP(IELEM,1).EQ. 0.0 .AND.
     @TEMP(IELEM,2) .EQ. 0.0)GOTO 90
C
C*** CALCULATE THE THERMAL RESTRAINT AXIAL AND
C    BENDING FORCES
C
      ALPHAT=PROPS(LPROP,NPROP+4)
      DEPTH=PROPS(LPROP,NPROP+5)
      YUP=PROPS(LPROP,NPROP+6)
      YDOWN=DEPTH-YUP
      AE=AREA*YOUNG
```

```
      TOPTEM=TEMP(IELEM,1)
      BOTTEM=TEMP(IELEM,2)
      RAXIAL=ALPHAT*AE/DEPTH*(TOPTEM*YDOWN+
     @BOTTEM*YUP)
      RMOMENT=ALPHAT*EI/DEPTH*(TOPTEM-BOTTEM)
C
C***STORE THE RESTRAINT FORCES AT THE ENDS IN A
C   TEMPORARY  ARRAY FORSTM(NEVAB).
C
      FORSTM(1)=-RAXIAL
      FORSTM(4)=-RAXIAL
      FORSTM(3)=RMOMENT
      FORSTM(6)=-RMOMENT
C
C***TOTAL FORCE = FORCES DUE TO (FIXED END +
C   DISPLACEMENT + TEMPERATURE)
C
   90 DO 70 IEVAB=1,NEVAB
      TOTFOR(IEVAB)=ELOAD(IEVAB)+DISFOR(IEVAB)+
     @FORSTM(IEVAB)
   70 CONTINUE
      WRITE(6,901)IELEM,IEND,JEND,(TOTFOR(IEVAB)
     @,IEVAB=1,NEVAB)
  901 FORMAT(I5,2(2X,I5),2X,E14.5,2(5X,E14.6)
     @3(2X,E14.6))
   10 CONTINUE
      RETURN
      END
```

```
C
C*********************************************
C
      SUBROUTINE STRESSGRID(NODFRE,LNODS,MATNO)
C
C***THIS SUBROUTINE CALCULATES THE SHEAR FORCE,
C   BENDING MOMENT AND TORQUE AT THE ENDS OF
C   THE ELEMENTS.
C
      COMMON/CONTRO/NPOIN,NELEM,NNODE,NDOFN,
     @NDIME,NPROP,NMATS,NRESND,NVFIX,NEVAB,
     @NFAIL,NLODEL,NEQNS,NBAND,ITEMP,IWARP
      COMMON/LGDATA/COORD(100,2),PROPS(10,12),
     @PRESC(80,4),TEMP(100,2),NDFFIX(80),
     @NDVFIX(80),IFPRE(80,4),FORSEL(80,2),
     @LODEL(80),ASDIS(800),GSTIF(100,30),
     @GLOAD(100)
      DIMENSION NODFRE(100,4),LNODS(100,2),
     @MATNO(100,2),DISFOR(8),TOTFOR(8),ELOAD(8),
     @FORSTM(8)
C .
C***STRESS CALCULATION IN THE MEMBERS
C
      DO 10 IELEM=1,NELEM
      LPROP=MATNO(IELEM,1)
      NTYPE=MATNO(IELEM,2)
      YOUNG=PROPS(LPROP,1)
      SHEARM=PROPS(LPROP,2)
      YINERA=PROPS(LPROP,3)
      TINERA=PROPS(LPROP,4)
      AREA=PROPS(LPROP,5)
```

```
        ZASF=PROPS(LPROP,6)
C
C***CALCULATE THE NODES CORRESPONDING TO THE TWO
C   ENDS OF THE ELEMENT
C
        IEND=LNODS(IELEM,1)
        JEND=LNODS(IELEM,2)
C
C***CALCULATE THE COORDINATES OF THE NODES AT THE
C   ENDS OF THE ELEMENT, THE LENGTH, DIRECTION
C   COSINES ETC.
C
        XPROJ=COORD(JEND,1)-COORD(IEND,1)
        YPROJ=COORD(JEND,2)-COORD(IEND,2)
        SPAN=SQRT(XPROJ*XPROJ+YPROJ*YPROJ)
        XL=XPROJ/SPAN
        YM=YPROJ/SPAN
        EI=YOUNG*YINERA
        GA=SHEARM*AREA
        EIBYL=EI/SPAN
        EIBYL2=EIBYL/SPAN
        EIBYL3=EIBYL2/SPAN
        BETA=12.0*EIBYL2/GA*ZASF
        GJBYL=TINERA*SHEARM/SPAN
C
C***ZERO THE VECTOR OF FIXED END FORCES
C
        DO 20 IEVAB=1,NEVAB
     20 ELOAD(IEVAB)=0.0
C
C***CALCULATE THE LATERAL LOADS ON THE ELEMENTS.
C   BUT INITIALLY ASSUME THAT THE ELEMENT IS
C   FREE OF LATERAL LOADS IE JLODEL=0.
C
        JLODEL=0
C
C***START THE CYCLE TO CHECK IF IELEM HAS ANY
C   LATERAL LOADS ON IT. IF IT HAS CALCULATE
C   WHAT THEY ARE.
C
        DO 50 ILODEL=1,NLODEL
        IF(LODEL(ILODEL).EQ.IELEM) JLODEL=ILODEL
        IF(JLODEL.NE.0) GOTO 55
     50 CONTINUE
        IF(JLODEL.EQ.0) GOTO 60
     55 UDLZT=FORSEL(JLODEL,1)*SPAN
        CONLZ=FORSEL(JLODEL,2)
     60 IF(NTYPE.EQ.2) GOTO 11
        IF(NTYPE.EQ.3) GOTO 12
        IF(NTYPE.EQ.4) GOTO 13
C
C***BENDING STIFFNESS COEFFICIENTS FOR AN
C   ORTHODOX BEAM ELEMENT
C
        S11=(4.0+BETA)/(1.0+BETA)
        S22=S11
        S12=(2.0-BETA)/(1.0+BETA)
        Q11=S11+S12
        Q22=Q11
        Q12=Q11
```

```
            Q21=Q11
            T11=2.0*Q11
            T22=T11
            T12=T11
C
C***IF THE ELEMENT IS FREE FROM LATERAL LOADS
C    SKIP THE CALCULATION OF FIXED END REACTIONS.
C
            IF(JLODEL.EQ.0) GOTO 14
C
C***CALCULATE THE FIXED END BM AND SF FOR AN
C    ORDINARY FIXED BEAM
C
            ELOAD(1)=-0.5*(UDLZT+CONLZ)
            ELOAD(NDOFN+1)=ELOAD(1)
            ELOAD(2)=SPAN*(UDLZT/12.0+CONLZ/8.0)
            ELOAD(NDOFN+2)=-ELOAD(2)
            GOTO 14
C
C***BENDING STIFFNESS COEFFICIENTS OF A BEAM ON
C    ELASTIC FOUNDATION
C
     11 SUBGRM=PROPS(LPROP,NPROP+1)
            AMDAL=SPAN*(SQRT(SQRT(SUBGRM/(4.0*EI))))
            SH=SINH(AMDAL)
            CH=COSH(AMDAL)
            S=SIN(AMDAL)
            C=COS(AMDAL)
            FACTO1=AMDAL/(SH*SH-S*S)
            FACTO2=FACTO1*AMDAL
            FACTO3=FACTO2*AMDAL
            S11=2.0*(SH*CH-S*C)*FACTO1
            S22=S11
            S12=2.0*(CH*S-SH*C)*FACTO1
            T11=4.0*(SH*CH+S*C)*FACTO3
            T22=T11
            T12=4.0*(CH*S+SH*C)*FACTO3
            Q11=2.0*(CH*CH-C*C)*FACTO2
            Q22=Q11
            Q12=4.0*(SH*S)*FACTO2
            Q21=Q12
C
C***IF THE ELEMENT IS FREE FROM LATERAL LOADS
C    SKIP THE CALCULATION OF FIXED END REACTIONS.
C
            IF(JLODEL.EQ.0) GOTO 14
C
C***CALCULATE THE FIXED END BM AND SF FOR A BEAM
C    ON ELASTIC SUBGRADE.   NOTE THAT PART
C    OF THE NORMAL LOAD ON THE BEAM IS ABSORBED
C    BY THE 'FOUNDATION'.
C
            AMDAL2=0.5*AMDAL
            SH2=SINH(AMDAL2)
            CH2=COSH(AMDAL2)
            S2=SIN(AMDAL2)
            C2=COS(AMDAL2)
C
C***CALCULATE THE MODIFICATION FACTORS DUE TO
C    THE LATERAL ELASTIC SUPPORT ON THE FIXED
```

```
C    END MOMENT AND SHEAR FORCES DUE TO UDL.
C
         FACTO1=6.0/AMDAL/AMDAL*(SH-S)/(SH+S)
         FACTO2=2.0/AMDAL*(CH-C)/(SH+S)
C
C***CALCULATE THE MODIFICATION FACTORS DUE TO
C    THE LATERAL ELASTIC SUPPORT ON THE THE FIXED
C    END MOMENT AND SHEAR FORCES DUE TO
C    CONCENTRATED LOAD AT MID-SPAN.
C
         FACTO3=8.0/AMDAL*(SH2*S2)/(SH+S)
         FACTO4=2.0*(SH2*C2+CH2*S2)/(SH+S)
         ELOAD(1)=-0.5*(UDLZT*FACTO2+CONLZ*FACTO4)
         ELOAD(NDOFN+1)=ELOAD(1)
         ELOAD(2)=SPAN*(UDLZT*FACTO1/12.0+CONLZ*
        @FACTO3/8.0)
         ELOAD(NDOFN+2)=-ELOAD(2)
         GOTO 14
C
C***BENDING STIFFNESS COEFFICIENTS FOR A
C    SEMI-RIGID BEAM
C
     12 SPRNG1=PROPS(LPROP,NPROP+2)
         SPRNG2=PROPS(LPROP,NPROP+3)
         ALPHA1=EIBYL/SPRNG1
         ALPHA2=EIBYL/SPRNG2
         DENOM=1.0+BETA+(4.0+BETA)*(ALPHA1+ALPHA2)
        @+12.0*ALPHA1*ALPHA2
         S11=(4.0+BETA+12.0*ALPHA2)/DENOM
         S12=(2.0-BETA)/DENOM
         S22=(4.0+BETA+12.0*ALPHA1)/DENOM
         Q11=S11+S12
         Q12=Q11
         Q22=S12+S22
         Q21=Q22
         T11=Q11+Q22
         T22=T11
         T12=T11
C
C***IF THE ELEMENT IS UNLOADED SKIP THE
C    CALCULATION OF FIXED END MOMENTS AND
C    VERTICAL REACTIONS.
C
         IF(JLODEL.EQ.0) GOTO 14
C
C***CALCULATE FIXED END BM AND SF FOR A SEMI-RIGID
C    BEAM
C
         D=(2.0+6.0*ALPHA1)*(2.0+6.0*ALPHA2)-1
C
C***CALCULATE THE MODIFICATION FACTORS DUE TO
C    THE SEMI-RIGID ENDS ON THE FIXED END MOMENT
C    AND VERTICAL REACTION.
C
         FACTO1=3.0*(1.0+6.0*ALPHA2)/D
         FACTO2=3.0*(1.0+6.0*ALPHA1)/D
         FEM1=SPAN*(UDLZT*FACTO1/12.0+CONLZ*2.0*
        @FACTO1/8.0)
         FEM2=SPAN*(UDLZT*FACTO2/12.0+CONLZ*2.0*
        @FACTO2/8.0)
```

```fortran
      ELOAD(2)=FEM1
      ELOAD(NDOFN+2)=-FEM2
      ELOAD(1)=-0.5*(UDLZT+CONLZ)-(FEM1-FEM2)/
     @SPAN
      ELOAD(NDOFN+1)=-0.5*(UDLZT+CONLZ)+(FEM1-FEM2)/
     @SPAN
      GOTO 14
C
C***BENDING STIFFNESS COEFFICIENTS FOR A
C   NONPRISMATIC ELEMENT FOR WHICH THE BENDING
C   STIFFNESS COEFFICIENTS COEFFICIENTS ARE READ
C   AS DATA.
C
   13 WRITE(6,910)
  910 FORMAT(//,'BENDING STIFFNESS COEFFICIENTS
     @FOR A NONPRISMATIC BEAM ELEMENT',//)
      READ(5,960)S11,S12,S22
      READ(5,960)T11,T12,T22
      READ(5,960)Q11,Q12,Q22,Q21
      WRITE(6,960)S11,S12,S22
      WRITE(6,960)T11,T12,T22
      WRITE(6,960)Q11,Q12,Q22,Q21
C
C***IF THE ELEMENT IS UNLOADED SKIP THE
C   CALCULATION OF FIXED END MOMENTS AND
C   VERTICAL REACTIONS.
C
      IF(JLODEL.EQ.0) GOTO 14
C
C***READ THE FIXED END SF, BM AND TORQUE FOR THE
C   NONPRISMATIC ELEMENT
C
      WRITE(6,930)
  930 FORMAT(//,'FIXED END SF BM & T FOR A NON
     @PRISMATIC BEAM ELEMENT')
      READ(5,960)(ELOAD(IEVAB),IEVAB=1,NEVAB)
      WRITE(6,960)(ELOAD(IEVAB),IEVAB=1,NEVAB)
  960 FORMAT(6F10.4)
   14 CONTINUE
      T11=T11*EIBYL3
      T22=T22*EIBYL3
      T12=T12*EIBYL3
      S11=S11*EIBYL
      S22=S22*EIBYL
      S12=S12*EIBYL
      Q11=Q11*EIBYL2
      Q22=Q22*EIBYL2
      Q12=Q12*EIBYL2
      Q21=Q21*EIBYL2
C
C***SET THE DISPLACEMENTS AT THE ENDS OF THE
C   BEAM AS ZERO.
C
      W1=0.0
      THETAX1=0.0
      THETAY1=0.0
      THETAZ1=0.0
      W2=0.0
      THETAX2=0.0
      THETAY2=0.0
```

```
      THETAZ2=0.0
C
C***CALCULATE THE NONZERO END DISPLACEMENTS.
C
      IF(NODFRE(IEND,1).NE.0)W1=
     @ASDIS(NODFRE(IEND,1))
      IF(NODFRE(IEND,2).NE.0)THETAX1=
     @ASDIS(NODFRE(IEND,2))
      IF(NODFRE(IEND,3).NE.0)THETAY1=
     @ASDIS(NODFRE(IEND,3))
      IF(NODFRE(JEND,1).NE.0)W2=
     @ASDIS(NODFRE(JEND,1))
      IF(NODFRE(JEND,2).NE.0)THETAX2=
     @ASDIS(NODFRE(JEND,2))
      IF(NODFRE(JEND,3).NE.0)THETAY2=
     @ASDIS(NODFRE(JEND,3))
C
C***CALCULATE THE TWIST AT THE ENDS.
C
      PSI1=THETAX1*XL+THETAY1*YM
      PSI2=THETAX2*XL+THETAY2*YM
C
C***CALCULATE THE BENDING ROTATIONS AT THE ENDS.
C
      THETA1=-THETAX1*YM+THETAY1*XL
      THETA2=-THETAX2*YM+THETAY2*XL
C
C***CALCULATE THE SHEAR FORCE AT THE ENDS DUE TO
C   DISPLACEMENTS.
C
      DISFOR(1)=T11*W1-T12*W2-Q11*THETA1-Q21*
     @THETA2
      DISFOR(NDOFN+1)=-T12*W1+T22*W2+Q12*THETA1
     @+Q22*THETA2
C
C***CALCULATE THE BENDING MOMENTS AT THE ENDS
C   DUE TO DISPLACEMENTS.
C
      DISFOR(2)=-Q11*W1+Q12*W2+S11*THETA1+S12*
     @THETA2
      DISFOR(NDOFN+2)=-Q21*W1+Q22*W2+S12*THETA1
     @+S22*THETA2
      IF(IWARP.EQ.0) GOTO 17
C
C***CALCULATE THE WARPING DISPLACEMENTS AND
C   OTHER PARAMETERS NEEDED TO CALCULATE THE
C   TORQUE AND BIMOMENT WHEN WARPING EFFECTS ARE
C   INCLUDED.
C
      IF(NODFRE(IEND,4).NE.0)THETAZ1=ASDIS
     @(NODFRE(IEND,4))
      IF(NODFRE(JEND,4).NE.0)THETAZ2=ASDIS
     @(NODFRE(JEND,4))
      WINERA=PROPS(LPROP,NPROP)
      GAMAL=SPAN*SQRT(SHEARM*TINERA/(YOUNG*
     @WINERA))
      SH=SINH(GAMAL)
      CH=COSH(GAMAL)
      D=SH*GAMAL-2.0*(CH - 1.0)
      GJBYLD=GJBYL/D
```

```fortran
      TW11=GJBYLD*SH*GAMAL
      TW12=GJBYLD*(CH-1.0)*SPAN
      TW22=GJBYLD*(CH-SH/GAMAL)*SPAN*SPAN
      TW24=GJBYLD*(SH/GAMAL-1.0)*SPAN*SPAN
C
C***CALCULATE THE TORQUE AND BIMOMENT AT THE
C    ENDS
C
      DISFOR(3)=TW11*(PSI1-PSI2)+TW12*(THETAZ1
     @+THETAZ2)
      DISFOR(3+NDOFN)=DISFOR(3)
      DISFOR(4)=TW12*(PSI1-PSI2)+TW22*THETAZ1
     @+TW24*THETAZ2
      DISFOR(8)=TW12*(PSI1-PSI2)+TW24*THETAZ1
     @+TW22*THETAZ2
      GOTO 18
C
C***CALCULATE THE TORQUE DUE TO THE
C   DISPLACEMENTS WHEN WARPING EFFECTS
C   ARE IGNORED.  THIS IS A CASE OF PURE TORQUE.
C
   17 DISFOR(3)=GJBYL*(PSI2-PSI1)
      DISFOR(NDOFN+3)=DISFOR(3)
C
C***ZERO THE VECTOR OF THERMAL RESTRAINT FORCES.
C
   18 DO 80 IEVAB=1,NEVAB
   80 FORSTM(IEVAB)=0.0
      IF(ITEMP.EQ.0) GOTO 90
      TOPTEM=TEMP(IELEM,1)
      BOTTEM=TEMP(IELEM,2)
C
C***IF THE THERMAL LOAD ON AN ELEMENT IS ZERO
C   SKIP THERMAL LOAD CALCULATIONS.
C
      IF(TOPTEM.EQ. 0.0 .AND.BOTTEM.EQ. 0.0
     @)GOTO 90
C
C*** CALCULATE THE BENDING MOMENT DUE TO THERMAL
C    RESTRAINT.
C
      ALPHAT=PROPS(LPROP,10)
      DEPTH=PROPS(LPROP,11)
      RMOMENT=ALPHAT*EI/DEPTH*(TOPTEM-BOTTEM)
      FORSTM(2)=RMOMENT
      FORSTM(NDOFN+2)=-RMOMENT
C
C***TOTAL FORCE = FORCE DUE TO (FIXED END +
C   DISPLACEMENT + TEMPERATURE.)
C
   90 DO 70 IEVAB=1,NEVAB
      TOTFOR(IEVAB)=ELOAD(IEVAB)+DISFOR(IEVAB)
     @+FORSTM(IEVAB)
   70 CONTINUE
      WRITE(6,900)IELEM,IEND,JEND,(TOTFOR(IEVAB)
     @,IEVAB=1,NEVAB)
   10 CONTINUE
  900 FORMAT(I4,4X,I4,6X,I4,6X,E10.4,7(1X,E10.4))
      RETURN
      END
```

```
C
C***********************************************
C
        SUBROUTINE PRESURRJ(NODFRE,LNODS,MATNO)
C
C***THIS SUBROUTINE CALCULATES THE BASE PRESSURE
C    ,BENDING MOMENT AND SHEAR FORCE AT TEN
C    SECTIONS FOR  ELEMENTS WITH ELASTIC
C    FOUNDATION.
C
        COMMON/CONTRO/NPOIN,NELEM,NNODE,NDOFN,
       @NDIME,NPROP,NMATS,NRESND,NVFIX,NEVAB,NFAIL,
       @NLODEL,NEQNS,NBAND,ITEMP
        COMMON/LGDATA/COORD(100,2),PROPS(20,11),
       @PRESC(80,3),TEMP(100,2),NDFFIX(80),
       @NDVFIX(80),IFPRE(80,3),FORSEL(80,4),
       @LODEL(80),ASDIS(800),GSTIF(100,30),
       @GLOAD(100)
        DIMENSION NODFRE(100,3),LNODS(100,2),
       @MATNO(100,2)
        DO 10 IELEM=1,NELEM
        NTYPE=MATNO(IELEM,2)
C
C***PRESSURE CALCULATION IS REQUIRED ONLY FOR
C    BEAMS ON ELASTIC FOUNDATION WHICH ARE OF
C    TYPE 2.
C
        IF(NTYPE.NE.2) GOTO 10
        WRITE(6,940)
  940 FORMAT(//,'PRESSURE, BM AND SF FOR BEAM ON
       @ELASTIC FOUNDATION')
        LPROP=MATNO(IELEM,1)
C
C***CALCULATE THE LENGTH, DIRECTION COSINES,
C    PROPERTIES ETC.
C
        IEND=LNODS(IELEM,1)
        JEND=LNODS(IELEM,2)
        XPROJ=COORD(JEND,1)-COORD(IEND,1)
        YPROJ=COORD(JEND,2)-COORD(IEND,2)
        SPAN=SQRT(XPROJ*XPROJ+YPROJ*YPROJ)
        XL=XPROJ/SPAN
        YM=YPROJ/SPAN
        SUBGRM=PROPS(LPROP,6)
        YOUNG=PROPS(LPROP,1)
        YINERA=PROPS(LPROP,3)
        EI=YOUNG*YINERA
        EIBYL=EI/SPAN
        EIBYL2=EIBYL/SPAN
        EIBYL3=EIBYL2/SPAN
        AMDAL=SPAN*(SQRT(SQRT(SUBGRM/(4.0*EI))))
        SH=SINH(AMDAL)
        CH=COSH(AMDAL)
        S=SIN(AMDAL)
        C=COS(AMDAL)
C
C***CALCULATE THE TRIGNOMETRIC AND HYPERBOLIC
C    FUNCTIONS FOR 0.5 X LAMDAL.  THIS IS
C    REQUIRED FOR CALCULATIONS IN CONNECTION
```

```
C     WITH MID-SPAN CONCENTRATED LOAD.
C
      AMDAL2=0.5*AMDAL
      SH2=SINH(AMDAL2)
      CH2=COSH(AMDAL2)
      S2=SIN(AMDAL2)
      C2=COS(AMDAL2)
C
C***ZERO ALL DISPLACEMENTS AT THE ENDS OF THE
C   ELEMENT.
C
      U1=0.0
      V1=0.0
      THETA1=0.0
      U2=0.0
      V2=0.0
      THETA2=0.0
C
C***CALCULATE NONZERO DISPLACEMENTS.
C
      IF(NODFRE(IEND,1).NE.0)U1=
     @ASDIS(NODFRE(IEND,1))
      IF(NODFRE(IEND,2).NE.0)V1=
     @ASDIS(NODFRE(IEND,2))
      IF(NODFRE(JEND,1).NE.0)U2=
     @ASDIS(NODFRE(JEND,1))
      IF(NODFRE(JEND,2).NE.0)V2=
     @ASDIS(NODFRE(JEND,2))
      IF(NODFRE(IEND,3).NE.0)THETA1=
     @ASDIS(NODFRE(IEND,3))
      IF(NODFRE(JEND,3).NE.0)THETA2=
     @ASDIS(NODFRE(JEND,3))
C
C***CALCULATE THE NORMAL COMPONENT OF THE
C   DISPLACEMENTS.
C
      DISPN1=-U1*YM+V1*XL
      DISPN2=-U2*YM+V2*XL
      FORUDL=0.0
      FORCON=0.0
C
C***CHECK IF THE ELEMENT CARRIES LATERAL
C   LOADS ON IT.
C
      JLODEL=0
      IF(NLODEL.EQ.0) GOTO 60
      DO 50 ILODEL=1,NLODEL
      IF(LODEL(ILODEL).EQ.IELEM) JLODEL=ILODEL
      IF(JLODEL.NE.0) GOTO 55
   50 CONTINUE
      IF(JLODEL.EQ.0) GOTO 60
C
C***CALCULATE THE TOTAL DISTRIBUTED AND
C   CONCENTRATED LOADS.
C
   55 UDLXT=FORSEL(JLODEL,1)*ABS(YPROJ)
      UDLYT=FORSEL(JLODEL,2)*ABS(XPROJ)
      CONLX=FORSEL(JLODEL,3)
      CONLY=FORSEL(JLODEL,4)
C
```

```
C***CALCULATE THE NORMAL COMPONENTS OF THE
C    APPLIED LOADS.
C
      FORUDL=UDLYT*XL-UDLXT*YM
      FORCON=CONLY*XL-CONLX*YM
C
C***CALCULATE THE VARIOUS FUNCTIONS NEEDED TO
C    EVALUATE V, V'' AND V''' DUE TO THE
C    DISPLACEMENTS AT THE ENDS OF THE ELEMENT.
C
   60 CONST1=1.0/(SH*SH-S*S)
      CONST2=CONST1/AMDAL*SPAN
      CONST3=2.0*AMDAL*AMDAL
      CONST4=CONST3*AMDAL
      S12=-(CH*CH-C*C)*CONST1
      S13=(SH*CH+S*C)*CONST1
      S14=-S13
      S22=(SH*CH-S*C)*CONST2
      S23=-SH*SH*CONST2
      S24=S*S*CONST2
      S32=2.0*SH*S*CONST1
      S33=-(CH*S+SH*C)*CONST1
      S34=-S33
      S42=(CH*S-SH*C)*CONST2
      S43=-SH*S*CONST2
      S44=-S43
      IF(JLODEL.EQ.0) GOTO 70
C
C***CALCULATE THE VARIOUS FUNCTIONS NEEDED TO
C    EVALUATE V, V'' AND V''' DUE TO DISTRIBUTED
C    AND CONCENTRATED LOAD.
C
      CONST5=1.0/(SH+S)
      UL1=(SH-S)*CONST5
      UL2=-(CH-C)*CONST5
      CL1=2.0*SH2*S2*AMDAL*CONST5
      CL2=-(SH2*C2+CH2*S2)*AMDAL*CONST5
   70 WRITE(6,910)IELEM,IEND,JEND
  910 FORMAT(//,' ELEMENT = ',I5,',',4X,'END 1
     @= NODE',I3,',','END 2 = NODE',I3)
      WRITE(6,920)
  920 FORMAT(//,6X,'X/L',8X,'PRESSURE',9X,'B.
     @MOMENT',6X,'SHEAR FORCE')
C
C***CALCULATE V, V'' AND V''' AT TEN SECTIONS
C    ALONG THE ELEMENT.
C
      DO 20 ISECTN=1,11
      XBYL=FLOAT(ISECTN-1)*0.1
      AMDAX=AMDAL*XBYL
      CHX=COSH(AMDAX)
      SHX=SINH(AMDAX)
      SX=SIN(AMDAX)
      CX=COS(AMDAX)
      CHCX=CHX*CX
      SHSX=SHX*SX
      CHSX=CHX*SX
      SHCX=SHX*CX
C
C***CALCULATE V, V'' AND V''' DUE TO UNIT
```

```
C    TRANSLATION AT END 1.
C
      SHAPE1=CHCX+S12*SHSX+S13*CHSX+S14*SHCX
      BM1=(-SHSX+S12*CHCX+S13*SHCX-S14*CHSX)*
     @CONST3
      SF1=(-(CHSX+SHCX)+S12*(SHCX-CHSX)+S13*
     @(CHCX-SHSX)-S14*(SHSX+CHCX))*CONST4
C
C***CALCULATE V, V'' AND V''' DUE TO UNIT
C    ROTATION AT END 1.
C
      SHAPE2=S22*SHSX+S23*CHSX+S24*SHCX
      BM2=(S22*CHCX+S23*SHCX-S24*CHSX)*CONST3
      SF2=(S22*(SHCX-CHSX)+S23*(CHCX-SHSX)-S24*
     @(SHSX+CHCX))*CONST4
C
C***CALCULATE V, V'' AND V''' DUE TO UNIT
C    TRANSLATION AT END 2.
C
      SHAPE3=S32*SHSX+S33*CHSX+S34*SHCX
      BM3=(S32*CHCX+S33*SHCX-S34*CHSX)*CONST3
      SF3=(S32*(SHCX-CHSX)+S33*(CHCX-SHSX)-S34*
     @(SHSX+CHCX))*CONST4
C
C***CALCULATE V, V'' AND V''' DUE TO UNIT
C    ROTATION AT END 2.
C
      SHAPE4=S42*SHSX+S43*CHSX+S44*SHCX
      BM4=(S42*CHCX+S43*SHCX-S44*CHSX)*CONST3
      SF4=(S42*(SHCX-CHSX)+S43*(CHCX-SHSX)-S44*
     @(SHSX+CHCX))*CONST4
C
C***ZERO THE V, V'' AND V''' DUE TO DISTRIBUTED
C    AND CONCENTRATED LOADS
C
      UDLDIS=0.0
      UDLBM=0.0
      UDLSF=0.0
      CONDIS=0.0
      CONBM=0.0
      CONSF=0.0
      IF(JLODEL.EQ.0) GOTO 90
      IF(FORUDL.EQ.0) GOTO 80
C
C***CALCULATE V, V'' AND V''' DUE TO DISTRIBUTED
C    LOAD.
C
      UDLDIS=(-CHCX+1.0)+UL1*SHSX+UL2*
     @(CHSX-SHCX)
      UDLBM=(SHSX +UL1*CHCX+UL2*(SHCX+CHSX))
     @*CONST3
      UDLSF=((CHSX+SHCX)+UL1*(SHCX-CHSX)+
     @UL2*2.0*CHCX)*CONST4
   80 IF(FORCON.EQ.0) GOTO 90
C
C***FOR CONCENTRATED LOAD AT MID-SPAN, THE
C    EXPRESSIONS ARE VALID ONLY FOR X/L =  OR
C    < 0.5. FOR X/L >  0.5 USE EXPRESSIONS FOR
C    (1-X/L)  BY MAKING USE OF 'SYMMETRY'.
C
```

```
      IF(ISECTN.LE.6) GOTO 100
      XBYL1=FLOAT(11-ISECTN)*0.1
      AMDAX=AMDAL*XBYL1
      CHX=COSH(AMDAX)
      SHX=SINH(AMDAX)
      SX=SIN(AMDAX)
      CX=COS(AMDAX)
      CHCX=CHX*CX
      SHSX=SHX*SX
      CHSX=CHX*SX
      SHCX=SHX*CX
C
C***CALCULATE V, V'' AND V''' DUE TO
C   CONCENTRATED LOAD AT MID-SPAN.
C
  100 CONDIS=CL1*SHSX+CL2*(CHSX-SHCX)
      CONBM=(CL1*CHCX+CL2*(SHCX+CHSX))*CONST3
      CONSF=(CL1*(SHCX-CHSX)+CL2*2.0*CHCX)*
     @CONST4
C
C***SHEAR FORCE DUE TO MID-SPAN CONCENTRATED
C   LOAD IS ANTI-SYMMETRIC WITH RESPECT TO
C   MID-SPAN.
C
      IF(ISECTN.GT.6) CONSF=-CONSF
   90 DISPLA=SHAPE1*DISPN1+SHAPE2*THETA1+
     @SHAPE3*DISPN2+SHAPE4*THETA2+(UDLDIS*
     @FORUDL+CONDIS*FORCON)/(SUBGRM*SPAN)
      PRESR=DISPLA*SUBGRM
      BM=(BM1*DISPN1+BM2*THETA1+BM3*DISPN2
     @+BM4*THETA2+(UDLBM*FORUDL+CONBM*
     @FORCON)/(SUBGRM*SPAN))*EIBYL2
      SF=(SF1*DISPN1+SF2*THETA1+SF3*DISPN2
     @+SF4*THETA2+(UDLSF*FORUDL+CONSF*
     @FORCON)/(SUBGRM*SPAN))*EIBYL3
      WRITE(6,930) XBYL,PRESR,BM,SF
  930 FORMAT(3X,4(F10.4,4X))
   20 CONTINUE
   10 CONTINUE
      RETURN
      END
```

```
C
C*********************************************
C
      SUBROUTINE PRESURGRID(NODFRE,LNODS,MATNO)
C
C***THIS SUBROUTINE CALCULATES THE BASE PRESSURE,
C   BENDING MOMENT AND SHEAR FORCE AT TEN
C   SECTIONS ALONG THE ELEMENTS WITH ELASTIC
C   FOUNDATION.
C
      COMMON/CONTRO/NPOIN,NELEM,NNODE,NDOFN,
     @NDIME,NPROP,NMATS,NRESND,NVFIX,NEVAB,
     @NFAIL,NLODEL,NEQNS,NBAND,ITEMP,IWARP
      COMMON/LGDATA/COORD(100,2),PROPS(10,12),
     @PRESC(80,4),TEMP(100,2),NDFFIX(80),
     @NDVFIX(80),IFPRE(80,4),FORSEL(80,2),
     @LODEL(80),ASDIS(800),GSTIF(100,30),
     @GLOAD(100)
```

```
      DIMENSION NODFRE(100,4),LNODS(100,2),
     @MATNO(100,2)
      DO 10 IELEM = 1,NELEM
      NTYPE=MATNO(IELEM,2)
C
C***PRESSURE CALCULATION IS NEEDED ONLY FOR BEAM
C   ON ELASTIC FOUNDATION.  THIS IS TYPE 2
C   ELEMENT.
C
      IF(NTYPE.NE.2) GOTO 10
      WRITE(6,940)
  940 FORMAT(//,' PRESSURE, BM AND SF FOR TEN
     @SECTIONS ALONG A BEF',//)
      LPROP=MATNO(IELEM,1)
      IEND=LNODS(IELEM,1)
      JEND=LNODS(IELEM,2)
      XPROJ=COORD(JEND,1)-COORD(IEND,1)
      YPROJ=COORD(JEND,2)-COORD(IEND,2)
      SPAN=SQRT(XPROJ*XPROJ+YPROJ*YPROJ)
      XL=XPROJ/SPAN
      YM=YPROJ/SPAN
      SUBGRM=PROPS(LPROP,7)
      YOUNG=PROPS(LPROP,1)
      YINERA=PROPS(LPROP,3)
      EI=YOUNG*YINERA
      EIBYL=EI/SPAN
      EIBYL2=EIBYL/SPAN
      EIBYL3=EIBYL2/SPAN
      AMDAL=SPAN*(SQRT(SQRT(SUBGRM/(4.0*EI))))
      SH=SINH(AMDAL)
      CH=COSH(AMDAL)
      S=SIN(AMDAL)
      C=COS(AMDAL)
C
C***CALCULATE THE TRIGNOMETRIC AND HYPERBOLIC
C   FUNCTIONS FOR 0.5 X LAMDAL.  THIS IS
C   REQUIRED FOR CALCULATIONS IN CONNECTION
C   WITH MID-SPAN CONCENTRATED LOAD.
C
      AMDAL2=0.5*AMDAL
      SH2=SINH(AMDAL2)
      CH2=COSH(AMDAL2)
      S2=SIN(AMDAL2)
      C2=COS(AMDAL2)
C
C***ZERO ALL DISPLACEMENTS AT THE ENDS OF THE
C   ELEMENT.
C
      W1=0.0
      THETAX1=0.0
      THETAY1=0.0
      W2=0.0
      THETAX2=0.0
      THETAY2=0.0
C
C***CALCULATE THE NON-ZERO DISPLACEMENTS AT THE
C   ENDS OF THE ELEMENT.
C
      IF(NODFRE(IEND,1).NE.0)W1=
     @ASDIS(NODFRE(IEND,1))
```

```
              IF(NODFRE(IEND,2).NE.0)THETAX1=
            @ASDIS(NODFRE(IEND,2))
              IF(NODFRE(IEND,3).NE.0)THETAY1=
            @ASDIS(NODFRE(IEND,3))
              IF(NODFRE(JEND,1).NE.0)W2=
            @ASDIS(NODFRE(JEND,1))
              IF(NODFRE(JEND,2).NE.0)THETAX2=
            @ASDIS(NODFRE(JEND,2))
              IF(NODFRE(JEND,3).NE.0)THETAY2=
            @ASDIS(NODFRE(JEND,3))
      C
      C***CALCULATE THE BENDING ROTATIONS AT THE ENDS.
      C
            THETA1=-THETAX1*YM+THETAY1*XL
            THETA2=-THETAX2*YM+THETAY2*XL
            FORUDL=0.0
            FORCON=0.0
      C
      C***CHECK IF THE ELEMENT CARRIES LATERAL LOADS
      C   ON IT.
      C
            JLODEL=0
            IF(NLODEL.EQ.0) GOTO 60
            DO 50 ILODEL=1,NLODEL
            IF(LODEL(ILODEL).EQ.IELEM) JLODEL=ILODEL
            IF(JLODEL.NE.0) GOTO 55
         50 CONTINUE
            IF(JLODEL.EQ.0) GOTO 60
      C
      C***CALCULATE THE TOTAL DISTRIBUTED AND
      C   CONCENTRATED LOADS.
      C
         55 FORUDL=FORSEL(JLODEL,1)*SPAN
            FORCON=FORSEL(JLODEL,2)
      C
      C***CALCULATE THE VARIOUS FUNCTIONS NEEDED TO
      C   CALCULATE V, V'' AND V''' DUE TO THE
      C   DISPLACEMENTS AT THE ENDS OF THE ELEMENT.
      C
         60 CONST1=1.0/(SH*SH-S*S)
            CONST2=CONST1/AMDAL*SPAN
            CONST3=2.0*AMDAL*AMDAL
            CONST4=CONST3*AMDAL
            S12=-(CH*CH-C*C)*CONST1
            S13=(SH*CH+S*C)*CONST1
            S14=-S13
            S22=(SH*CH-S*C)*CONST2
            S23=-SH*SH*CONST2
            S24=S*S*CONST2
            S32=2.0*SH*S*CONST1
            S33=-(CH*S+SH*C)*CONST1
            S34=-S33
            S42=(CH*S-SH*C)*CONST2
            S43=-SH*S*CONST2
            S44=-S43
            IF(JLODEL.EQ.0) GOTO 70
      C
      C***CALCULATE THE VARIOUS FUNCTIONS NEEDED TO
      C   EVALUATE V, V'' AND V''' DUE TO THE LATERAL
      C   DISTRIBUTED AND CONCENTRATED LOADS.
```

```
C
      CONST5=1.0/(SH+S)
      UL1=(SH-S)*CONST5
      UL2=-(CH-C)*CONST5
      CL1=2.0*SH2*S2*AMDAL*CONST5
      CL2=-(SH2*C2+CH2*S2)*AMDAL*CONST5
   70 WRITE(6,910)IELEM,IEND,JEND
  910 FORMAT(//,' ELEMENT = ',I3,',',4X,'END 1
     @= NODE',I3,',','END 2 = NODE',I3)
      WRITE(6,920)
  920 FORMAT(//,7X,'X/L',8X,'PRESSURE',9X,'B.
     @MOMENT',6X,'SHEAR FORCE')
C
C***CALCULATE V, V'' AND V''' AT TEN SECTIONS
C   ALONG THE ELEMENT.
C
      DO 20 ISECTN=1,11
      XBYL=FLOAT(ISECTN-1)*0.1
      AMDAX=AMDAL*XBYL
      CHX=COSH(AMDAX)
      SHX=SINH(AMDAX)
      SX=SIN(AMDAX)
      CX=COS(AMDAX)
      CHCX=CHX*CX
      SHSX=SHX*SX
      CHSX=CHX*SX
      SHCX=SHX*CX
C
C***CALCULATE V, V'' AND V''' DUE TO TRANSLATION
C   AT END 1.
C
      SHAPE1=CHCX+S12*SHSX+S13*CHSX+S14*SHCX
      BM1=(-SHSX+S12*CHCX+S13*SHCX-S14*CHSX)*
     @CONST3
      SF1=(-(CHSX+SHCX)+S12*(SHCX-CHSX)+S13*
     @(CHCX-SHSX)-S14*(SHSX+CHCX))*CONST4
C
C***CALCULATE V, V'' AND V''' DUE TO ROTATION
C   AT END 1.
C
      SHAPE2=S22*SHSX+S23*CHSX+S24*SHCX
      BM2=(S22*CHCX+S23*SHCX-S24*CHSX)*CONST3
      SF2=(S22*(SHCX-CHSX)+S23*(CHCX-SHSX)-S24*
     @(SHSX+CHCX))*CONST4
C
C***CALCULATE V, V'' AND V''' DUE TO TRANSLATION
C   AT END 2.
C
      SHAPE3=S32*SHSX+S33*CHSX+S34*SHCX
      BM3=(S32*CHCX+S33*SHCX-S34*CHSX)*CONST3
      SF3=(S32*(SHCX-CHSX)+S33*(CHCX-SHSX)-
     @S34*(SHSX+CHCX))*CONST4
C
C***CALCULATE V, V'' AND V''' DUE TO ROTATION
C   AT END 2.
C
      SHAPE4=S42*SHSX+S43*CHSX+S44*SHCX
      BM4=(S42*CHCX+S43*SHCX-S44*CHSX)*CONST3
      SF4=(S42*(SHCX-CHSX)+S43*(CHCX-SHSX)-
     @S44*(SHSX+CHCX))*CONST4
```

```
C
C***ZERO V, V'' AND V''' DUE TO LATERAL LOADS ON
C    THE ELEMENT.
C
      UDLDIS=0.0
      UDLBM=0.0
      UDLSF=0.0
      CONDIS=0.0
      CONBM=0.0
      CONSF=0.0
      IF(JLODEL.EQ.0) GOTO 90
      IF(FORUDL.EQ.0) GOTO 80
C
C***CALCULATE V, V'' AND V''' DUE TO
C    DISTRIBUTED LOAD.
C
      UDLDIS=(-CHCX+1.0)+UL1*SHSX+
     @UL2*(CHSX-SHCX)
      UDLBM=(SHSX+UL1*CHCX+UL2*(SHCX+CHSX))
     @*CONST3
      UDLSF=((CHSX+SHCX)+UL1*(SHCX-CHSX)+
     @UL2*2.0*CHCX)*CONST4
   80 IF(FORCON .EQ. 0) GOTO 90
C
C***FOR CONCENTRATED LOAD AT MID-SPAN, THE
C    EXPRESSIONS ARE VALID ONLY FOR X/L = OR
C    < 0.5.  FOR X/L > 0.5 USE EXPRESSIONS FOR
C    (1-X/L) BY MAKING USE OF 'SYMMETRY'.
C
      IF(ISECTN.LE.6) GOTO 100
      XBYL1=FLOAT(11-ISECTN)*0.1
      AMDAX=AMDAL*XBYL1
      CHX=COSH(AMDAX)
      SHX=SINH(AMDAX)
      SX=SIN(AMDAX)
      CX=COS(AMDAX)
      CHCX=CHX*CX
      SHSX=SHX*SX
      CHSX=CHX*SX
      SHCX=SHX*CX
C
C***CALCULATE V, V'' AND V''' DUE TO
C    DISTRIBUTED LOAD.
C
  100 CONDIS=CL1*SHSX+CL2*(CHSX-SHCX)
      CONBM=(CL1*CHCX+CL2*(SHCX+CHSX))*CONST3
      CONSF=(CL1*(SHCX-CHSX)+CL2*2.0*CHCX)*
     @CONST4
C
C***SHEAR FORCE DUE TO MID-SPAN CONCENTRATED
C    LOAD IS ANTI-SYMMETRIC WITH RESPECT TO
C    MID-SPAN.
C
      IF(ISECTN.GT.6) CONSF=-CONSF
   90 DISPLA=SHAPE1*W1+SHAPE2*THETA1+SHAPE3*W2
     @+SHAPE4*THETA2 +(UDLDIS*FORUDL+CONDIS*
     @FORCON)/(SUBGRM*SPAN)
      PRESR=DISPLA*SUBGRM
      BM=(BM1*W1+BM2*THETA1+BM3*W2+
     @BM4*THETA2+(UDLBM*FORUDL+CONBM*FORCON)
```

```
      @/(SUBGRM*SPAN))*EIBYL2
       SF=(SF1*W1+SF2*THETA1+SF3* W2+
      @SF4*THETA2+(UDLSF*FORUDL+CONSF*FORCON)
      @/(SUBGRM*SPAN))*EIBYL3
       WRITE(6,930) XBYL,PRESR,BM,SF
  930 FORMAT(3X,4(F10.4,4X))
   20 CONTINUE
   10 CONTINUE
      RETURN
      END
```

CHAPTER 7   Elastic stability of 2-D rigid-jointed structures

Elastic instability of structures results from the reduction in the elastic rotational and translational stiffness coefficients of beam elements due to the effect of axial compressive force in the members. If the level of the axial compressive force in the elements is high enough to reduce the 'stiffness of the structure' to zero, then the structure will be unable to resist any disturbing force. In other words, the structure becomes unstable even though the structure is assumed to remain elastic. In the stiffness method, the structure is assumed to reach its limit of elastic stability, when the determinant of the structural stiffness matrix tends to zero.

7.1 THE ELASTIC STABILITY STIFFNESS MATRIX FOR AN ORTHODOX BEAM ELEMENT

The bending stiffness coefficients are a function of the axial load in the element as was derived in Chapter 1. Three cases have to be considered as follows.

(i) *Positive (i.e. tensile) axial load P.* In this case the stiffness coefficients, as defined in Fig. 3.2, are expressed in terms of the axial load P as follows:

$$S_{11} = S_{22} = S, \quad S_{12} = SC$$

$$Q_{11} = Q_{12} = Q_{22} = Q_{21} = S(1 + C)$$

$$T_{11} = T_{12} = T_{22} = 2Q_{11} + \frac{PL^2}{EI}$$

The stability functions S and C are defined as follows:

$$S = \frac{1 - kL \coth kL}{(2/kL) \tanh 0.5kL - 1}$$

$$C = \frac{kL - \sinh kL}{\sinh kL - kL \cosh kL}$$

$$kL = \pi\sqrt{\rho}, \rho = \frac{P}{P_{\text{Euler}}}, P_{\text{Euler}} = \frac{\pi^2 EI}{L^2}$$

(ii) *Negative (i.e. compressive) axial load P.*   In this case the equations given for the stiffness coefficients are valid except for the fact that the stability function S and C are given by

$$S = \frac{1 - kL \cot kL}{(2/kL) \tan 0.5kL - 1}$$

$$C = \frac{kL - \sin kL}{\sin kL - kL \cos kL}, \quad k = \pi\sqrt{(-\rho)}$$

(iii) *Zero axial load P.*   In this case $S = 4.0$ and $C = 0.5$.

Table 1.8 gives the elastic stability stiffness matrix for an orthodox beam element.

7.2   FIXED END REACTIONS DUE TO THE LATERAL LOADS ON THE BEAM ELEMENT CARRYING THE AXIAL LOAD

There are only two types of loading considered in the programs presented in this book. The fixed end reactions are, like the stiffness coefficients, a function of the axial load in the element. Referring to Figure 4.2, the fixed end reactions are given by the following equations:

$$M_1 = -M_2 = L\left(f_1 \frac{\text{UDLN}}{12.0} + f_2 \frac{\text{CN}}{8.0} \right), \quad R_1 = R_2 = 0.5(\text{UDLN} + \text{CN})$$

where UDLN is the total uniformly distributed normal load on the element and CN is the total concentrated normal load on the element. The functions f_1 and f_2 depend on the axial load on the element and are given by the following equations.

(i) *Tensile axial load.*   In this case,

$$f_1 = \frac{12EI}{PL^2} (0.5kL \coth 0.5kL - 1), \quad f_2 = \frac{0.5Q_{11}}{T_{11}}$$

(ii) *Compressive axial load.*   In this case,

$$f_1 = \frac{12EI}{PL^2} (0.5kL \cot 0.5kL - 1), \quad f_2 = \frac{0.5Q_{11}}{T_{11}}$$

(iii) *Zero axial load.*   In this case, $f_1 = f_2 = 1.0$

The stiffness coefficients Q_{11} and T_{11} in the above equations are evaluated using the formulae given in section 7.1 but using $0.5L$ instead of L.

7.3   THE PROCEDURE FOR THE CALCULATION OF THE ELASTIC STABILITY LOAD FACTOR

The determination of the load factor to cause the elastic instability of the structure involves iterative calculations. The structural stiffness matrix, which is a function of the applied load level, is calculated at increasingly higher load levels until the determinant of the structural stiffness matrix reaches a zero value. This represents the outermost iteration cycle. In addition to the interations on the applied load, there is an inner cycle of iterations. This arises because, at any given external load level, the element stiffness matrix is a function of the axial load in the member. Unfortunately the exact axial load in the member is not known at the beginning of the element stiffness calculations. Therefore the calculations are done on the assumption that the axial load in the element is the same as that determined from a previous set of calculations. After solving for the axial load in the member, the assumed and calculated axial loads are compared. If the difference is within the limits of acceptable error, then the inner cycle of iterations are terminated. If the error is unacceptable, then the calculated axial load is assumed to be the axial load in the element and the inner cycle of iterations are repeated until acceptable agreement is obtained between the assumed and calculated axial loads in the members.

The steps involved in the determination of the load factor to cause elastic instability can be summarized as follows.

(1)   Input the basic parameters of the structure and its elements.

(2)   Input the loads on the elements and at the nodes. Also input the starting value of the load factor on the applied loads and the percentage increment in the initial load factor for each outermost iteration cycle.

(3)   Start the outermost cycle. Choose the starting load factor.

(4)   Assume the axial loads in the elements. For the first cycle it is convenient to assume that they are zero. Calculate the element stiffness matrix and assemble the structural stiffness matrix. This starts the 'innermost' iteration cycle.

(5)   Calculate the load vector at the current load factor.

(6)   Calculate the displacements of the joints of the structure by solving the equations of equilibrium using the Gaussian elimination method.

(7)   From the displacements, calculate the axial loads in the elements.

(8)   Compare the calculated and assumed axial loads in the elements. If the differences are within acceptable limits, then go to the next step. If they are not acceptable, then go back to (4) after setting the assumed axial load as equal to the values calculated in (7).

(9)   Calculate the determinant of the structural stiffness matrix. The determinant is simply calculated as the continued product of the diagonal terms of the stiffness matrix after it has been reduced by the Gaussian elimination procedure. If the determinant is positive, then go back to (3) because the structure is still stable. If the determinant is negative, then evidently the load factor is greater than the load factor at which instability sets in. The load factor at which the determinant is zero is obtained by linear interpolation between the previous load factor at which the determinant was positive and the current load factor at which the determinant is negative.

7.4 ISOLATION OF A PARTICULAR EIGENVALUE

As explained in Chapter 1, the determination of the load factor at which the structure becomes unstable as well as the determination of the natural frequencies of the structure involve the calculation of a certain parameter (either the load factor or the frequency) at which the determinant of the structural stiffness matrix attains a zero value. In applied mathematics, these problems are known as eigenvalue problems and the particular parameter as an eigenvalue. An eigenvalue problem has several eigenvalues at which the determinant is zero, as shown in Fig. 7.1. Since a particular eigenvalue is calculated by interpolating between two values of the parameter such that at one value of the parameter the determinant is positive and at the other the determinant is negative, it is important to ensure that, between the two values of the parameter, only one eigenvalue is present. If this is not ensured, then the interpolated value will not be correct as shown in Fig. 7.1. Fortunately this can be easily ensured by using the following well-known result from linear algebra (see Peters and Wilkinson [5]).

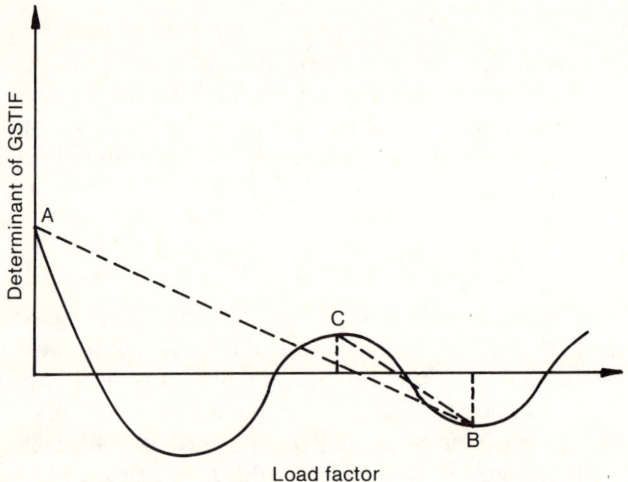

Fig. 7.1   Correct (CB) and incorrect (AB) method for isolating an eigenvalue.

Let the structural stiffness matrix GSTIF be formed at a value of the parameter equal to λ. Then the number of eigenvalues smaller than λ is equal to the number of agreements in sign between the consecutive members of the sequence—the determinant of K_m, where $m = 0, 1, 2, \ldots, N$. The determinant of $K_0 = 1$ by definition and the matrix K_r is the leading principal submatrix of order r of the structutral stiffness matrix GSTIF. In practice the above theorem is used as follows.

(1) Form the structural stiffness matrix at a value of the parameter equal to λ.
(2) Use the Gaussian elimination method to reduce it to the upper triangular form.

(3) Calculate the total number of agreements in sign between the consecutive members S_m and S_{m+1}:

determinant of

$$K_m = S_m = (-1)^m \cdot \text{GSTIF}(1,1) \cdot \text{GSTIF}(2,2) \cdot \ldots \cdot \text{GSTIF}(m,m)$$

determinant of $K_{m+1} = S_{m+1} = -S_m \cdot \text{GSTIF}(m+1,m+1)$ where $m = 0$ to (number of unknowns $- 1$) and $S_0 = 1$ by definition.

(4) The total number of agreements in sign indicates how many eigenvalues are smaller than λ.

(5) Interpolation is done only between those values of the parameter between which lies only one eigenvalue.

7.5 NOMENCLATURE OF NAMES OF VARIABLES

The program for the elastic stability analysis of 2-D rigid-jointed structures uses variable names additional to these used in the previous chapters. The additional names of variables and their meanings are as follows.

DETNEW	Determinant of the structural stiffness matrix from the current calculation. The actual value of the determinant is equal to $\text{DETNEW} \cdot 10^{(50 \cdot \text{NNEW})}$.
DETOLD	Determinant from a previous calculation. The actual value of the determinant is equal to $\text{DETOLD} \cdot 10^{(50 \cdot \text{NOLD})}$.
FACTLD	Multiplication **FACT**or on **L**oa**D**s.
FORSND(NLODND,NDOFN)	This array stores the concentrated nodal loads at NLODND nodes in the NDOFN freedoms.
ICONV	**I**ndication of **CONV**ergence. If ICONV = 1, it indicates that the assumed and calculated axial loads in the elements are within acceptable margin of error.
INCREM	Percentage **INCREM**ent in the initial load factor for each 'outermost' iteration cycle on loads.
LODPT(NLODPT)	This array stores the node numbers of the NLODPT nodes where concentrated loads are applied. LODPT stands for **L**Oa**D**ed **P**oin**T**.
NITER	Maximum **N**umber of **ITER**ations for the axial forces in the elements to converge to the required accuracy.
NNEW	See DETNEW.
NOLD	See DETOLD.

NSINAG	<u>N</u>umber of <u>SIgN</u> <u>AG</u>reements between the consecutive members of the sequence S_m *and* S_{m+1} as explained in section 7.4.
PNEW(NELEM)	This array stores the value of the axial loads in the NELEM elements calculated from the current displacements.
POLD(NELEM)	Similar to PNEW except that the axial loads from a previous calculation are stored. Convergence of the axial loads is checked by comparing the vectors POLD and PNEW. See section 7.6 under subroutine CONVER for details of the check used.
STRFAC	<u>STaR</u>ting value of the load <u>FAC</u>tor. This is the value of the load factor at which the 'outermost' iteration cycle starts.

7.6 SUBROUTINES STIFFRJS, LOADRJS, STRESSRJS, READLOD, CONVER AND STURM

There are three 'replacement' subroutines which are essentially similar to the subroutines used in the linear elastic analysis and three new subroutines specific to the elastic stability analysis. They are as follows.

(1) STIFFRJS. This subroutine is similar to STIFFRJ except for the following aspects.

    (a) Only prismatic beam elements are included.

    (b) The stiffness matrix is a function of the axial load in the element. The axial load in the element is stored in the array POLD-(NELEM).

(2) LOADRJS. This is similar to LOADRJ except for the following aspects.

    (a) There is no data input about the loads at the nodes and on the elements. This is done in the new subroutine READLOD.

    (b) The fixed end reactions are a function of the axial loads in the elements which are stored in the array POLD(NELEM).

    (c) The external lateral loads on the elements stored in the arrays FORSND(NLODND) and FORSEL(NLODEL) have to be multiplied by the load factor appropriate to the iteration cycle on the loads.

(3) STRESSRJS. This is very similar to STRESSRJ except that only prismatic beam elements are considered and thermal stresses are not calculated.

(4) READLOD. This subroutine is used to input external loads and to echo it back. The loads acting at the nodes and the corresponding node

numbers are stored in the arrays FORSND(NLODND,NDOFN) and LODPT(NLODND) respectively. The data about the uniformly distributed loads on the elements are stored in the arrays FORSEL-(NLODEL) and FORSEL(NLODEL,4) as in the subroutine LOADRJ.

(5) CONVER. The object of this subroutine is to calculate whether the assumed axial loads in the elements stored in the array POLD-(NELEM) and the calculated values of the axial loads stored in the array PNEW(NELEM) are close enough or not. The criterion for convergence is the force norm

$$\text{FACTOR} = \sqrt{\left\{ \frac{\sum [\text{POLD(IELEM)} - \text{PNEW(IELEM)}]^2}{\sum [\text{POLD(IELEM)}^2]} \right\}}$$

where the summation $\sum$ is on IELEM = 1 to NELEM. If FACTOR is less than 0.01 (i.e. 1%), then the difference between the assumed and actual axial loads in the elements are assumed to be close enough for all practical purposes. It should be noted that the subroutine CONVER calculates only the axial forces in the members from the calculated values of displacements. Only when convergence in the values of the axial loads has taken place is the subroutine STRESSRJS called to calculate the 'stresses' in the elements. If one is not interested in the nonlinear stress analysis but only in the load factor to cause instability, then the call to the subroutine STRESSSRJS can be omitted. This can be done by setting the parameter ISTRES to zero if no stress output is required or to unity if stress output is required.

(6) STURM. This subroutine calculates the value of the determinant of the structural stiffness matrix and the number of agreements in sign between the consecutive members of the sequence S_m and S_{m+1} as explained in section 7.4. Note that the determinant is obtained as the continued product of the diagonal terms of the structural stiffness matrix after Gaussian reduction. In the rectangular format in which the stiffness matrix is stored, the diagonal elements are in the first column. Because the value of the determinant can be very large, it is scaled down by dividing the intermediate value of the determinant by 10^{50} whenever its value exceeds 10^{50} and a record is kept of how many times the division is done. The current value of the determinant is given by

$$\text{determinant} = \text{DETNEW} \cdot 10^{(50 \cdot \text{NNEW})}$$

A record is kept of both DETNEW and NNEW.

7.7 REFERENCES

[1] P. Bhatt, *Problems in Structural Analysis by Matrix Methods*, Construction Press, 1981.

[2] A. K. Chugh, Stiffness matrix for a beam element including transverse shear and axial force effects, *International Journal for Numerical Methods in Engineering*, Vol. 11, 1977, pp. 1681–1697.

[3]   M. R. Horne and W. Merchant, *The Stability of Frames*, Pergamon Press, 1965.

[4]   G. D. Manolis and D. E. Beskos, Internal force distribution effect of framework stability, *Journal of Structural Engineering, Proceedings of the American Society of Civil Engineers*, Vol. 109, No. 1, 1983, pp. 250–257.

[5]   G. Peters and J. H. Wilkinson, Eigenvalues of $Ax = \lambda Bx$ with band symmetric matrices A and B, *Computer Journal*, Vol. 12, 1969, pp. 398–404.

```
C
C**********************************************
C
      SUBROUTINE STIFFRJS(NODFRE,LNODS,MATNO)
C
C***THIS SUBROUTINE COMPUTES THE ELEMENT
C    STIFFNESS MATRIX AND ASSEMBLES THE
C    STRUCTURAL STIFFNESS MATRIX IN BANDED FORMAT
C    WITH THE DIAGONAL TERM IN THE FIRST COLUMN.
C
      COMMON/CONTRO/NPOIN,NELEM,NNODE,NDOFN,
     @NDIME,NPROP,NMATS,NRESND,NEVAB,NFAIL,
     @NLODEL,NEQNS,NBAND,NLODND,NSINAG,ICONV,
     @DETNEW,NNEW,FACTLD,CONFAC
      COMMON/LGDATA/COORD(100,3),PROPS(10,3),
     @PNEW(100),POLD(100),NDFFIX(80),
     @IFPRE(80,3),FORSEL(80,4),LODEL(80),
     @FORSND(80,3),LODPT(80),ASDIS(800),
     @GSTIF(100,30),GLOAD(100)
      DIMENSION ESTIF(6,6),MEMDIS(6),
     @NODFRE(100,3),LNODS(100,2),MATNO(100,2)
C
C*** CYCLE ON ALL ELEMENTS  AND CALCULATE
C    ELEMENT STIFFNESS MATRIX AND ADD TO THE
C    STRUCTURAL STIFFNESS MATRIX.
C
      DO 10 IELEM=1,NELEM
      LPROP=MATNO(IELEM,1)
C
C*** CALCULATE THE PROPERTIES OF ELEMENTS.
C
      YOUNG=PROPS(LPROP,1)
      YINERA=PROPS(LPROP,2)
      AREA=PROPS(LPROP,3)
C
C***CALCULATE THE NODES CORRESPONDING TO THE TWO
C    ENDS OF THE ELEMENT, THE ELEMENT LENGTH AND
C    THE DIRECTION COSINES.
C
      IEND=LNODS(IELEM,1)
      JEND=LNODS(IELEM,2)
      XPROJ=COORD(JEND,1)-COORD(IEND,1)
      YPROJ=COORD(JEND,2)-COORD(IEND,2)
      SPAN=SQRT(XPROJ*XPROJ+YPROJ*YPROJ)
      XL=XPROJ/SPAN
      YM=YPROJ/SPAN
      EI=YOUNG*YINERA
      EIBYL=EI/SPAN
      EIBYL2=EIBYL/SPAN
      EIBYL3=EIBYL2/SPAN
      AEBYL=AREA*YOUNG/SPAN
C
C***CALCULATE THE P EULER AND RATIO OF AXIAL
C    LOAD TO P EULER.
C
      PEULER=9.8696044*EIBYL2
      ROH=POLD(IELEM)/PEULER
      IF(ROH.EQ.0) GOTO 11
      IF(ROH.LT.0) GOTO 12
C
```

```
C***CALCULATE THE STABILITY FUNCTIONS S AND C
C    AND THE BENDING STIFFNESS WHEN AXIAL LOAD
C    IS TENSILE.
C
      AMDA=3.141592654*SQRT(ROH)
      HSIN=SINH(AMDA)
      HCOS=COSH(AMDA)
      HTAN2=SINH(0.5*AMDA)/COSH(0.5*AMDA)
      S=(1.0-AMDA*HCOS/HSIN)/(2.0*HTAN2/AMDA
     @-1.0)
      C=(AMDA-HSIN)/(HSIN-AMDA*HCOS)
      S11=S
      S12=S*C
      Q11=S11+S12
      T11=2.0*Q11+AMDA*AMDA
      GOTO 13
C
C***BENDING STIFFNESS COEFFICIENTS FOR ZERO
C    AXIAL LOAD.
C
   11 S11=4.0
      S12=2.0
      Q11=6.0
      T11=12.0
      GOTO 13
C
C***CALCULATE THE STABILITY FUNCTIONS S AND C
C    AND THE BENDING STIFFNESS COEFFICIENTS WHEN
C    AXIAL LOAD IS COMPRESSIVE.
C
   12 AMDA=3.141592654*SQRT(-ROH)
      TSIN=SIN(AMDA)
      TCOS=COS(AMDA)
      TTAN2=SIN(0.5*AMDA)/COS(0.5*AMDA)
      S=(1.0-AMDA*TCOS/TSIN)/(2.0*TTAN2/AMDA
     @-1.0)
      C=(AMDA-TSIN)/(TSIN-AMDA*TCOS)
      S11=S
      S12=S*C
      Q11=S11+S12
      T11=2.0*Q11-AMDA*AMDA
   13 T11=T11*EIBYL3
      S11=S11*EIBYL
      S12=S12*EIBYL
      Q11=Q11*EIBYL2
C
C***CALCULATE THE ELEMENT STIFFNESS MATRIX
C
      ESTIF(1,1)=T11*YM*YM+AEBYL*XL*XL
      ESTIF(1,2)=-T11*XL*YM+AEBYL*XL*YM
      ESTIF(1,3)=Q11*YM
      ESTIF(1,4)=-ESTIF(1,1)
      ESTIF(1,5)=-ESTIF(1,2)
      ESTIF(1,6)=ESTIF(1,3)
      ESTIF(2,2)=T11*XL*XL+AEBYL*YM*YM
      ESTIF(2,3)=-Q11*XL
      ESTIF(2,4)=ESTIF(1,5)
      ESTIF(2,5)=-ESTIF(2,2)
      ESTIF(2,6)=ESTIF(2,3)
      ESTIF(3,3)=S11
```

```
      ESTIF(3,4)=-ESTIF(1,3)
      ESTIF(3,5)=-ESTIF(2,3)
      ESTIF(3,6)=S12
      ESTIF(4,4)=ESTIF(1,1)
      ESTIF(4,5)=ESTIF(1,2)
      ESTIF(4,6)=ESTIF(3,4)
      ESTIF(5,5)=ESTIF(2,2)
      ESTIF(5,6)=ESTIF(3,5)
      ESTIF(6,6)=S11
C
C***FILL UP THE SYMMETRICAL LOWER HALF OF THE
C   ESTIF MATRIX
C
      DO 20 IEVAB=2,NEVAB
      IEVAB1=IEVAB-1
      DO 20 JEVAB=1,IEVAB1
   20 ESTIF(IEVAB,JEVAB)=ESTIF(JEVAB,IEVAB)
C
C***CALCULATE THE NODE FREEDOMS CORRESPONDING TO
C   THE NODES AT THE ENDS OF THE ELEMENT.  THIS
C   IS STORED IN ARRAY MEMDIS(NDOFN).
C
      DO 30 IDOFN=1,NDOFN
      MEMDIS(IDOFN)=NODFRE(IEND,IDOFN)
   30 MEMDIS(IDOFN+NDOFN)=NODFRE(JEND,IDOFN)
C
C***ASSEMBLE THE STRUCTURAL STIFFNESS MATRIX IN
C   THE BANDED FORM WITH THE DIAGONAL ELEMENTS
C   IN THE FIRST COLUMN
C
      DO 40 IEVAB=1,NEVAB
      IF(MEMDIS(IEVAB).EQ.0) GOTO 40
      DO 40 JEVAB=1,NEVAB
      IF(MEMDIS(JEVAB).EQ.0) GOTO 40
C
C***CALCULATE THE COLUMN POSITION IN THE BANDED
C   FORMAT.
C
      NEWCOL=MEMDIS(JEVAB)-MEMDIS(IEVAB)+1
      IF(NEWCOL.LE.0) GOTO 40
      GSTIF(MEMDIS(IEVAB),NEWCOL)=GSTIF(MEMDIS
     @(IEVAB),NEWCOL)+ESTIF(IEVAB,JEVAB)
   40 CONTINUE
   10 CONTINUE
      RETURN
      END
C
C*******************************************
C
      SUBROUTINE LOADRJS(NODFRE,LNODS,MATNO)
C
C***THIS SUBROUTINE CALCULATES THE GLOBAL LOAD
C   VECTOR
C
      COMMON/CONTRO/NPOIN,NELEM,NNODE,NDOFN,
     @NDIME,NPROP,NMATS,NRESND,NEVAB,NFAIL,
     @NLODEL,NEQNS,NBAND,NLODND,NSINAG,ICONV,
     @DETNEW,NNEW,FACTLD,CONFAC
      COMMON/LGDATA/COORD(100,3),PROPS(10,3),
```

```
      @PNEW(100),POLD(100),NDFFIX(80),IFPRE(80,3)
      @,FORSEL(80,4),LODEL(80),FORSND(80,3),
      @LODPT(80),ASDIS(800),GSTIF(100,30),
      @GLOAD(100)
       DIMENSION ELOAD(6),MEMDIS(6),NODFRE(100,3)
      @,LNODS(100,2),MATNO(100,2)
C
C***ZERO LOAD VECTOR
C
      DO 10 IEQNS=1,NEQNS
   10 GLOAD(IEQNS)=0.0
C
C***ADD THE LOADS AT JOINTS TO THE APPROPRIATE
C   POSITIONS IN THE GLOBAL LOAD VECTOR.
C
      IF(NLODND.EQ.0) GOTO 40
      DO 20 ILODND=1,NLODND
      DO 30 IDOFN=1,NDOFN
      IF(NODFRE(LODPT(ILODND),IDOFN).EQ.0)
     @GOTO 30
      GLOAD(NODFRE(LODPT(ILODND),IDOFN))=
     @FORSND(ILODND,IDOFN)*FACTLD
   30 CONTINUE
   20 CONTINUE
C
C***CALCULATE THE FIXED END FORCES ALONG X,Y AND
C   THETAZ DIRECTIONS DUE TO THE LATERAL LOADS
C   ON ELEMENTS.
C
   40 IF(NLODEL.EQ.0) GOTO 80
      DO 50 ILODEL=1,NLODEL
      IELEM=LODEL(ILODEL)
      IEND=LNODS(IELEM,1)
      JEND=LNODS(IELEM,2)
      XPROJ=COORD(JEND,1)-COORD(IEND,1)
      YPROJ=COORD(JEND,2)-COORD(IEND,2)
      SPAN=SQRT(XPROJ*XPROJ+YPROJ*YPROJ)
      XL=XPROJ/SPAN
      YM=YPROJ/SPAN
C
C*** CALCULATE THE PROPERTIES OF ELEMENTS.
C
      LPROP=MATNO(IELEM,1)
      YOUNG=PROPS(LPROP,1)
      YINERA=PROPS(LPROP,2)
      EI=YOUNG*YINERA
      PEULER=9.8696044*EI/SPAN/SPAN
      ROH=POLD(IELEM)/PEULER
      IF(ROH.EQ.0) GOTO 11
      IF(ROH.LT.0) GOTO 12
C
C***FOR TENSILE AXIAL LOAD CALCULATE THE
C   MODIFICATION FACTORS FOR THE FIXED END
C   MOMENTS DUE TO UDL AND MID-SPAN CONCENTRATED
C   LOADS.
C
      AMDA=1.570796*SQRT(ROH)
      HSIN=SINH(AMDA)
      HCOS=COSH(AMDA)
      FACTO1=3.0/AMDA/AMDA*(AMDA*HCOS/HSIN-1.0)
```

```
C
C***MID-SPAN CONCENTRATED LOAD FIXED END MOMENT
C   MODIFICATION FACTOR.
C
      AMDA=3.141592654*SQRT(0.25*ROH)
      HSIN=SINH(AMDA)
      HCOS=COSH(AMDA)
      HTAN2=SINH(0.5*AMDA)/COSH(0.5*AMDA)
      S=(1.0-AMDA*HCOS/HSIN)/(2.0*HTAN2/AMDA
     @-1.0)
      C=(AMDA-HSIN)/(HSIN-AMDA*HCOS)
      S11=S
      S12=S*C
      Q11=S11+S12
      T11=2.0*Q11+AMDA*AMDA
      FACTO2=2.0*Q11/T11
      GOTO 13
C
C***NO MODIFICATION OF FIXED END MOMENTS IF
C   AXIAL LOAD IS ZERO.
C
   11 FACTO1=1.0
      FACTO2=1.0
      GOTO 13
C
C***FOR COMPRESSIVE AXIAL LOAD CALCULATE THE
C   MODIFICATION FACTORS FOR THE FIXED END
C   MOMENTS DUE TO UDL AND MID-SPAN CONCENTRATED
C   LOADS.
   12 AMDA=1.570796*SQRT(-ROH)
      TSIN=SIN(AMDA)
      TCOS=COS(AMDA)
      FACTO1=3.0/AMDA/AMDA*(1.0-AMDA*TCOS/TSIN)
C
C***MID-SPAN CONCENTRATED LOAD FIXED END MOMENT
C   MODIFICATION FACTOR.
C
      AMDA=3.141592654*SQRT(-0.25*ROH)
      TSIN=SIN(AMDA)
      TCOS=COS(AMDA)
      TTAN2=SIN(0.5*AMDA)/COS(0.5*AMDA)
      S=(1.0-AMDA*TCOS/TSIN)/(2.0*TTAN2/AMDA
     @-1.0)
      C=(AMDA-TSIN)/(TSIN-AMDA*TCOS)
      S11=S
      S12=S*C
      Q11=S11+S12
      T11=2.0*Q11-AMDA*AMDA
      FACTO2=2.0*Q11/T11
C
C***CALCULATE THE TOTAL DISTRIBUTED LOAD AND
C   CONCENTRATED LOAD ON THE ELEMENT.
C
   13 UDLXT=FORSEL(ILODEL,1)*ABS(YPROJ)*FACTLD
      UDLYT=FORSEL(ILODEL,2)*ABS(XPROJ)*FACTLD
      CONLX=FORSEL(ILODEL,3)*FACTLD
      CONLY=FORSEL(ILODEL,4)*FACTLD
C
C***CALCULATE THE FIXED END FORCES ALONG X,Y
C   AND THETAZ DIRECTIONS FOR AN ORDINARY FIXED
```

```
C    BEAM.
C
      ELOAD(1)=-0.5*(UDLXT+CONLX)
      ELOAD(4)=ELOAD(1)
      ELOAD(2)=-0.5*(UDLYT+CONLY)
      ELOAD(5)=ELOAD(2)
      ELOAD(3)=SPAN/12.0*FACTO1*(UDLYT*XL-
     @UDLXT*YM)+SPAN/8.0*FACTO2*(CONLY*XL-
     @CONLX*YM)
      ELOAD(6)=-ELOAD(3)
C
C***CALCULATE THE NODE FREEDOMS AT THE TWO NODES
C   AT THE ENDS OF THE BEAM AND STORE IT IN THE
C   ARRAY MEMDIS(NDOFN).
C
      DO 60 IDOFN=1,NDOFN
      MEMDIS(IDOFN)=NODFRE(IEND,IDOFN)
      MEMDIS(IDOFN+NDOFN)=NODFRE(JEND,IDOFN)
   60 CONTINUE
C
C***ADD THE FIXED END FORCES TO THE APPROPRIATE
C   LOAD VECTOR
C
      DO 70 IEVAB=1,NEVAB
      IF(MEMDIS(IEVAB).EQ.0) GOTO 70
      GLOAD(MEMDIS(IEVAB))=GLOAD(MEMDIS(IEVAB))
     @-ELOAD(IEVAB)
   70 CONTINUE
   50 CONTINUE
   80 RETURN
      END
```

```
C
C***********************************************
C
      SUBROUTINE STRESSRJS(NODFRE,LNODS,MATNO)
C
C***THIS SUBROUTINE CALCULATES THE FINAL AXIAL,
C   BENDING AND SHEAR FORCES AT THE ENDS OF
C   ELEMENTS.
C
      COMMON/CONTRO/NPOIN,NELEM,NNODE,NDOFN,
     @NDIME,NPROP,NMATS,NRESND,NEVAB,NFAIL,
     @NLODEL,NEQNS,NBAND,NLODND,NSINAG,ICONV,
     @DETNEW,NNEW,FACTLD,CONFAC
      COMMON/LGDATA/COORD(100,3),PROPS(10,3),
     @PNEW(100),POLD(100),NDFFIX(80),IFPRE(80,3)
     @,FORSEL(80,4),LODEL(80),FORSND(80,3),
     @LODPT(80),ASDIS(800),GSTIF(100,30),
     @GLOAD(100)
      DIMENSION NODFRE(100,3),LNODS(100,2),
     @MATNO(100,2),DISFOR(6),TOTFOR(6),ELOAD(6)
C
C***STRESS CALCULATION IN THE MEMBERS
C
      DO 10 IELEM=1,NELEM
      LPROP=MATNO(IELEM,1)
C
C***CALCULATE THE PROPERTIES OF THE ELEMENT.
C
```

```
      YOUNG=PROPS(LPROP,1)
      YINERA=PROPS(LPROP,2)
      AREA=PROPS(LPROP,3)
C
C***CALCULATE THE NODES CORRESPONDING TO THE TWO
C    ENDS OF THE ELEMENT
C
      IEND=LNODS(IELEM,1)
      JEND=LNODS(IELEM,2)
C
C***CALCULATE THE COORDINATES OF THE NODES AT THE
C    ENDS OF THE ELEMENT, THE LENGTH AND THE
C    DIRECTION COSINES.
C
      XPROJ=COORD(JEND,1)-COORD(IEND,1)
      YPROJ=COORD(JEND,2)-COORD(IEND,2)
      SPAN=SQRT(XPROJ*XPROJ+YPROJ*YPROJ)
      XL=XPROJ/SPAN
      YM=YPROJ/SPAN
      EI=YOUNG*YINERA
      EIBYL=EI/SPAN
      EIBYL2=EIBYL/SPAN
      EIBYL3=EIBYL2/SPAN
      AEBYL=AREA*YOUNG/SPAN
C
C***BENDING STIFFNESS COEFFICIENTS FOR AN
C    ORTHODOX BEAM ELEMENT
C
      PEULER=9.8696044*EIBYL2
      ROH=POLD(IELEM)/PEULER
      IF(ROH.EQ.0) GOTO 11
      IF(ROH.LT.0) GOTO 12
C
C***CALCULATE THE STABILITY FUNCTIONS S AND C
C    AND THE BENDING STIFFNESS WHEN AXIAL LOAD IS
C    TENSILE.
C
      AMDA=3.141592654*SQRT(ROH)
      HSIN=SINH(AMDA)
      HCOS=COSH(AMDA)
      HTAN2=SINH(0.5*AMDA)/COSH(0.5*AMDA)
      S=(1.0-AMDA*HCOS/HSIN)/(2.0*HTAN2/AMDA
     @-1.0)
      C=(AMDA-HSIN)/(HSIN-AMDA*HCOS)
      S11=S
      S12=S*C
      Q11=S11+S12
      T11=2.0*Q11+AMDA*AMDA
      GOTO 13
C
C***BENDING STIFFNESS COEFFICIENTS FOR ZERO
C    AXIAL LOAD.
C
   11 S11=4.0
      S12=2.0
      Q11=6.0
      T11=12.0
      GOTO 13
C
C***CALCULATE THE STABILITY FUNCTIONS S AND C
```

```
C     AND THE BENDING STIFFNESS COEFFICIENTS WHEN
C     AXIAL LOAD IS COMPRESSIVE.
C
   12 AMDA=3.141592654*SQRT(-ROH)
      TSIN=SIN(AMDA)
      TCOS=COS(AMDA)
      TTAN2=SIN(0.5*AMDA)/COS(0.5*AMDA)
      S=(1.0-AMDA*TCOS/TSIN)/(2.0*TTAN2/AMDA
     @-1.0)
      C=(AMDA-TSIN)/(TSIN-AMDA*TCOS)
      S11=S
      S12=S*C
      Q11=S11+S12
      T11=2.0*Q11-AMDA*AMDA
   13 T11=T11*EIBYL3
      S11=S11*EIBYL
      S12=S12*EIBYL
      Q11=Q11*EIBYL2
C
C***SET THE VALUES OF THE END DISPLACEMENTS OF
C   THE ELEMENT AS ZERO.
C
      U1=0.0
      V1=0.0
      THETA1=0.0
      U2=0.0
      V2=0.0
      THETA2=0.0
C
C***CALCULATE THE NONZERO END DISPLACEMENTS OF
C   THE ELEMENT.
C
      IF(NODFRE(IEND,1).NE.0)U1=
     @ASDIS(NODFRE(IEND,1))
      IF(NODFRE(IEND,2).NE.0)V1=
     @ASDIS(NODFRE(IEND,2))
      IF(NODFRE(IEND,3).NE.0)THETA1=
     @ASDIS(NODFRE(IEND,3))
      IF(NODFRE(JEND,1).NE.0)U2=
     @ASDIS(NODFRE(JEND,1))
      IF(NODFRE(JEND,2).NE.0)V2=
     @ASDIS(NODFRE(JEND,2))
      IF(NODFRE(JEND,3).NE.0)THETA2=
     @ASDIS(NODFRE(JEND,3))
C
C***CALCULATE THE AXIAL DISPLACEMENTS.
C
      DISPAX1=U1*XL+V1*YM
      DISPAX2=U2*XL+V2*YM
C
C***CALCULATE THE NORMAL DISPLACEMENTS.
C
      DISPN1=-U1*YM+V1*XL
      DISPN2=-U2*YM+V2*XL
C
C***CALCULATE THE AXIAL FORCE DUE TO
C   DISPLACEMENTS.
C
      DISFOR(1)=AEBYL*(DISPAX2-DISPAX1)
      DISFOR(4)=DISFOR(1)
```

```
C
C***CALCULATE THE BENDING MOMENTS AT THE ENDS
C   DUE TO DISPLACEMENTS.
C
      DISFOR(3)=S11*THETA1+S12*THETA2-Q11*DISPN1
     @+Q11*DISPN2
      DISFOR(6)=S12*THETA1+S11*THETA2-Q11*DISPN1
     @+Q11*DISPN2
C
C***CALCULATE THE SHEAR FORCES DUE TO
C   DISPLACEMENTS.
C
      DISFOR(2)=-Q11*THETA1-Q11*THETA2+T11*
     @DISPN1-T11*DISPN2
      DISFOR(5)=+Q11*THETA1+Q11*THETA2-T11*
     @DISPN1+T11*DISPN2
C
C***ZERO THE VECTOR OF FIXED END FORCES
C
      DO 20 IEVAB=1,NEVAB
   20 ELOAD(IEVAB)=0.0
C
C***CALCULATE THE FIXED END AXIAL, MOMENT AND
C   SHEAR FORCES DUE TO THE LATERAL LOADS ON THE
C   ELEMENTS.
C
      JLODEL=0
C
C***CALCULATE THE LATERAL LOADS ON THE ELEMENTS.
C   IF NO LATERAL LOADS ACT THEN THERE IS NO
C   NEED TO CALCULATE THE FIXED END FORCES.
C
      IF(NLODEL.EQ.0) GOTO 60
      DO 50 ILODEL=1,NLODEL
      IF(LODEL(ILODEL).EQ.IELEM) JLODEL=ILODEL
      IF(JLODEL.NE.0) GOTO 55
   50 CONTINUE
      IF(JLODEL.EQ.0) GOTO 60
   55 UDLXT=FORSEL(JLODEL,1)*ABS(YPROJ)*FACTLD
      UDLYT=FORSEL(JLODEL,2)*ABS(XPROJ)*FACTLD
      CONLX=FORSEL(JLODEL,3)*FACTLD
      CONLY=FORSEL(JLODEL,4)*FACTLD
C
C***CALCULATE THE FIXED END AXIAL, BENDING AND
C   SHEAR FORCES FOR  AN ORDINARY BEAM.
C
      IF(ROH.EQ.0) GOTO 61
      IF(ROH.LT.0) GOTO 62
C
C***FOR TENSILE AXIAL LOAD CALCULATE THE
C   MODIFICATION FACTORS FOR THE FIXED END
C   MOMENTS DUE TO UDL AND MID-SPAN CONCENTRATED
C   LOADS.
C
      AMDA=1.570796*SQRT(ROH)
      HSIN=SINH(AMDA)
      HCOS=COSH(AMDA)
      FACTO1=3.0/AMDA/AMDA*(AMDA*HCOS/HSIN-1.0)
C
C***MID-SPAN CONCENTRATED LOAD FIXED END MOMENT
```

```
C    MODIFICATION FACTOR.
C
      AMDA=3.141592654*SQRT(0.25*ROH)
      HSIN=SINH(AMDA)
      HCOS=COSH(AMDA)
      HTAN2=SINH(0.5*AMDA)/COSH(0.5*AMDA)
      S=(1.0-AMDA*HCOS/HSIN)/(2.0*HTAN2/AMDA
     @- 1.0)
      C=(AMDA-HSIN)/(HSIN-AMDA*HCOS)
      S11=S
      S12=S*C
      Q11=S11+S12
      T11=2.0*Q11+AMDA*AMDA
      FACTO2=2.0*Q11/T11
      GOTO 63
C
C***NO MODIFICATION OF FIXED END MOMENTS IF AXIAL
C    LOAD IS ZERO.
C
   61 FACTO1=1.0
      FACTO2=1.0
      GOTO 63
C
C***FOR COMPRESSIVE AXIAL LOAD CALCULATE THE
C    MODIFICATION FACTORS FOR THE FIXED END
C    MOMENTS DUE TO UDL AND MID-SPAN CONCENTRATED
C    LOADS.
   62 AMDA=1.570796*SQRT(-ROH)
      TSIN=SIN(AMDA)
      TCOS=COS(AMDA)
      FACTO1=3.0/AMDA/AMDA*(1.0-AMDA*TCOS/TSIN)
C
C***MID-SPAN CONCENTRATED LOAD FIXED END MOMENT
C    MODIFICATION FACTOR.
C
      AMDA=3.141592654*SQRT(-0.25*ROH)
      TSIN=SIN(AMDA)
      TCOS=COS(AMDA)
      TTAN2=SIN(0.5*AMDA)/COS(0.5*AMDA)
      S=(1.0-AMDA*TCOS/TSIN)/(2.0*TTAN2/AMDA
     @- 1.0)
      C=(AMDA-TSIN)/(TSIN-AMDA*TCOS)
      S11=S
      S12=S*C
      Q11=S11+S12
      T11=2.0*Q11-AMDA*AMDA
      FACTO2=2.0*Q11/T11
   63 ELOAD(1)=0.5*((UDLXT+CONLX)*XL+(UDLYT+
     @CONLY)*YM)
      ELOAD(4)=-ELOAD(1)
      ELOAD(2)=-0.5*((UDLYT+CONLY)*XL-(UDLXT+
     @CONLX)*YM)
      ELOAD(5)=ELOAD(2)
      ELOAD(3)=SPAN/12.0*FACTO1*(UDLYT*XL-
     @UDLXT*YM)+SPAN/8.0*FACTO2*(CONLY*XL-
     @CONLX*YM)
      ELOAD(6)=-ELOAD(3)
C
C***TOTAL FORCE = FORCES DUE TO (FIXED END +
C    DISPLACEMENT)
```

```
C
   60 DO 70 IEVAB=1,NEVAB
      TOTFOR(IEVAB)=ELOAD(IEVAB)+DISFOR(IEVAB)
   70 CONTINUE
      WRITE(6,900)IELEM,IEND,JEND,(TOTFOR(IEVAB)
     @,IEVAB=1,NEVAB)
   10 CONTINUE
  900 FORMAT(4X,3(I5),6E14.6)
      RETURN
      END
```

```
C
C**********************************************
C
      SUBROUTINE CONVER(NODFRE,LNODS,MATNO)
C
C***THIS SUBROUTINE CHECKS IF THE AXIAL LOADS
C   ASSUMED IN THE CALCULATION OF THE STIFFNESS
C   MATRIX ARE ACCURATE OR NOT.
C
      COMMON/CONTRO/NPOIN,NELEM,NNODE,NDOFN,
     @NDIME,NPROP,NMATS,NRESND,NEVAB,NFAIL,
     @NLODEL,NEQNS,NBAND,NLODND,NSINAG,ICONV,
     @DETNEW,NNEW,FACTLD,CONFAC
      COMMON/LGDATA/COORD(100,3),PROPS(10,3),
     @PNEW(100),POLD(100),NDFFIX(80),
     @IFPRE(80,3),FORSEL(80,4),LODEL(80)
     @,FORSND(80,3),LODPT(80),ASDIS(800),
     @GSTIF(100,30),GLOAD(100)
      DIMENSION NODFRE(100,3),LNODS(100,2),
     @MATNO(100,2)
      SUMDIF=0.0
      SUMOLD=0.0
      ICONV=0
C
C***AXIAL FORCE  CALCULATION IN THE MEMBERS
C
      DO 10 IELEM=1,NELEM
      LPROP=MATNO(IELEM,1)
C
C***CALCULATE THE PROPERTIES OF THE ELEMENT.
C
      YOUNG=PROPS(LPROP,1)
      AREA=PROPS(LPROP,3)
C
C***CALCULATE THE NODES CORRESPONDING TO THE TWO
C   ENDS OF THE ELEMENT
C
      IEND=LNODS(IELEM,1)
      JEND=LNODS(IELEM,2)
C
C***CALCULATE THE COORDINATES OF THE NODES AT THE
C   ENDS OF THE ELEMENT, THE LENGTH AND THE
C   DIRECTION COSINES.
C
      XPROJ=COORD(JEND,1)-COORD(IEND,1)
      YPROJ=COORD(JEND,2)-COORD(IEND,2)
      SPAN=SQRT(XPROJ*XPROJ+YPROJ*YPROJ)
      XL=XPROJ/SPAN
      YM=YPROJ/SPAN
```

```
      AEBYL=AREA*YOUNG/SPAN
C
C***SET THE VALUES OF THE END DISPLACEMENTS OF
C   THE ELEMENT AS ZERO.
C
      U1=0.0
      V1=0.0
      U2=0.0
      V2=0.0
C
C***CALCULATE THE NONZERO END DISPLACEMENTS OF
C   THE ELEMENT.
C
      IF(NODFRE(IEND,1).NE.0)U1=
     @ASDIS(NODFRE(IEND,1))
      IF(NODFRE(IEND,2).NE.0)V1=
     @ASDIS(NODFRE(IEND,2))
      IF(NODFRE(JEND,1).NE.0)U2=
     @ASDIS(NODFRE(JEND,1))
      IF(NODFRE(JEND,2).NE.0)V2=
     @ASDIS(NODFRE(JEND,2))
C
C***CALCULATE THE AXIAL DISPLACEMENTS.
C
      DISPAX1=U1*XL+V1*YM
      DISPAX2=U2*XL+V2*YM
C
C***CALCULATE THE AXIAL FORCE DUE TO
C   DISPLACEMENTS.
C
      FORAX1D=AEBYL*(DISPAX2-DISPAX1)
C
C***CALCULATE THE FIXED END AXIAL FORCE DUE TO
C   THE LATERAL LOADS ON THE ELEMENTS.
C
      FORAX1=0.0
      JLODEL=0
      IF(NLODEL.EQ.0) GOTO 60
C
C***CALCULATE THE LATERAL LOADS ON THE ELEMENTS.
C   IF NO LATERAL LOADS ACT THEN THERE IS NO
C   NEED TO CALCULATE THE FIXED END FORCES.
C
      DO 50 ILODEL=1,NLODEL
      IF(LODEL(ILODEL).EQ.IELEM) JLODEL=ILODEL
      IF(JLODEL.NE.0) GOTO 55
   50 CONTINUE
      IF(JLODEL.EQ.0) GOTO 60
   55 UDLXT=FORSEL(JLODEL,1)*ABS(YPROJ)*FACTLD
      UDLYT=FORSEL(JLODEL,2)*ABS(XPROJ)*FACTLD
      CONLX=FORSEL(JLODEL,3)*FACTLD
      CONLY=FORSEL(JLODEL,4)*FACTLD
C
C***CALCULATE THE FIXED END AXIAL  FORCES FOR
C   AN  ORDINARY BEAM.
C
      FORAX1=0.5*((UDLXT+CONLX)*XL+(UDLYT+
     @CONLY)*YM)
   60 PEND1=FORAX1+FORAX1D
      PEND2=-FORAX1+FORAX1D
```

```
      PNEW(IELEM)=PEND1
      IF(PEND2.LT.PEND1) PNEW(IELEM)=PEND2
      DIF=PNEW(IELEM)-POLD(IELEM)
      SUMDIF=SUMDIF+DIF*DIF
      SUMOLD=SUMOLD+POLD(IELEM)*POLD(IELEM)
      POLD(IELEM)=PNEW(IELEM)
   10 CONTINUE
      CONFAC=100*SQRT(SUMDIF/SUMOLD)
      IF(CONFAC.LT. 1.0)ICONV=1
      RETURN
      END
```

```
C
C*********************************************
C
      SUBROUTINE STURM
C
C***THIS SUBROUTINE CALCULATES THE VALUE OF THE
C   DETERMINANT AND NUMBER OF AGREEMENTS IN THE
C   SIGNS OF THE DETERMINANTS OF THE SUBMATRICES.
C
      COMMON/CONTRO/NPOIN,NELEM,NNODE,NDOFN,
     @NDIME,NPROP,NMATS,NRESND,NEVAB,NFAIL,
     @NLODEL,NEQNS,NBAND,NLODND,NSINAG,ICONV,
     @DETNEW,NNEW,FACTLD,CONFAC
      COMMON/LGDATA/COORD(100,3),PROPS(10,3),
     @PNEW(100),POLD(100),NDFFIX(80),
     @IFPRE(80,3),FORSEL(80,4),LODEL(80),
     @FORSND(80,3),LODPT(80),ASDIS(800),
     @GSTIF(100,30),GLOAD(100)
      NNEW=0
      DETNEW=1.0
      NSINAG=0
      I1=1
      DO 10 IEQNS=1,NEQNS
      DETNEW=DETNEW*GSTIF(IEQNS,1)
      ISIGN=GSTIF(IEQNS,1)/ABS(GSTIF(IEQNS,1))
      I2=-I1*ISIGN
      IF(I2.EQ.I1)NSINAG=NSINAG+1
      I1=I2
      IF(ABS(DETNEW).LT.1.0E50) GOTO 10
      DETNEW=DETNEW*1.0E-50
      NNEW=NNEW+1
   10 CONTINUE
      RETURN
      END
```

```
C
C**********************************************
C
      SUBROUTINE READLOD
C
C***THIS SUBROUTINE READS LOADS ACTING AT THE
C   NODES AND ON THE ELEMENTS.
C
      COMMON/CONTRO/NPOIN,NELEM,NNODE,NDOFN,
     @NDIME,NPROP,NMATS,NRESND,NEVAB,NFAIL,
     @NLODEL,NEQNS,NBAND,NLODND,NSINAG,ICONV,
     @DETNEW,NNEW,FACTLD,CONFAC
      COMMON/LGDATA/COORD(100,3),PROPS(10,3),
```

```
      @PNEW(100),POLD(100),NDFFIX(80),
      @IFPRE(80,3),FORSEL(80,4),LODEL(80),
      @FORSND(80,3),LODPT(80),ASDIS(800),
      @GSTIF(100,30),GLOAD(100)
C
C***READ AND WRITE LOADS ACTING AT THE JOINTS
C    ONLY.
C
      READ(5,900)NLODND
  900 FORMAT(I5)
      IF(NLODND.EQ.0) GOTO 25
      WRITE(6,910)NLODND
  910 FORMAT(/, ' NUMBER OF LOADED NODES = ',I3
      @,/,'-------------------------',//)
      WRITE(6,920)
  920 FORMAT(//,' NODE',6X,'X-LOAD',11X,'Y-LOAD'
      @,8X,'THETAZ-LOAD',/,'-------------------
      @----------------------------------------
      @--------',//)
      DO 20 ILODND=1,NLODND
      READ(5,930)LODPT(ILODND),(FORSND(ILODND,
      @IDOFN),IDOFN=1,NDOFN)
      WRITE(6,931)LODPT(ILODND),(FORSND(ILODND,
      @IDOFN),IDOFN=1,NDOFN)
  931 FORMAT(I3,5X,E11.4,5X,E11.4,4X,E11.4,
      @4X,E11.4)
   20 CONTINUE
C
C***READ AND WRITE THE LATERAL DISTRIBUTED AND
C    CONCENTRATED LOADS ON THE ELEMENTS
C
   25 READ(5,900)NLODEL
      IF(NLODEL.EQ.0) GOTO 80
      WRITE(6,940)NLODEL
  940 FORMAT(//,' NO. OF ELEMENTS CARRYING
      @DISTRIBUTED OR CONCENTRATED LATERAL LOAD =
      @',I5,/,'---------------------------------
      @----------------------------------------
      @--',//)
      WRITE(6,950)
  950 FORMAT(//,' ELEMENT',6X,'UDL-X',9X,'UDL-Y'
      @,8X,'CONLOAD-X',6X,'CONLOAD-Y')
      DO 40 ILODEL=1,NLODEL
      READ(5,930)LODEL(ILODEL),(FORSEL(ILODEL,J)
      @,J=1,4)
   40 WRITE(6,931)LODEL(ILODEL),(FORSEL(ILODEL,J)
      @,J=1,4)
  930 FORMAT(I5,4E11.4)
   80 RETURN
      END
```

CHAPTER 8   Frequency Analysis of 2-D Rigid-jointed Structures and Plane Grids

If a structure is vibrated by an external force whose frequency is equal to one of the natural frequencies of the structure, then the structure will resonate. In the absence of damping, resonance implies that the displacements of the structure tend to infinity. As explained in Chapter 1, for displacements to tend to infinity, the determinant of the dynamic stiffness matrix must tend to zero. In fact the problem of the determination of the natural frequencies of a structure and the problem of the determination of the load to cause elastic instability of a structure are very similar and require the solution of an eigenvalue problem. However, there are some differences as follows.

(1) In an elastic stability analysis, one is mainly interested in the smallest load to cause elastic instability. In a frequency analysis, one is interested in all the natural frequencies in a given frequency band which is of interest to the designer.

(2) In an instability analysis, one is sometimes interested in the stress in the nonlinear range. In a frequency analysis, 'stress' has no real meaning as it is proportional to the amplitude of a chosen displacement.

8.1 DYNAMIC ELEMENT STIFFNESS MATRIX FOR A PRISMATIC BEAM ELEMENT WITH UNIFORMLY DISTRIBUTED MASS

The dynamic stiffness coefficients for flexural as well as for longitudinal vibrations were derived in Chapter 1. They are repeated below for easy reference. The dynamic element stiffness matrix for a beam element in a 2-D rigid-jointed structure is given in Table 1.9.

(1) The bending stiffness coefficients are

$$S_{11} = (Cs\text{-}Sc)D, \quad S_{12} = (S\text{-}s)D$$
$$Q_{11} = \lambda LSsD, \quad Q_{12} = \lambda L(C\text{-}c)D$$
$$T_{11} = (\lambda L)^2(Sc + Cs)D, \quad T_{12} = (\lambda L)^2(S + s)D$$

where $D = \lambda L/(1 - Cc)$, $\lambda = \sqrt{\sqrt{(m\omega^2/EI)}}$, $S = \sinh \lambda L$, $C = \cosh \lambda L$, $s = \sin \lambda L$, $c = \cos \lambda L$, ω is the frequency of vibration in radians per second, m is the uniformly distributed mass on the beam element, EI is the flexural rigidity and L is the span. Note that, if $m = 0$, then $S_{11} = 4.0$, $S_{12} = 2.0$, $Q_{11} = Q_{12} = 6.0$ and $T_{11} = T_{12} = 12.0$.

(2) The axial stiffness coefficients are

$$A_1 = \beta L \cot \beta L, \quad A_2 = -\beta L \operatorname{cosec} \beta L$$

where $\beta = \sqrt{(m\omega^2/AE)}$ and AE is the axial rigidity. Note that, if $m = 0$, then $A_1 = 1.0$ and $A_2 = -1.0$. In calculating the values of λ and β it is important to ensure that the units of mass, force and length are all consistent. In other words, care has to be taken that the units of acceleration due to gravity are consistent with the units of mass and force.

8.2 DYNAMIC STIFFNESS MATRIX FOR A GRID ELEMENT CARRYING UNIFORMLY DISTRIBUTED MASS

The dynamic stiffness coefficients for flexural vibration were given in the previous section. The dynamic stiffness coefficients for torsional vibration are given below. The dynamic stiffness matrix for a plane grid element is given in Table 1.10.

The torsional stiffness coefficients are

$$A_1 = \beta L \cot \beta L, \quad A_2 = -\beta L \operatorname{cosec} \beta L$$

where $\beta = \sqrt{(m\omega^2 I_\mathrm{p}/GJA)}$, GJ is the torsional rigidity, I_p is the polar second moment of area, A is the area of cross section, m is the uniformly distributed mass and ω is the frequency of vibration. Note that, if $m = 0$, then $A_1 = 1.0$ and $A_2 = -1.0$.

Occasionally cases do arise where one has to deal with superimposed mass in addition to the self-mass as shown in Fig. 8.1. In deriving the equations of torsional stiffness coefficients given in section 8.2, it was assumed (as explained in section 1.36) that the mass m is wholly due to the self-mass of the element. The best procedure in such nonstandard cases is to use m to stand for the total uniformly distributed mass but to adjust I_p to reflect the nonstandard condition.

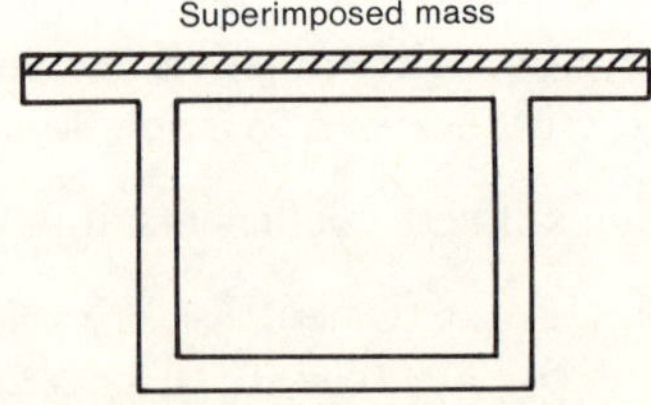

Fig. 8.1   A member with superimposed mass.

8.3   CONCENTRATED MASS AT THE JOINTS

If a structure carries a concentrated mass M at a joint, then the amplitude of the translational intertial force associated with the mass is equal to $M\omega^2$ multiplied by the amplitude of translational displacement. It should be noted that the intertial force due to rotation is ignored because the polar inertia of the mass is unlikely to be sufficiently large in most common situations to induce significant rotational inertial force. If the mass occurs on a beam element, it is necessary to treat the point where the mass is situated as a joint because the dynamic stiffness coefficients given in section 8.1 include the inertial forces associated with the uniformly distributed mass only.

8.4   THE PROCEDURE FOR THE DETERMINATION
OF ALL THE NATURAL FREQUENCIES
IN A GIVEN FREQUENCY BAND

As in a stability analysis, the determination of natural frequencies requires iterative calculations because it is necessary to determine those frequencies at which the determinant of the dynamic structural stiffness matrix is zero. The steps involved can be summarized as follows.

(1)   Input the basic parameters of the structure and its elements.
(2)   Input the uniformly distributed mass on the elements and the concentrated mass at the joints. If a concentrated mass is situated in between the acutual joints of the structure, then it is necessary to treat that point as an additional joint.
(3)   Input the lower and upper limits of the frequency band within which it is desired to determine all the natural frequencies which lie in that band.
(4)   Start the calculations at the lower limit of the frequency band. Calculate the element stiffness matrices and assemble the structural stiffness matrix. Note that, if there is a concentrated mass M at the joint, then the diagonal terms of the stiffness matrix corresponding to the translational degrees of freedom at the joint are reduced by $M\omega^2$ (see section 1.40). Apply the Gaussian elimination procedure and determine the determinant and the number of agreements in sign between the consecutive members of the sequence S_m and S_{m+1} as described in section 7.3. Let the number of agreements in sign be NSNSTR. NSNSTR stands for Number of SigN agreements at the STaRt of the frequency band.
(5)   Repeat step (4) at the upper limit of frequency band. Let the number of agreements in sign be NSNEND. NSNEND stands for Number of SigN agreements at the END of frequency band.
(6)   The difference between the number of agreements in sign calculated in steps (5) and (4) shows how many natural frequencies lie in the given frequency band.
(7)   Adopt the root bisection procedure to determine the values of all the natural frequencies lying in the given frequency band. The root bisection procedure is explained in the next section.

8.5 THE ROOT BISECTION PROCEDURE

Let the number of frequencies in the given frequency band be NFREQ. As a crude approximation, assume that the lower limit and upper limit for all the NFREQ frequencies are the lower and upper limits of the frequency band. The frequencies of vibration of NFREQ frequencies are isolated as follows.

(1) Start at the first natural frequency in the given band. Start the calculations at a frequency equal to the average of the upper and lower limits of the first frequency. Repeat step (4) of section 8.4. Let the mumber of agreements in sign be NSINAG. NSINAG stands for Number of SigN AGreements.

(2) Calculate   IBELOW = NSINAG − NSNSTR   and   IABOVE = NSNEND − NSINAG. Evidently, IBELOW frequencies are below the frequency at which the calculation was done and IABOVE frequencies are above the frequency at which the calculation was done. Therefore the upper limit of IBELOW frequencies and the lower limits of IABOVE frequencies can be reset if the new values are better than the previously calculated values. Similar adjustments can be made to the corresponding determinants as well.

(3) Check whether the difference between the upper and lower limits is within acceptable error. If the accuracy is outside the limits, then go back to (1).

(4) If the limit is within acceptable limits, then repeat the calculations (1) to (3) starting with the next frequency.

(5) When all the frequencies are determined to within acceptable limits, then the exact value can be determined by linear intrerpolation between the upper and lower limits of that particular frequency.

8.6 NOMENCLATURE OF NAMES OF VARIABLES

In addition to the basic nomenclature described in the previous chapters, the following variable names are used in the subroutines for the frequency analysis of 2-D rigid-jointed structures and grids.

ACURCY	Limit of ACcURaCY to which the difference between the upper and lower limits of any frequency must be determined before linear interpolation can be done to determine its value.
ENDFRQ	END or upper limit of FReQency band.
FREQ	FREQuency (in hertz) at which a set of calculations are being done.
NFREQ	Number of natural FREQuencies to be determined.
OMEGA	Frequency in radians per second.
STRFRQ	STaRting or lower limit of FReQuency band.

FRQTAB(NFREQ,6)   This array stores for all the NFREQ natural frequencies the following information: in columns one and two the lower and upper limits of frequency; in columns three and four the information needed to calculate the value of the determinant at the lower limit; similarly, in columns five and six the information needed to calculate the value of the determinant at the upper limit of frequency. For example at the lower limit of frequency the value of determinant is given by

$$\text{determinant} = \text{value in column three} \cdot 10^{(50 \cdot \text{value in column four})}$$

8.7   SUBROUTINES STIFRJV, STIFGRV AND READMAS

There are three new subroutines used in the frequency analysis of structures. They are as follows.

(1)   STIFRJV. This is similar to STIFFRJ except for the following aspects.

    (a)   Only prismatic beam elements are used.

    (b)   The stiffness coefficients are a function of the uniformly distributed mass on the element and frequency of vibration. The uniformly distributed mass on the element is stored in the array FORSEL-(NLODEL) where NLODEL stands for the number of elements carrying uniformly distributed mass.

    (c)   The diagonal elements of the structural stiffness matrix are modified to take account of the inertial force due to the translation only of concentrated mass at the joints as described in section 8.3.

(2)   STIFGRV. This is similar to STIFRJV except that it is used for the analysis of plane grid structures.

(3)   READMAS. This subroutine is similar to READLOD described in section 7.5 under stability analysis except that, since the mass is a scalar quantity, the uniformly distributed mass on the elements is stored as a column vector FORSEL(NLODEL) and the concentrated mass at the joints are stored in the column vector FORSND(NLODND).

8.8   REFERENCES

[1]   P. Bhatt, *Problems in Structural Analysis by Matrix Methods*, Construction Press, 1981.

[2]   F. Y. Cheng, Vibration of Timoshenko beams and frameworks, *Journal of the Structural Division, Proceedings of the American Society of Civil Engineers*, Vol. 96, ST3, March 1970, pp. 351–371.

[3]   R. W. Clough and J. Penzien, *Dynamics of Structures*, McGraw-Hill, 1975, Chapter 20, pp. 345–360.

[4]   G. Peters and J. H. Wilkinson, Eigenvalues of $Ax = \lambda Bx$ with band symmetric A and B, *Computer Journal*, Vol. 12, 1969, pp. 398–404.

```
C
C**********************************************
C
      SUBROUTINE STIFFRJV(NODFRE,LNODS,MATNO)
C
C***THIS SUBROUTINE COMPUTES THE ELEMENT
C   STIFFNESS MATRIX AND ASSEMBLES THE
C   STRUCTURAL STIFFNESS MATRIX IN BANDED FORMAT
C   WITH THE DIAGONAL TERM IN THE FIRST COLUMN.
C   THIS IS DONE FOR A 2-D RIGID-JOINTED
C   STRUCTURE.
C
      COMMON/CONTRO/NPOIN,NELEM,NNODE,NDOFN,
     @NDIME,NPROP,NMATS,NRESND,NEVAB,NFAIL,
     @NLODEL,NEQNS,NBAND,NLODND,NSINAG,OMEGA,
     @DETNEW,NNEW
      COMMON/LGDATA/COORD(100,2),PROPS(10,6),
     @NDFFIX(80),IFPRE(80,3),FORSEL(80),
     @LODEL(80),FORSND(80),LODPT(80),
     @GSTIF(100,30)
      DIMENSION ESTIF(6,6),MEMDIS(6),
     @NODFRE(100,3),LNODS(100,2),MATNO(100,2)
C
C*** CYCLE ON ALL ELEMENTS AND CALCULATE ELEMENT
C    STIFFNESS MATRIX AND ADD TO THE STRUCTURAL
C    STIFFNESS MATRIX.
C
      DO 10 IELEM=1,NELEM
      LPROP=MATNO(IELEM,1)
C
C*** CALCULATE THE PROPERTIES OF ELEMENTS.
C
      YOUNG=PROPS(LPROP,1)
      YINERA=PROPS(LPROP,2)
      AREA=PROPS(LPROP,3)
C
C***CALCULATE THE NODES CORRESPONDING TO THE TWO
C   ENDS OF THE ELEMENT, THE ELEMENT LENGTH AND
C   THE DIRECTION COSINES.
C
      IEND=LNODS(IELEM,1)
      JEND=LNODS(IELEM,2)
      XPROJ=COORD(JEND,1)-COORD(IEND,1)
      YPROJ=COORD(JEND,2)-COORD(IEND,2)
      SPAN=SQRT(XPROJ*XPROJ+YPROJ*YPROJ)
      XL=XPROJ/SPAN
      YM=YPROJ/SPAN
      EI=YOUNG*YINERA
      EIBYL=EI/SPAN
      EIBYL2=EIBYL/SPAN
      EIBYL3=EIBYL2/SPAN
      AEBYL=AREA*YOUNG/SPAN
C
C***DETERMINE THE UDL MASS ON THE ELEMENTS WHICH
C   IS STORED IN THE ARRAY FORSEL(NLODEL).
C
      UDLMASS=0.0
      IF(NLODEL.EQ.0) GOTO 11
      JLODEL=0
C
```

```
C***SEARCH THROUGH THE NLODEL ELEMENTS TO SEE
C   IF THE ELEMENT HAS  UDL MASS ON IT.
C
      DO 15 ILODEL=1,NLODEL
      IF(LODEL(ILODEL).EQ.IELEM) JLODEL=ILODEL
      IF(JLODEL.NE.0) GOTO 25
   15 CONTINUE
      IF(JLODEL.EQ.0) GOTO 11
   25 UDLMASS=FORSEL(JLODEL)
C
C***CALCULATE THE STIFFNESS COEFFICIENTS WHEN
C   UDL MASS IS PRESENT.
C
      AMDAL=SPAN*SQRT(SQRT(UDLMASS*OMEGA*OMEGA
     @/EI))
      BETAL=SPAN*SQRT(UDLMASS*OMEGA*OMEGA/
     @(YOUNG*AREA))
      A1=BETAL*COS(BETAL)/SIN(BETAL)
      A2=-BETAL/SIN(BETAL)
      TSIN=SIN(AMDAL)
      TCOS=COS(AMDAL)
      HSIN=SINH(AMDAL)
      HCOS=COSH(AMDAL)
      D=AMDAL/(1.0-HCOS*TCOS)
      S11=D*(HCOS*TSIN-HSIN*TCOS)
      S12=D*(HSIN-TSIN)
      Q11=D*AMDAL*HSIN*TSIN
      Q12=D*AMDAL*(HCOS-TCOS)
      T11=D*AMDAL*AMDAL*(HCOS*TSIN+HSIN*TCOS)
      T12=D*AMDAL*AMDAL*(HSIN+TSIN)
      GOTO 13
C
C***BENDING STIFFNESS COEFFICIENTS FOR ZERO
C   UNIFORMLY DISTRIBUTED MASS.
C
   11 A1=1.0
      A2=-1.0
      S11=4.0
      S12=2.0
      Q11=6.0
      Q12=6.0
      T11=12.0
      T12=12.0
   13 T11=T11*EIBYL3
      T12=T12*EIBYL3
      S11=S11*EIBYL
      S12=S12*EIBYL
      Q11=Q11*EIBYL2
      Q12=Q12*EIBYL2
      A1=A1*AEBYL
      A2=A2*AEBYL
C
C***CALCULATE THE ELEMENT STIFFNESS MATRIX
C
      ESTIF(1,1)=T11*YM*YM+A1*XL*XL
      ESTIF(1,2)=(-T11 + A1)*XL*YM
      ESTIF(1,3)=Q11*YM
      ESTIF(1,4)=-T12*YM*YM+A2*XL*XL
      ESTIF(1,5)=(T12 + A2)*XL*YM
      ESTIF(1,6)=Q12*YM
```

```fortran
      ESTIF(2,2)=T11*XL*XL+A1*YM*YM
      ESTIF(2,3)=-Q11*XL
      ESTIF(2,4)=ESTIF(1,5)
      ESTIF(2,5)=-T12*XL*XL+A2*YM*YM
      ESTIF(2,6)=-Q12*XL
      ESTIF(3,3)=S11
      ESTIF(3,4)=-ESTIF(1,6)
      ESTIF(3,5)=-ESTIF(2,6)
      ESTIF(3,6)=S12
      ESTIF(4,4)=ESTIF(1,1)
      ESTIF(4,5)=ESTIF(1,2)
      ESTIF(4,6)=-ESTIF(1,3)
      ESTIF(5,5)=ESTIF(2,2)
      ESTIF(5,6)=-ESTIF(2,3)
      ESTIF(6,6)=S11
C
C***FILL UP THE SYMMETRICAL LOWER HALF OF THE
C   ESTIF MATRIX
C
      DO 20 IEVAB=2,NEVAB
      IEVAB1=IEVAB-1
      DO 20 JEVAB=1,IEVAB1
   20 ESTIF(IEVAB,JEVAB)=ESTIF(JEVAB,IEVAB)
C
C***CALCULATE THE NODE FREEDOMS CORRESPONDING
C   TO THE NODES AT THE ENDS OF THE ELEMENT.
C   THIS IS STORED IN ARRAY MEMDIS(NDOFN).
C
      DO 30 IDOFN=1,NDOFN
      MEMDIS(IDOFN)=NODFRE(IEND,IDOFN)
   30 MEMDIS(IDOFN+NDOFN)=NODFRE(JEND,IDOFN)
C
C***ASSEMBLE THE STRUCTURAL STIFFNESS MATRIX IN
C   THE BANDED FORM WITH THE DIAGONAL ELEMENTS
C   IN THE FIRST COLUMN
C
      DO 40 IEVAB=1,NEVAB
      IF(MEMDIS(IEVAB).EQ.0) GOTO 40
      DO 40 JEVAB=1,NEVAB
      IF(MEMDIS(JEVAB).EQ.0) GOTO 40
C
C***CALCULATE THE COLUMN POSITION IN THE BANDED
C   FORMAT.
C
      NEWCOL=MEMDIS(JEVAB)-MEMDIS(IEVAB)+1
      IF(NEWCOL.LE.0) GOTO 40
      GSTIF(MEMDIS(IEVAB),NEWCOL)=GSTIF(MEMDIS
     @(IEVAB),NEWCOL)+ESTIF(IEVAB,JEVAB)
   40 CONTINUE
   10 CONTINUE
C
C***MODIFY THE STIFFNESS MATRIX FOR THE EFFECT
C   OF CONCENTRATED MASSES.  IT IS ASSUMED THAT
C   THE CONCENTRATED MASS EXERTS INERTIAL FORCE
C   WITH RESPECT TO TRANSLATIONS ONLY IE. IDOFN
C   = 1 AND 2 ONLY.
C
      IF(NLODND.EQ.0) GOTO 100
      DO 70 ILODND=1,NLODND
      DO 80 IDOFN=1,2
```

```
      IF(NODFRE(LODPT(ILODND),IDOFN).EQ.0)
     @GOTO 80
      GSTIF(NODFRE(LODPT(ILODND),IDOFN),1)=
     @GSTIF(NODFRE(LODPT(ILODND),IDOFN),1)-
     @FORSND(ILODND)*OMEGA*OMEGA
   80 CONTINUE
   70 CONTINUE
  100 RETURN
      END
```

```
C
C*********************************************
C
      SUBROUTINE STIFFGRV(NODFRE,LNODS,MATNO)
C
C***THIS SUBROUTINE COMPUTES THE ELEMENT
C    STIFFNESS MATRIX AND ASSEMBLES THE
C    STRUCTURAL STIFFNESS MATRIX IN BANDED FORMAT
C    WITH THE DIAGONAL TERM IN THE FIRST COLUMN.
C    THIS IS DONE FOR A PLANE GRID STRUCTURE.
C
      COMMON/CONTRO/NPOIN,NELEM,NNODE,NDOFN,
     @NDIME,NPROP,NMATS,NRESND,NEVAB,NFAIL,
     @NLODEL,NEQNS,NBAND,NLODND,NSINAG,OMEGA,
     @DETNEW,NNEW
      COMMON/LGDATA/COORD(100,2),
     @PROPS(10,6),NDFFIX(80),IFPRE(80,3),
     @FORSEL(80),LODEL(80),FORSND(80),LODPT(80),
     @GSTIF(100,30)
      DIMENSION ESTIF(6,6),MEMDIS(6),
     @NODFRE(100,3),LNODS(100,2),MATNO(100,2)
C
C*** CYCLE ON ALL ELEMENTS AND CALCULATE ELEMENT
C    STIFFNESS MATRIX AND ADD TO THE STRUCTURAL
C    STIFFNESS MATRIX.
C
      DO 10 IELEM=1,NELEM
      LPROP=MATNO(IELEM,1)
C
C*** CALCULATE THE PROPERTIES OF ELEMENTS.
C
      YOUNG=PROPS(LPROP,1)
      SHEARM=PROPS(LPROP,2)
      YINERA=PROPS(LPROP,3)
      TINERA=PROPS(LPROP,4)
      AREA=PROPS(LPROP,5)
      POLAR=PROPS(LPROP,6)
C
C***CALCULATE THE NODES CORRESPONDING TO THE TWO
C    ENDS OF THE ELEMENT, THE ELEMENT LENGTH AND
C    THE DIRECTION COSINES.
C
      IEND=LNODS(IELEM,1)
      JEND=LNODS(IELEM,2)
      XPROJ=COORD(JEND,1)-COORD(IEND,1)
      YPROJ=COORD(JEND,2)-COORD(IEND,2)
      SPAN=SQRT(XPROJ*XPROJ+YPROJ*YPROJ)
      XL=XPROJ/SPAN
      YM=YPROJ/SPAN
      EI=YOUNG*YINERA
```

```
      EIBYL=EI/SPAN
      EIBYL2=EIBYL/SPAN
      EIBYL3=EIBYL2/SPAN
      GJBYL=TINERA*SHEARM/SPAN
C
C***DETERMINE THE UDL MASS ON THE ELEMENTS WHICH
C   IS STORED IN THE ARRAY FORSEL(NLODEL).
C
      UDLMASS=0.0
      IF(NLODEL.EQ.0) GOTO 11
      JLODEL=0
C
C***SEARCH THROUGH THE NLODEL ELEMENTS TO SEE IF
C   THE ELEMENT HAS  UD MASS ON IT.
C
      DO 15 ILODEL=1,NLODEL
      IF(LODEL(ILODEL).EQ.IELEM) JLODEL=ILODEL
      IF(JLODEL.NE.0) GOTO 25
   15 CONTINUE
      IF(JLODEL.EQ.0) GOTO 11
   25 UDLMASS=FORSEL(JLODEL)
C
C***CALCULATE THE STIFFNESS COEFFICIENTS WHEN
C   UD MASS IS PRESENT.
C
      AMDAL=SPAN*SQRT(SQRT(UDLMASS*OMEGA*OMEGA
     @/EI))
      BETAL=SPAN*SQRT(UDLMASS*OMEGA*OMEGA*POLAR
     @/(SHEARM*TINERA*AREA))
      G1=BETAL*COS(BETAL)/SIN(BETAL)
      G2=-BETAL/SIN(BETAL)
      TSIN=SIN(AMDAL)
      TCOS=COS(AMDAL)
      HSIN=SINH(AMDAL)
      HCOS=COSH(AMDAL)
      D=AMDAL/(1.0-HCOS*TCOS)
      S11=D*(HCOS*TSIN-HSIN*TCOS)
      S12=D*(HSIN-TSIN)
      Q11=D*AMDAL*HSIN*TSIN
      Q12=D*AMDAL*(HCOS-TCOS)
      T11=D*AMDAL*AMDAL*(HCOS*TSIN+HSIN*TCOS)
      T12=D*AMDAL*AMDAL*(HSIN+TSIN)
      GOTO 13
C
C***BENDING STIFFNESS COEFFICIENTS FOR ZERO
C   UD MASS.
C
   11 G1=1.0
      G2=-1.0
      S11=4.0
      S12=2.0
      Q11=6.0
      Q12=6.0
      T11=12.0
      T12=12.0
   13 T11=T11*EIBYL3
      T12=T12*EIBYL3
      S11=S11*EIBYL
      S12=S12*EIBYL
      Q11=Q11*EIBYL2
```

```
      Q12=Q12*EIBYL2
      G1=G1*GJBYL
      G2=G2*GJBYL
C
C***CALCULATE THE ELEMENT STIFFNESS MATRIX
C
      ESTIF(1,1)=T11
      ESTIF(1,2)=Q11*YM
      ESTIF(1,3)=-Q11*XL
      ESTIF(1,4)=-T12
      ESTIF(1,5)=Q12*YM
      ESTIF(1,6)=-Q12*XL
      ESTIF(2,2)=S11*YM*YM+G1*XL*XL
      ESTIF(2,3)=-(S11-G1)*XL*YM
      ESTIF(2,4)=-ESTIF(1,5)
      ESTIF(2,5)=S12*YM*YM+G2*XL*XL
      ESTIF(2,6)=-(S12-G2)*XL*YM
      ESTIF(3,3)=S11*XL*XL+G1*YM*YM
      ESTIF(3,4)=-ESTIF(1,6)
      ESTIF(3,5)=ESTIF(2,6)
      ESTIF(3,6)=S12*XL*XL+G2*YM*YM
      ESTIF(4,4)=ESTIF(1,1)
      ESTIF(4,5)=-ESTIF(1,2)
      ESTIF(4,6)=-ESTIF(1,3)
      ESTIF(5,5)=ESTIF(2,2)
      ESTIF(5,6)=ESTIF(2,3)
      ESTIF(6,6)=ESTIF(3,3)
C
C***FILL UP THE SYMMETRICAL LOWER HALF OF THE
C   ESTIF MATRIX
C
      DO 20 IEVAB=2,NEVAB
      IEVAB1=IEVAB-1
      DO 20 JEVAB=1,IEVAB1
   20 ESTIF(IEVAB,JEVAB)=ESTIF(JEVAB,IEVAB)
C
C***CALCULATE THE NODE FREEDOMS CORRESPONDING
C   TO THE NODES AT THE ENDS OF THE ELEMENT.
C   THIS IS STORED IN ARRAY MEMDIS(NDOFN).
C
      DO 30 IDOFN=1,NDOFN
      MEMDIS(IDOFN)=NODFRE(IEND,IDOFN)
   30 MEMDIS(IDOFN+NDOFN)=NODFRE(JEND,IDOFN)
C
C***ASSEMBLE THE STRUCTURAL STIFFNESS MATRIX IN
C   THE BANDED FORM WITH THE DIAGONAL ELEMENTS
C   IN THE FIRST COLUMN
C
      DO 40 IEVAB=1,NEVAB
      IF(MEMDIS(IEVAB).EQ.0) GOTO 40
      DO 40 JEVAB=1,NEVAB
      IF(MEMDIS(JEVAB).EQ.0) GOTO 40
C
C***CALCULATE THE COLUMN POSITION IN THE BANDED
C   FORMAT.
C
      NEWCOL=MEMDIS(JEVAB)-MEMDIS(IEVAB)+1
      IF(NEWCOL.LE.0) GOTO 40
      GSTIF(MEMDIS(IEVAB),NEWCOL)=GSTIF(MEMDIS
     @(IEVAB),NEWCOL)+ESTIF(IEVAB,JEVAB)
```

```
   40 CONTINUE
   10 CONTINUE
C
C***MODIFY THE STIFFNESS MATRIX FOR THE EFFECT
C   OF CONCENTRATED MASSES.  IT IS ASSUMED THAT
C   THE CONCENTRATED MASS EXERTS INERTIAL FORCE
C   WITH RESPECT TO TRANSLATIONS ONLY IE. IDOFN
C   = 1 ONLY.
C
      IF(NLODND.EQ.0) GOTO 100
      DO 70 ILODND=1,NLODND
      IF(NODFRE(LODPT(ILODND),1).EQ.0)
     @GOTO 70
      GSTIF(NODFRE(LODPT(ILODND),1),1)=GSTIF
     @(NODFRE(LODPT(ILODND),1),1)-FORSND
     @(ILODND)*OMEGA*OMEGA
   70 CONTINUE
  100 RETURN
      END
```

```
C
C***********************************************
C
      SUBROUTINE READMAS
C
C***THIS SUBROUTINE READS CONCENTRATED MASS AT
C   THE NODES AND UDL MASS ON THE ELEMENTS.
C
      COMMON/CONTRO/NPOIN,NELEM,NNODE,NDOFN,
     @NDIME,NPROP,NMATS,NRESND,NEVAB,NFAIL,
     @NLODEL,NEQNS,NBAND,NLODND,NSINAG,OMEGA,
     @DETNEW,NNEW
      COMMON/LGDATA/COORD(100,2),PROPS(10,6),
     @NDFFIX(80),IFPRE(80,3),FORSEL(80),
     @LODEL(80),FORSND(80),LODPT(80),
     @GSTIF(100,30)
C
C***READ AND WRITE MASS AT THE JOINTS ONLY.
C
      READ(5,900)NLODND
  900 FORMAT(I5)
      IF(NLODND.EQ.0) GOTO 25
      WRITE(6,910)NLODND
  910 FORMAT(/, ' NUMBER OF LOADED NODES = ',I3,/
     @,'----------------------------',//)
      WRITE(6,920)
  920 FORMAT(//,' NODE',6X,'CONC. MASS',/,'----
     @----------------------------------------
     @----------------------------------------
     @----',//)
      DO 20 ILODND=1,NLODND
      READ(5,930)LODPT(ILODND),FORSND(ILODND)
      WRITE(6,931)LODPT(ILODND),FORSND(ILODND)
  931 FORMAT(I3,5X,E14.6)
   20 CONTINUE
C
C***READ AND WRITE THE DISTRIBUTED MASS ON THE
C   ELEMENTS.
C
   25 READ(5,900)NLODEL
```

```
      IF(NLODEL.EQ.0) GOTO 80
      WRITE(6,940)NLODEL
  940 FORMAT(//,' NO. OF ELEMENTS CARRYING
     @DISTRIBUTED MASS = ',I5,/,'---------------
     @-------------------------------------------
     @-------------------------------------------
     @',//)
      WRITE(6,950)
  950 FORMAT(//,' ELEMENT',3X,'UDL MASS')
      DO 40 ILODEL=1,NLODEL
      READ(5,930)LODEL(ILODEL),FORSEL(ILODEL)
   40 WRITE(6,931)LODEL(ILODEL),FORSEL(ILODEL)
  930 FORMAT(I5,E14.6)
   80 RETURN
      END
```

CHAPTER 9   Main Programs and Data Preparation

The previous chapters have described the subroutines used to accomplish the various tasks involved in the analysis of skeletal structures by the stiffness method. In this chapter, the master programs which use the subroutines are described. Also included are instructions for data preparation.

9.1   MASTER PROGRAMS TRUSS, 2-DFRAME, GRID, STABIL AND VIBES

There are five master programs called TRUSS, 2-DFRAME, GRID, STABIL and VIBES in this book. The first three programs, i.e. TRUSS, 2-DFRAME and GRID, are used for the linear elastic analysis of 2-D and 3-D pin-jointed structures, 2-D rigid-jointed frames and plane grids respectively. They can handle external loads applied at joints as well as on members, thermal loads and loads due to settlement of supports. The program STABIL is used for the calculation of the elastic stability load factor of 2-D rigid-jointed frames, and the program VIBES is used for calculating the natural frequencies of 2-D rigid-jointed frames and plane grids. A brief description of the programs are given below.

(1)   TRUSS. This program carries out the linear elastic analysis of 2-D and 3-D pin-jointed structures. External loads must be applied at the joints only.

(2)   2-DFRAME. This program includes four types of bending element as described in chapter 3. There is facility for including any nonstandard beam element provided that the corresponding element stiffness matrix in general coordinate directions is known.

(3)   GRID. This is similar to 2-DFRAME. The program has two types of torsion element, i.e. free warping and restrained warping torsion cases.

(4)   STABIL. In this program, only simple prismatic bending elements are included. There is the option either to carry out a complete nonlinear analysis or to calculate only the load factor to induce instability.

(5) VIBES. This program includes only simple prismatic beam elements. The mass can be both distributed on elements as well as concentrated at the nodes. The effects of shear deformation are excluded from the analysis.

9.1.1 Structure of the linear elastic stress analysis programs

The structure of the three programs TRUSS, 2-DFRAME and GRID is very similar and can be summarized as follows.

(1) Input how many structures to be analysed.
(2) For each structure, do the following.

    (a) Input basic parameters such as NPOIN, NELEM, NRESND, NVFIX, NMATS, NCASE, etc.

    (b) Use subroutine CORDGEN to read and, if necessary, to generate the coordinates of all the NPOIN nodes. Similarly, use subroutine INTGEN to read and, if necessary, to generate the node numbers at the ends of elements and the group number of all the elements.

    (c) Input NOPTN. If NOPTN (Number of the OPTioN) = 1, generate the node freedom array NODFRE using subroutine INTGEN. If NOPTN = 0, read data about the nodes where one or more freedoms are restrained and then use subroutine NFGEN to calculate the NODFRE array.

    (d) Input the basic geometrical and material properties of all the NMATS group of elements.

    (e) If there are nodes where displacements are required to have a fixed nonzero value, read the data about these and store the appropriate information.

    (f) Use subroutines MAXEQNS and BANDWIDTH to calculate the number of simultaneous equations to be solved and their half-bandwidth respectively.

    (g) Use subroutine NULLGSTIF to zero the structural stiffness matrix. Use the appropriate subroutines such as STIFFPJ, STIFFRJ or STIFFGRID to calculate the element stiffness matrix ESTIF and to add it to the structural stiffness matrix GSTIF. When all the elements have been considered, the assembling of the structural stiffness matrix is complete.

    (h) If there are fixed values of displacements, then modify the matrix GSTIF using the subroutine STIFSPRNG as explained in section 5.6 under stiff spring approach.

    (i) Using the subroutine GAUSRED, reduce the matrix GSTIF to the upper triangular form.

    (j) Use the appropriate subroutine LOADPJ, LOADRJ or LOAD-GRID to read the applied load at the nodes and on the elements and assemble the GLOAD vector.

    (k) Input ITEMP. If ITEMP = 1, then use the appropriate subroutines THERMPJ, THERMRJ or THERMGRID to read the temperature change in the elements and the material and geometrical properties needed for thermal load analysis. Calculate the contribution to the vector GLOAD from the elements due to restraining the thermal expansion.

(l)   Use the subroutine FIXDISP to modify the vector GLOAD to ensure that the GSTIF modification in (h) is consistent with the modification of GLOAD to ensure that the fixed displacement case is correctly solved.

(m)   Use subroutines BACKSUB to reduce the load vector GLOAD and to calculate the displacements and to store them in the array ASDIS.

(n)   Use the appropriate subroutines STRESSPJ, STRESSRJ or STRESSGRID to calculate the final stresses at the ends of the elements.

(o)   If there are beams on elastic foundations, then use the subroutines PRESURRJ or PRESURGRID to calculate the displacements and stresses at ten intermediate sections along the element.

The above steps can be easily followed in the program printouts given. It should be appreciated that a complete program consists of a master program and the associated subroutines. It is important to remember that all the subroutines should have the appropriate COMMON BLOCKS at the beginning of the subroutines. The COMMON BLOCKS are exactly the same as those at the beginning of the corresponding main programs.

9.1.2   Data input

The data input for the programs are formatted. In many systems, it is possible to use 'free format', where the individual pieces of data are separated by commas. This is very convenient to use and the reader should investigate this possibility when using the programs.

9.2   DATA INPUT FOR TRUSS

The order in which the data should be input is as follows.

(a)   NPROB in I5 formats, where NPROB is the number of problems to be solved.
       Now follows NPROB sets of data from (b) to (k).

(b)   TITLE (any description of the problem) with a maximum of 60 characters.

(c)   NPOIN, NELEM, NRESND, NVFIX, NMATS, NCASE, NDIME in 7(I5) format. If NDIME = 2, then omit all reference to the z coordinate in the following.

(d)   Data about the coordinates of nodes. Since the subroutine CORDGEN will be used to generate intermediate data, it is necessary to input only the 'basic' data as explained in Chapter 2. It is important that the last node for which the data are input is NPOIN.

IPOIN	NSTEP	Coordinates of IPOIN		
		x	y	z
.	.	.	.	.
.	.	.	.	.
NPOIN	1	.	.	.

See section 2.2.1 and figure 2.4 for an explanation of NSTEP. Use the format (2I5, 3F10.4) for data input.

(e)   Data regarding the node numbers at the ends of elements. Since the subroutine INTGEN will be used to generate the intermediate data, only 'basic' data need to be input. The last data must be those of the NELEMth element. Use 4(I5) format.

IELEM	NSTEP	Node numbers at	
		End 1	End 2
.	.	.	.
.	.	.	.
NELEM	1	.	.

(f)   Data regarding the group number to which the NELEM elements belong. Here also the subroutine INTGEN will be used to generate 'intermediate' data. Therefore, only 'basic' data need to be input. Use 3(I5) format.

IELEM	NSTEP	Group number
.	.	.
.	.	.
NELEM	1	.

(g)   NOPTN in I5 format. The number of option NOPTN is interpreted as follows.

(1)   If NOPTN = 1, it means that the node freedom numbers will be input as data and subroutine INTGEN will be used to calculate intermediate data. Use 5I5 format. The last data must be those of the NPOINth node. Skip this section if NOPTN = 0.

		Node freedom number for		
IPOIN	NSTEP	u	v	w
.	.	.	.	.
.	.	.	.	.
NPOIN	1	.	.	.

(2)   If NOPTN = 0, it means that the node freedom numbers will be
      generated from the data about the restrained nodes using the
      subroutine NFGEN. Data to be input about the restraint state of
      NDOFN freedoms at NRESND nodes. There are NRESND sets
      of data to follow. Use 4(I5) format. Use the code free = 0 and
      restrained = 1. Skip this section if NOPTN = 1.

	Fixity state w.r.t.		
Node number	u	v	w
.	.	.	.
.	.	.	.
.	.	.	.

(h)   Data about the material and cross-sectional geometric properties of the
      NMATS group of elements. There are NMATS sets of data to follow.
      Use 2(E14.6) format.

Young's modulus	Area
.	.
.	.
.	.

(j)   Data about nodes where one or more displacements are required to
      have a fixed value. If a particular freedom is free, input the 'fixed'
      displacement as 0!. There are NVFIX sets of data to follow. Use (I5,
      3 F10.4) format. Skip this section if NVFIX = 0.

Node number	Fixed value of displacement		
	u	v	w
.	.	.	.
.	.	.	.
.	.	.	.

(k) Data regarding the loads on the structure. NCASE sets of data to follow.

    (1) TITLE (any description of load case) with a maximum of 60 characters.

    (2) NLODPT in I5 format. NLODPT is the number of nodes at which concentrated loads are applied.

    (3) NLODPT sets of data to follow. Use (I5, 3F10.4) format. If NLODPT = 0, skip this section.

Node number	x load	y load	z load
.	.	.	.
.	.	.	.
.	.	.	.

    (4) ITEMP in I5 format. If ITEMP = 0, then thermal effects need not be considered. Skip the sections (5) and (6).

    (5) Data about linear thermal expansion coefficient α_t for all the NMATS group of elements. There are NMATS sets of data to follow. Use E14.6 format.

α_t
.
.
.

    (6) Data regarding the temperature changes in elements. It is necessary to input data about only those elements for which there is a change in temperature. The data about the last element must be those of NELEM even if it does not suffer any change in temperature.

Element number	$T(+$ is a rise)
.	.
.	.
.	.
NELEM	.

9.3 DATA INPUT FOR 2-DFRAME

The order in which the data should be input is as follows.

(a) NPROB in I5 format, where NPROB is the number of problems to be solved. Now follows NPROB sets of data from (b) to (q).

(b) TITLE (any description of the problem) with a maximum of 60 characters.

(c) NPOIN, NELEM, NRESND, NVFIX, NMATS, NCASE, in 6(I5) format.

(d) Data about the coordinates of nodes. Since the subroutine CORDGEN will be used to generate intermediate data, it is necessary to input only the 'basic' data as explained in Chapter 2. It is important that the last node for which the data are input is NPOIN.

		Coordinates of IPOIN	
IPOIN	NSTEP	x	y
.	.	.	.
.	.	.	.
NPOIN	1	.	.

See section 2.2.1 and Fig. 2.4 for an explanation of NSTEP. Use the (2I5, 2F10.4) format.

(e) Data regarding the node numbers at the ends of elements. Since the subroutine INTGEN will be used to generate the intermediate data, only 'basic' data need to be input. The last data must be those of the NELEMth element. Use 4(I5) format.

		Node numbers at	
IELEM	NSTEP	End 1	End 2
.	.	.	.
.	.	.	.
NELEM	1	.	.

(f)   Data regarding the group number and type number to which the NELEM elements belong. Here also the subroutine INTGEN will be used to generate 'intermediate' data. Therefore, only 'basic' data need to be input. Use 4(I5) format. The type numbers of the element are as follows: type 1, an ordinary prismatic beam; type 2, a beam on an elastic foundation; type 3, a semi-rigid beam; type 4. a nonstandard beam. The last data must be that of the NELEMth element.

IELEM	NSTEP	Group number	Type number
.	.	.	.
.	.	.	.
NELEM	1	.	.

(g)   NOPTN in I5 format. The number of option NOPTN is interpreted as follows.

(1)   If NOPTN = 1, it means that the node freedom numbers will be input as data and subroutine INTGEN will be used to calculate intermediate data. The last data must be of the NPOINth node. Use 5I5 format. Skip this section if NOPTN = 0.

		Node freedom number for		
IPOIN	NSTEP	u	v	θ
.	.	.	.	.
.	.	.	.	.
NPOIN	1	.	.	.

(2)   If NOPTN = 0, it means that the node freedom numbers will be generated from the data about the restrained nodes using the subroutine NFGEN. Data to be input about the restraint state of NDOFN freedoms at NRESND nodes. There are NRESND sets of data to follow. Use 4(I5) format. Use the code free = 0 and restrained = 1. Skip this section if NOPTN = 1.

	Fixity state w.r.t. displacements		
Node number	u	v	θ
.	.	.	.
.	.	.	.
.	.	.	.

(h)   Data about the material and cross-sectional geometric properties of the NMATS group of elements. There are NMATS sets of data to follow. Use 5(E14.6) format.

Young's modulus E	Shear modulus G	Inertia I	Area A	ZASF
.	.	.	.	.
.	.	.	.	.
.	.	.	.	.

(j)   Additional data concerning the subgrade modulus and the stiffness of the springs at the ends of semi-rigid elements. NMATS sets of data to follow. Use 3(E14.6) format.

	Spring stiffness at	
Subgrade modulus	End 1	End 2
.	.	.
.	.	.
.	.	.

(k)   Data about nodes where one or more displacements are required to have a fixed value. If a particular freedom is free, input the 'fixed' displacement as 0!. There are NVFIX sets of data to follow. Use (I5, 3 F10.4) format. Skip this section if NVFIX = 0.

	Fixed value of displacement		
Node number	u	v	θ
.	.	.	.
.	.	.	.
.	.	.	.

(m)   Data concerning the stiffness coefficients of nonstandard elements. Skip this section if there are no type 4 elements. Otherwise, for each type 4 element, input the following stiffness factors:

        S11, S12, S22, the rotational stiffness factors

        T11, T12, T22, the translational stiffness factors

        Q11, Q12, Q21, Q22, the cross-stiffness factors

See Fig. 3.2 for additional explanations. Use 6E14.6 format.

(n)　Data regarding the loads on the structure. NCASE sets of data to follow.

    (1)　TITLE (any description of load case) with a maximum of 60 characters.

    (2)　NLODPT in I5 format. NLODPT is the number of nodes at which concentrated loads are applied.

    (3)　NLODPT sets of data to follow. Use (I5, 3F10.4) format. If NLODPT = 0, skip this section.

Node number	x load	y load	θ load
.	.	.	.
.	.	.	.
.	.	.	.

    (4)　NLODEL in I5 format. This shows how many elements carry lateral loads.

    (5)　If NLODEL = 0, skip this section. There are NLODEL sets of data to follow. Use (I5, 4F10.4) format.

	UDL intensity in direction		Mid-span load in direction	
Element number	x	y	x	y
.	.	.	.	.
.	.	.	.	.
.	.	.	.	.

    (6)　Data concerning the fixed end reactions in the x, y and θ directions for the nonstandard element. For each type 4 element, input

$$R_{x1}, R_{y1}, M_1, R_{x2}, R_{y2}, M_2$$

See Figure 4.3 for notation. Use 6(E14.6) format.

    (7)　ITEMP in I5 format. If ITEMP = 0, then thermal effects need not be considered. Skip the sections (8) and (9).

    (8)　Data about the linear thermal expansion coefficient α_t, the depth d and the distance Y_u from neutral axis to the top fibre for all the NMATS group of elements. If the element is assumed to be orientated horizontally with end 1 on the left-hand side, the upper face is the 'natural' top face. There are NMATS sets of data to follow. Use (E14.6, 2F10.4) format.

α_t	Depth d	Y_u
.	.	.
.	.	.
.	.	.

(9)   Data regarding the temperature changes in elements. It is necessary to input data about only those elements for which there is a change in temperature. The data about the last element must be those of NELEM even if it does not suffer any change in temperature.

Element number	$T(+$ is a rise)	
	Top	Bottom
.	.	.
.	.	.
.	.	.
NELEM	.	.

(p)   Repeat (m). If there are no type 4 elements, then skip this section.

(q)   Input fixed end axial force, shear force, moment at end 1 and fixed end axial force, shear force and moment at end 2 for all type 4 elements. See Fig. 4.2 for sign convention. If there are no type 4 elements, skip this section.

9.4   DATA INPUT FOR GRID

The order in which the data should be input is as follows.

(a)   NPROB in I5 format, where NPROB is the number of problems to be solved. Now follows NPROB sets of data from (b) to (r).

(b)   TITLE (any description of the problem) with a maximum of 60 characters.

(c)   IWARP in I5 format. IWARP = 0, if the effects of warping restraint is to be ignored. IWARP = 1, if the effects of warping restraint are to be included.

(d)   NPOIN, NELEM, NRESND, NVFIX, NMATS, NCASE, in 6(I5) format.

(e)   Data about the coordinates of nodes. Since the subroutine CORDGEN will be used to generate intermediate data, it is necessary to input only the 'basic' data as explained in Chapter 2. It is important that the last node for which the data are input is NPOIN.

		Coordinates of IPOIN	
IPOIN	NSTEP	x	y
.	.	.	.
.	.	.	.
NPOIN	1	.	.

See section 2.2.1 and Fig. 2.4 for an explanation of NSTEP. Use the (2I5, 2F10.4) format.

(f)   Data regarding the node numbers at the ends of elements. Since the subroutine INTGEN will be used to generate the intermediate data, only 'basic' data need to be input. The last data must be those of the NELEMth element. Use 4(I5) format.

		Node numbers at	
IELEM	NSTEP	End 1	End 2
.	.	.	.
.	.	.	.
NELEM	1	.	.

(g)   Data regarding the group number and type number to which the NELEM elements belong. Here also the subroutine INTGEN will be used to generate 'intermediate' data. Therefore only 'basic' data need to be input. Use 4(I5) format. The type numbers of the element are as follows: type 1, an ordinary prismatic beam; type 2, a beam on an elastic foundation; type 3, a semi-rigid beam; type 4, a nonstandard beam. The last data must be of the NELEMth element.

IELEM	NSTEP	Group number	Type number
.	.	.	.
.	.	.	.
NELEM	1	.	.

(h)   NOPTN in I5 format. The number of option NOPTN is interpreted as follows.

   (1)   If NOPTN = 1, it means that the node freedom numbers will be input as data and subroutine INTGEN will be used to calculate intermediate data. Use 6I5 format. If IWARP = 0, ignore reference to θ_z freedom. The last data must be those of the NPOINth node. Skip this section if NOPTN = 0.

		Node freedom number for			
IPOIN	NSTEP	w	θ_x	θ_y	θ_z
.	.	.	.	.	.
.	.	.	.	.	.
NPOIN	1	.	.	.	.

(2) If NOPTN = 0, it means that the node freedom numbers will be generated from the data about the restrained nodes using the subroutine NFGEN. Data to be input about the restraint state of NDOFN freedoms at NRESND nodes. There are NRESND sets of data to follow. Use 5(I5) format. Use the code free = 0, and restrained = 1. If IWARP = 0, ignore reference to θ_z freedom. Skip this section if NOPTN = 1.

	Fixity state w.r.t.			
Node number	w	θ_x	θ_y	θ_z
.	.	.	.	.
.	.	.	.	.
.	.	.	.	.

(j) Data about the material and cross-sectional and geometric properties of the NMATS group of elements. There are NMATS sets of data to follow. Use 7(E14.6) format. If IWARP = 0, ignore data for I_w.

E	G	I	J	A	ZASF	I_w
.	.	.	.	.	.	.
.	.	.	.	.	.	.
.	.	.	.	.	.	.

(k) Additional data concerning the subgrade modulus and the stiffness of the springs at the ends of semi-rigid elements. NMATS sets of data to follow. Use 3(E14.6) format.

	Spring stiffness at	
Subgrade modulus	End 1	End 2
.	.	.
.	.	.
.	.	.

(m)   Data about nodes where one or more displacements are required to have a fixed value. If a particular freedom is free, input the 'fixed' displacement as 0!. There are NVFIX sets of data to follow. Use (I5, 4 F10.4) format. If IWARP = 0, ignore reference to θ_z. Skip this section if NVFIX = 0.

	Fixed value of displacement			
Node number	w	θ_x	θ_y	θ_z
.	.	.	.	.
.	.	.	.	.
.	.	.	.	.

(n)   Data concerning the stiffness coefficients of nonstandard elements. Skip this section if there are no type 4 elements. Otherwise for each type 4 element input the following stiffness factors:

S11, S12, S22, the rotational stiffness factors

T11, T12, T22, the translation stiffness factors

Q11, Q12, Q21, Q22, the cross-stiffness factors

See Fig. 3.2 for additional explanations. Use 6E14.6 format.

(p)   Data regarding the loads on the structure. NCASE sets of data to follow.

(1)   TITLE (any description of load case) with a maximum of 60 characters.

(2)   NLODPT in I5 format. NLODPT is the number of nodes at which concentrated loads are applied.

(3)   NLODPT sets of data to follow. Use (I5, 4F10.4) format. Ignore reference to θ_z if IWARP = 0. If NLODPT = 0, skip this section.

Node number	w load	θ_x load	θ_y load	θ_z load
.	.	.	.	.
.	.	.	.	.
.	.	.	.	.

(4) NLODEL in I5 format. This shows how many elements carry lateral loads.

(5) If NLODEL = 0, skip this section. There are NLODEL sets of data to follow. Use (I5, 2F10.4) format.

Element number	Intensity of uniformly distributed load	Mid-span load
.	.	.
.	.	.
.	.	.

(6) Data concerning the fixed end reactions in the w, θ_x, θ_y and θ_z directions for the nonstandard element. For each type 4 element, input

$$R_{z1}, M_{x1}, M_{y1}, M_{z1}, R_{z2}, M_{x2}, M_{y2}, M_{z2}$$

See Figure 4.3 for notation. Use 8(E14.6) format. If IWARP = 0, ignore reference to M_z forces. Note that for the type of loads on the element considered in the program M_z is always equal to zero.

(7) ITEMP in I5 format. If ITEMP = 0, then thermal effects need not be considered. Skip sections (8) and (9).

(8) Data about the linear thermal expansion coefficient α_t, the depth d and the distance Y_u from the neutral axis to the top face for all the NMATS group of elements. If the element is assumed to be orientated horizontally with end 1 on the left-hand side, the upper face is the 'natural' top face. There are NMATS sets of data to follow. Use (E14.6, 2F10.4) format.

α_t	Depth d	Y_u
.	.	.
.	.	.
.	.	.

(9) Data regarding the temperature changes in elements. It is necessary to input data about only those elements for which there is a change in temperature. The data about the last element must be those of NELEM even if it does not suffer any change in temperature.

Element number	$T(+$ is a rise)	
	Top	Bottom
.	.	.
.	.	.
.	.	.
NELEM	.	.

(q)   Repeat (n). If there are no type 4 elements, then skip this section.
(r)   Input the fixed end axial force, the shear force, the moment at end 1 and the fixed end axial force, the shear force and the moment at end 2 respectively for all type 4 elements. If there are no type 4 elements, skip this section.

9.5   DATA INPUT FOR STABIL

The order in which the data should be input is as follows.

(a)   NPROB in I5 format, where NPROB is the number of problems to be solved. Now follows NPROB sets of data from (b) to (k).
(b)   TITLE (any description of the problem) with a maximum of 60 characters.
(c)   NPOIN, NELEM, NRESND, NMATS, ISTRES, in 5(I5) format. If ISTRES = 0, then only the load factor to cause instability is required. If ISTRES = 1, then stresses and displacements at different load factors are required.
(d)   Data about the coordinates of nodes. Since the subroutine CORDGEN will be used to generate intermediate data, it is necessary to input only the 'basic' data as explained in Chapter 2. It is important that the last node for which the data are input is NPOIN.

IPOIN	NSTEP	Coordinates of IPOIN	
		x	y
.	.	.	.
.	.	.	.
NPOIN	1	.	.

See section 2.2.1 and Fig. 2.4 for an explanation of NSTEP. Use the format (2I5, 2F10.4).

(e)　Data regarding the node numbers at the ends of elements. Since the subroutine INTGEN will be used to generate the intermediate data, only 'basic' data need to be input. The last data must be those of the NELEMth element. Use 4(I5) format.

		Node numbers at	
IELEM	NSTEP	End 1	End 2
.	.	.	.
.	.	.	.
NELEM	1	.	.

(f)　Data regarding the group number to which the NELEM elements belong. Here also the subroutine INTGEN will be used to generate 'intermediate' data. Therefore, only 'basic' data need to be input. Use 3(I5) format. The last data must be of the NELEMth element.

IELEM	NSTEP	Group number
.	.	.
.	.	.
NELEM	1	.

(g)　NOPTN in I5 format. The number of option NOPTN is interpreted as follows.

(1)　If NOPTN = 1, it means that the node freedom numbers will be input as data and subroutine INTGEN will be used to calculate intermediate data. Use 5I5 format. The last data must be those of the NPOINth node. Skip this section if NOPTN = 0.

		Node freedom number for		
IPOIN	NSTEP	u	v	θ
.	.	.	.	.
.	.	.	.	.
NPOIN	1	.	.	.

(2)　If NOPTN = 0, it means that the node freedom numbers will be generated from the data about the restrained nodes using the subroutine NFGEN. Data to be input about the restraint state of

NDOFN freedoms at NRESND nodes. There are NRESND sets of data to follow. Use 4(I5) format. Use the code free = 0 and restrained = 1. Skip this section if NOPTN = 1.

Node number	Fixity state w.r.t.		
	u	v	θ
	.	.	.
	.	.	.
	.	.	.

(h)   Data about the material and cross-sectional and geometric properties of the NMATS group of elements. There are NMATS sets of data to follow. Use 3(E14.6) format.

Young's modulus E	Inertia I	Area A
.	.	.
.	.	.
.	.	.

(j)   Data regarding the loads on the structure.

(1)   TITLE (any description of load case) with a maximum of 60 characters.

(2)   NLODPT in I5 format. NLODPT is the number of nodes at which concentrated loads are applied.

(3)   NLODPT sets of data to follow. Use (I5, 3F10.4) format. If NLODPT = 0, skip this section.

Node number	x load	y load	θ load
.	.	.	.
.	.	.	.
.	.	.	.

(4)   NLODEL in I5 format. This shows how many elements carry lateral loads.

(5)   If NLODEL = 0, skip this section. There are NLODEL sets of data to follow. Use (I5, 4F10.4) format.

Element number	Intensity of uniformly distributed load in direction		Mid-span load in direction	
	x	y	x	y
.	.	.	.	.
.	.	.	.	.
.	.	.	.	.

(k)   Input STRFAC, NITER, INCREM in (F10.4, 2I5) format.

9.6   DATA INPUT FOR VIBES

The order in which the data should be input is as follows.

(a)   NPROB in I5 format, where NPROB is the number of problems to be solved. Now follows NPROB sets of data from (b) to (I).

(b)   TITLE (any description of the problem) with a maximum of 60 characters.

(c)   ISTRUCT in I5 format. If ISTRUCT = 1, then the structure is a 2-D frame. If STRUCT = 2, then it is a plane grid.

(d)   NPOIN, NELEM, NRESND, NMATS in 4(I5) format.

(e)   Data about the coordinates of nodes. Since the subroutine CORDGEN will be used to generate intermediate data, it is necessary to input only the 'basic' data as explained in Chapter 2. It is important that the last node for which the data are input is NPOIN.

IPOIN	NSTEP	Coordinates of IPOIN	
		x	y
.	.	.	.
.	.	.	.
NPOIN	1	.	.

See section 2.2.1 and Fig. 2.4 for an explanation of NSTEP. Use the format (2I5, 2F10.4) for data input.

(f)   Data regarding the node numbers at the ends of elements. Since the subroutine INTGEN will be used to generate the intermediate data, only 'basic' data need to be input. The last data must be those of the NELEMth element. Use 4(I5) format.

IELEM	NSTEP	Node numbers at	
		End 1	End 2
.	.	.	.
.	.	.	.
NELEM	1	.	.

(g) Data regarding the group number to which the NELEM elements belong. Here also the subroutine INTGEN will be used to generate 'intermediate' data. Therefore only 'basic' data need to be input. Use 3(I5) format. The last data must be those of the NELEMth element.

IELEM	NSTEP	Group number
.	.	.
.	.	.
NELEM	1	.

(h) NOPTN in I5 format. The number of option NOPTN is interpreted as follows.

(1) If NOPTN = 1, it means that the node freedom numbers will be input as data and subroutine INTGEN will be used to calculate intermediate data. Use 5I5 format. The last data must be those of the NPOINth node. Skip this section if NOPTN = 0.

IPOIN	NSTEP	Node freedom number for		
		u	v	θ
		or		
		w	θ_x	θ_y
.	.	.	.	.
.	.	.	.	.
NPOIN	1	.	.	.

(2) If NOPTN = 0, it means that the node freedom numbers will be generated from the data about the restrained nodes using the subroutine NFGEN. Data to be input about the restraint state of

NDOFN freedoms at NRESND nodes. There are NRESND sets of data to follow. Use 4(I5) format. Use the code free = 0 and restrained = 1. Skip this section if NOPTN = 1.

	Fixity state w.r.t.		
	u	v	θ
	or		
Node number	w	θ_x	θ_y
	.	.	.
	.	.	.
	.	.	.

(j)   Data about the material and cross-sectional and geometric properties of the NMATS group of elements. There are NMATS sets of data to follow. Use 3(E14.6) format.

Young's modulus E	Inertia I	Area A			
or					
E	G	I	J	A	I_p
.	.	.	.	.	.
.	.	.	.	.	.
.	.	.	.	.	.

(k)   Data regarding the mass on the structure.

   (1)   TITLE (any description of the particular load case) with a maximum of 60 characters.
   (2)   NLODPT in I5 format. NLODPT is the number of nodes at which the concentrated mass acts.
   (3)   NLODPT sets of data to follow. Use (I5, F10.4) format. If NLODPT = 0, skip this section.

Node number	Mass
.	.
.	.
.	.

(4) NLODEL in I5 format. This shows how many elements carry uniformly distributed mass.

(5) If NLODEL = 0, skip this section. There are NLODEL sets of data to follow. Use (I5, F10.4) format.

Element number	UD mass intensity
.	.
.	.
.	.

(l) STRFRQ, ENDFRQ, ACURCY in 3(F10.4) format.

```
                    PROGRAM TRUSS

      C
      C***THIS PROGRAM CALLED TRUSS CAN BE USED FOR
      C    THE ANALYSIS OF 2-D AND 3-D PIN-JOINTED
      C    STRUCTURES ACTED ON BY EXTERNAL LOADS AT
      C    JOINTS AND THERMAL LOADS.  IT CAN ALSO BE
      C    USED TO ANALYSE CASES OF SELF-STRAINING DUE
      C    TO SETTLEMENT OF SUPPORTS.
      C
            COMMON/CONTRO/NPOIN,NELEM,NNODE,NDOFN,
           @NDIME,NPROP,NMATS,NRESND,NVFIX,NEVAB,
           @NFAIL,NOPTN,NEQNS,NBAND,ITEMP
            COMMON/LGDATA/COORD(100,3),PROPS(10,3),
           @PRESC(80,3),TEMP(100),NDFFIX(80),
           @NDVFIX(80),IFPRE(80,3),ASDIS(400),
           @GSTIF(100,30),GLOAD(100)
            DIMENSION DISP(3),NODFRE(100,3),
           @LNODS(100,2),MATNO(100,2)
            READ(5,900)NPROB
        900 FORMAT(8I5)
            WRITE(6,905)NPROB
        905 FORMAT(//,5X,' TOTAL NUMBER OF PROBLEMS =
           @ ',I3)
        908 FORMAT('--------------------------------
           @----------------',//)
            WRITE (6,908)
            WRITE (6,908)
      C
      C***START THE CYCLE ON THE NUMBER OF PROBLEMS.
      C
            DO 1 IPROB=1,NPROB
            READ(5,'(A60)')TITLE
            WRITE(6,910)IPROB
        910 FORMAT(/////,6X,' PROBLEM NO ',I3)
            WRITE(6,'(A60)')TITLE
            WRITE(6,908)
      C
      C***SET THE BASIC PARAMETERS OF THE 2-D PIN-
      C    JOINTED STRUCTURE
      C
            NNODE=2
            NPROP=2
      C
      C***READ AND PRINT THE BASIC PARAMETERS OF THE
      C    PARTICULAR PROBLEM
      C
            READ(5,900)NPOIN,NELEM,NRESND,NVFIX,NMATS
           @,NCASE,NDIME
            WRITE(6,915)NPOIN,NELEM,NRESND,NVFIX,NMATS
           @,NCASE,NDIME
        915 FORMAT(//,' NUMBER OF NODES =
           @                ',I3,/,' NUMBER OF MEMBERS =
           @                      ',I3,/,' NUMBER OF
           @ RESTRAINED NODES =               ',I3,/,
           @' NUMBER OF NODES WITH FIXED DISPLACEMENTS = ',I3,/,
           @' NUMBER OF DIFFERENT TYPES OF MATERIALS =   ',I3,/,
           @' NUMBER OF LOAD CASES =                     ',I3,/,
           @' NUMBER OF DIMENSIONS =                     ',I3)
            NDOFN=NDIME
```

```
      NEVAB=NNODE*NDOFN
C
C***READ THE COORDINATES OF SOME OF THE JOINTS,
C    GENERATE THE OMITTED DATA AND PRINT THE
C    COORDINATES OF ALL THE NODES.  IF NFAIL = 1,
C    THEN IT MEANS ERROR IN DATA.  THIS STOPS
C    ALL COMPUTATIONS.
C
      CALL CORDGEN
      IF(NFAIL.EQ.1) GOTO 140
      WRITE(6,920)
  920 FORMAT(//,' COORDINATES OF NODES')
      WRITE (6,908)
      WRITE(6,925)
  925 FORMAT(//,' NODE',7X,'X-CORD',6X,'Y-CORD'
     @,6X,'Z-CORD')
      WRITE (6,908)
      DO 10 IPOIN=1,NPOIN
   10 WRITE(6,935)IPOIN,(COORD(IPOIN,JDIME),
     @JDIME=1,NDIME)
  935 FORMAT(1X,I3,4X,3F11.2)
C
C***READ THE NODE NUMBERS DEFINING THE ENDS OF
C    SOME OF THE ELEMENTS AND GENERATE THE
C    OMITTED DATA.  NFAIL = 1 MEANS ERROR IN DATA
C    WHICH STOPS ALL CALCULATIONS.
C
      CALL INTGEN(LNODS,NNODE,NELEM)
      IF(NFAIL.EQ.1) GOTO 140
C
C***READ THE MATERIAL NUMBERS OF SOME OF THE
C    ELEMENTS AND GENERATE THE OMITTED DATA.
C    NFAIL = 1 INDICATES ERROR IN DATA AND
C    THIS STOPS ALL CALCULATIONS.
C
      CALL INTGEN(MATNO,1,NELEM)
      IF(NFAIL.EQ.1) GOTO 140
C
C***PRINT THE MATERIAL NUMBERS AND THE NODE
C    NUMBERS AT THE ENDS FOR ALL THE ELEMENTS
C
      WRITE(6,940)
  940 FORMAT(///,' ELEMENT NO.',3X,' MATERIAL NO
     @.',6X,'NODE 1',6X,' NODE-2')
      WRITE (6,908)
      DO 20 IELEM = 1, NELEM
   20 WRITE(6,945)IELEM,MATNO(IELEM,1),(LNODS(
     @IELEM,JNODE),JNODE=1,NNODE)
  945 FORMAT(1X,I5,7X,I9,11X,I5,7X,I5)
C
C***READ NOPTN.  THIS TELLS WHETHER THE NODE
C    FREEDOM ARRAY IS EITHER INPUT AS DATA
C    (FOR NOPTN = 1) OR TO BE GENERATED FROM
C    THE DATA ABOUT THE NODES WHERE ONE MORE
C    DISPLACEMENTS ARE ZERO(FOR NOPTN = 0)
C
      READ(5,900)NOPTN
C
C***READ NODE FREEDOMS OF SOME OF THE NODES AND
C    GENERATE THE OMITTED DATA.  NFAIL = 1 MEANS
```

```fortran
C    THAT THERE IS AN ERROR IN DATA AND THEREFORE
C    ALL CALCULATIONS ARE HALTED.
C
      IF(NOPTN.EQ.1) CALL INTGEN(NODFRE,NDOFN,
     @NPOIN)
      IF(NFAIL.EQ.1) GOTO 140
      IF(NOPTN.EQ.1) GOTO 45
C
C***READ DATA ABOUT THE NODE NUMBERS AND THE
C    FIXITY STATE OF NODES WHERE ONE OR MORE
C    FREEDOMS ARE RESTRAINED TO ZERO VALUE.
C
      DO 30 IRESND=1,NRESND
   30 READ(5,900)NDFFIX(IRESND),(IFPRE(IRESND,
     @JDOFN),JDOFN=1,NDOFN)
      WRITE(6,950)
  950 FORMAT(///,' DEGREES OF FREEDOM AT
     @RESTRAINED NODES',/,'------------------
     @---------------------------------------
     @--',/,'1 = FIXED AND 0 = FREE',//,'NODE
     @NO.',6X,'X-FREEDOM',6X,'Y-FREEDOM',6X,
     @'Z-FREEDOM',/,'------------------------
     @---------------------------------',/)
      DO 40 IRESND=1,NRESND
   40 WRITE(6,955)NDFFIX(IRESND),(IFPRE(IRESND,
     @JDOFN),JDOFN=1,NDOFN)
  955 FORMAT(1X,I3,11X,I4,2(10X,I4))
C
C***CALL THE SUBROUTINE TO GENERATE THE NODE
C    FREEDOM ARRAY FROM THE DATA ABOUT THE
C    RESTRAINED NODES
C
      CALL NFGEN(NODFRE)
   45 CONTINUE
      WRITE(6,960)
  960 FORMAT(//,' DEGREES OF FREEDOM AT NODES',/,
     @'--------------------------------------
     @--------------------',//,'NODES',6X,'X-
     @FRDOM',5X,' Y-FRDOM.',6X,' Z-FRDOM.',/,
     @'--------------------------------------
     @--------------------',/)
      DO 50 IPOIN=1,NPOIN
   50 WRITE(6,18)IPOIN,(NODFRE(IPOIN,IDOFN),
     @IDOFN=1,NDOFN)
   18 FORMAT(I5,3(6X,I5))
C
C***READ AND PRINT THE PROPERTIES IE. AREA,
C    YOUNG'S MODULUS ETC. ABOUT THE DIFFERENT
C    MATERIALS
C
      DO 60 IMATS=1,NMATS
   60 READ(5,970)(PROPS(IMATS,IPROP),IPROP=1,
     @NPROP)
  970 FORMAT(5E14.6)
      WRITE(6,975)
  975 FORMAT(///,' PROPERTIES OF DIFFERENT
     @MATERIALS' )
      WRITE (6,908)
      WRITE(6,980)
  980 FORMAT(//,' MATERIAL NO.',6X,'YOUNGS MOD.
```

```
      @',6X,'AREA')
        WRITE (6,908)
        DO 70 IMATS=1,NMATS
     70 WRITE(6,985)IMATS,(PROPS(IMATS,IPROP),
        @IPROP=1,NPROP)
    985 FORMAT(2X,I5,9X,E14.6,4X,F10.4)
        WRITE (6,'(/)')
C
C***READ THE DATA ABOUT NODE NUMBERS AND THE
C   FIXED DISPLACEMENT VALUES OF NODES WITH
C   PRESCRIBED DISPLACEMENTS.
C
        IF(NVFIX.EQ.0) GOTO 95
        DO 80 IVFIX=1,NVFIX
     80 READ(5,987)NDVFIX(IVFIX),(PRESC(IVFIX,
        @IDOFN),IDOFN=1,NDOFN)
    987 FORMAT(I5,4E14.6)
        WRITE(6,988)
    988 FORMAT(//,' NODES AT WHICH THE VALUE OF
        @DISPLACEMENTS ARE FIXED',//,'NODE',6X,'
        @X-DISP.',6X,'Y-DISP.',6X,'Z-DISP',/)
        DO 90 IVFIX=1,NVFIX
     90 WRITE(6,989)NDVFIX(IVFIX),(PRESC(IVFIX,
        @IDOFN),IDOFN=1,NDOFN)
    989 FORMAT(2X,I5,3F10.4)
C
C***CALCULATE THE MAXIMUM NUMBER OF EQUATIONS
C
     95 CALL MAXEQNS(NODFRE)
C
C***CALCULATE THE HALF-BANDWIDTH
C
        CALL BANDWIDTH(NODFRE,LNODS)
C
C***FIRST ZERO AND THEN FORM THE STRUCTURAL
C   STIFFNESS MATRIX
C
        CALL NULLGSTIF
        CALL STIFFPJ(NODFRE,LNODS,MATNO)
C
C***IF NEED BE MODIFY THE DIAGONAL ELEMENTS OF
C   THE STRUCTURAL STIFFNESS MATRIX TO
C   INCORPORATE STIFF SPRING APPROACH.
C
        IF(NVFIX.EQ.0) GOTO 105
        CALL STFSPRG(NODFRE)
C
C***REDUCE THE STIFFNESS MATRIX TO UPPER
C   TRIANGULAR FORM.  IF NFAIL = 1, THEN THE
C   STRUCTURAL STIFFNESS MATRIX IS SINGULAR
C   INDICATING EITHER THERE IS AN ERROR IN DATA
C   OR THAT THE STRUCTURE IS NOT PROPERLY BRACED
C   MAKING IT UNSTABLE.
C
    105 CALL GAUSRED
        IF(NFAIL.EQ.1) GOTO 140
C
C***START THE LOAD CASE CYCLE
C
        DO 110 ICASE=1,NCASE
```

```fortran
      READ(5,'(A60)')TITLE
C
C***FORM THE LOAD VECTOR DUE TO EXTERNAL LOADS
C
      CALL LOADPJ(NODFRE)
C
C***DECIDE IF THERMAL LOADS ARE TO BE INCLUDED.
C   IF SO MODIFY THE LOAD VECTOR TO INCLUDE THE
C   EFFECTS OF THERMAL RESTRAINT FORCES.
C
      READ(5,900)ITEMP
      WRITE(6,993)ITEMP
  993 FORMAT(//,' THERMAL STRESSES TO BE INCLU
     @DED ? ( YES = 1 AND NO = 0) =',I2)
      IF(ITEMP.EQ.1) CALL THERMPJ(NODFRE,LNODS,MATNO)
C
C***IF NEED BE MODIFY THE LOAD VECTOR TO
C   INCORPORATE THE STIFF SPRING APPROACH.
C
      IF(NVFIX.EQ.0) GOTO 115
      CALL FIXDISP(NODFRE)
C
C***CALCULATE THE DISPLACEMENTS
C
  115 CALL BACKSUB
      WRITE(6,994)ICASE
  994 FORMAT(//,' DISPLACEMENTS OF JOINTS FOR
     @LOAD CASE = ',I3,/,'----------------------
     @---------------------------------',//)
      WRITE(6,'(A60)')TITLE
      WRITE(6,995)
  995 FORMAT(//,' NODE',6X,'X-DISP.',6X,'Y-DISP
     @.',6X,'Z-DISP.',/,'---------------------
     @-----------------------------------'
     @,/)
      DO 120 IPOIN=1,NPOIN
      DO 130 IDOFN=1,NDOFN
      DISP(IDOFN)=0.0
      IF(NODFRE(IPOIN,IDOFN).NE.0)DISP(IDOFN)=
     @ASDIS(NODFRE(IPOIN,IDOFN))
  130 CONTINUE
      WRITE(6,996)IPOIN,(DISP(IDOFN),IDOFN=1,
     @NDOFN)
  120 CONTINUE
  996 FORMAT(I3,3X,3(E14.6))
      WRITE(6,997)ICASE
  997 FORMAT(//,' FORCES IN THE MEMBERS FOR
     @LOAD CASE =',I3,/,'----------------------
     @-------------------------------------'
     @,//' ELEMENT',6X,'END 1',6X,'END 2',6X,
     @'AXIAL FORCE (+ IS TENSION)',/,'---------
     @------------------------------------
     @-----------','-----------',/)
C
C***CALCULATE THE AXIAL FORCE IN THE ELEMENTS.
C
      CALL STRESSPJ(NODFRE,LNODS,MATNO)
  110 CONTINUE
    1 CONTINUE
      STOP
```

```
  140 END
--------------------------------------------------------


                     PROGRAM 2-D FRAME

C
C***THIS PROGRAM CALLED 2-DFRAME IS USED FOR THE
C    ANALYSIS OF 2-D RIGID-JOINTED STRUCTURES.
C    THIS CAN HANDLE EXTERNAL LOADS, THERMAL
C    LOADS AND SELF-STRAINING LOADS DUE TO
C    SETTLEMENT OF SUPPORTS.  THIS HAS A CHOICE
C    OF DIFFERENT TYPES OF ELEMENTS INCLUDING
C    BEF AND SEMI-RIGID ELEMENTS.
C
      COMMON/CONTRO/NPOIN,NELEM,NNODE,NDOFN,
     @NDIME,NPROP,NMATS,NRESND,NVFIX,NEVAB,
     @NFAIL,NLODEL,NEQNS,NBAND,ITEMP
      COMMON/LGDATA/COORD(100,2),PROPS(20,11),
     @PRESC(80,3),TEMP(100,2),NDFFIX(80),
     @NDVFIX(80),IFPRE(80,3),FORSEL(80,4),
     @LODEL(80),ASDIS(800),GSTIF(100,30),
     @GLOAD(100)
      DIMENSION DISP(3),NODFRE(100,3),
     @LNODS(100,2),MATNO(100,2)
      READ(5,900)NPROB
  900 FORMAT(8I5)
      WRITE(6,905)NPROB
  905 FORMAT(//,5X,' TOTAL NUMBER OF PROBLEMS
     @= ',I5)
  908 FORMAT ('---------------------------------
     @-------------------',/)
      WRITE (6,908)
C
C***START THE CYCLE ON THE NUMBER OF PROBLEMS.
C
      DO 1 IPROB=1,NPROB
      READ(5,'(A60)')TITLE
      WRITE(6,910)IPROB
  910 FORMAT(/////,6X,' PROBLEM NO ',I3)
      WRITE(6,'(A60)')TITLE
      WRITE(6,908)
C
C***SET THE BASIC PARAMETERS OF THE 2-D RIGID-
C    JOINTED STRUCTURE
C
      NNODE=2
      NDOFN=3
      NDIME=2
      NPROP=5
      NEVAB=NNODE*NDOFN
C
C***READ AND PRINT THE BASIC PARAMETERS OF THE
C    PARTICULAR PROBLEM
C
      READ(5,900)NPOIN,NELEM,NRESND,NVFIX,
     @NMATS,NCASE
      WRITE(6,915)NPOIN,NELEM,NRESND,NVFIX,
     @NMATS,NCASE
  915 FORMAT(//,' NUMBER OF NODES =
     @                 ',I3,/,' NUMBER OF MEMBERS =
```

```fortran
      @                              ',I3,/,' NUMBER
      @OF RESTRAINED NODES =                  ',I3,/,
      @' NUMBER OF NODES WITH FIXED DISPLACEMENTS
      @ = ',I3,/,' NUMBER OF DIFFERENT TYPES OF
      @MATERIALS =     ',I3,/,' NUMBER OF LOAD
      @CASES =                      ',I3)
C
C***READ THE COORDINATES OF SOME OF THE JOINTS,
C    GENERATE THE OMITTED DATA AND PRINT THE
C    COORDINATES OF ALL THE NODES.  NFAIL = 1
C    INDICATES ERROR IN DATA AND THEREFORE
C    TERMINATION OF ALL FURTHER CALCULATIONS.
C
      CALL CORDGEN
      IF(NFAIL.EQ.1) GOTO 140
      WRITE(6,920)
  920 FORMAT(//,' COORDINATES OF NODES')
      WRITE (6,908)
      WRITE(6,925)
  925 FORMAT(//,'    NODE',5X,'X-CORD',7X,'Y-
      @CORD')
      WRITE (6,908)
      DO 10 IPOIN=1,NPOIN
   10 WRITE(6,935)IPOIN,(COORD(IPOIN,JDIME),
      @JDIME=1,NDIME)
  935 FORMAT(2X,I5,3(3X,F10.3))
C
C***READ THE NODE NUMBERS DEFINING THE ENDS OF
C    SOME OF THE ELEMENTS AND GENERATE THE
C    OMITTED DATA.  NFAIL = 1 INDICATES ERROR IN
C    DATA AND THEREFORE TERMINATION OF ALL
C    CALCULATIONS.
C
      CALL INTGEN(LNODS,NNODE,NELEM)
      IF(NFAIL.EQ.1) GOTO 140
C
C***READ THE MATERIAL NUMBERS OF SOME OF THE
C    ELEMENTS AND GENERATE THE OMITTED DATA.
C    NFAIL = 1 INDICATES ERROR IN DATA
C    AND THEREFORE TERMINATION OF ALL CALCULATIONS.
C
      CALL INTGEN(MATNO,2,NELEM)
      IF(NFAIL.EQ.1) GOTO 140
C
C***PRINT THE MATERIAL NUMBERS AND THE NODE
C    NUMBERS AT THE ENDS FOR ALL THE ELEMENTS
C
      WRITE(6,940)
  940 FORMAT(///,' ELEMENT NO.',3X,' MATERIAL
      @NO.',4X,'TYPE',6X,'NODE-1',6X,' NODE-2',/,
      @'------------------------------------------
      @------------------------------------------
      @--------',//)
      DO 20 IELEM =1,NELEM
   20 WRITE(6,945)IELEM,MATNO(IELEM,1),MATNO(
      @IELEM,2),(LNODS(IELEM,JNODE),JNODE=1,NNODE)
  945 FORMAT(1X,I5,12X,I5,3(7X,I5))
C
C***READ NOPTN. THIS TELLS WHETHER THE NODE
C    FREEDOM ARRAY IS EITHER INPUT AS DATA
```

```
C     (FOR NOPTN=1) OR TO BE GENERATED FROM
C     THE DATA ABOUT THE NODES WHERE ONE MORE
C     DISPLACEMENTS ARE ZERO (FOR NOPTN = 0)
C
      READ(5,900)NOPTN
C
C***READ NODE FREEDOMS OF SOME OF THE NODES AND
C     GENERATE THE OMITTED DATA.  NFAIL=1 MEANS
C     ERROR IN DATA AND THIS STOPS ALL CALCULATIONS.
C
      IF(NOPTN.EQ.1) CALL INTGEN(NODFRE,NDOFN,
     @NPOIN)
      IF(NFAIL.EQ.1) GOTO 140
      IF(NOPTN.EQ.1) GOTO 45
C
C***READ DATA ABOUT  THE NODE NUMBERS AND THE
C     FIXITY CONDITION AT NODES WHERE ONE OR MORE
C     FREEDOMS ARE EQUAL TO ZERO.
C
      DO 30 IRESND=1,NRESND
   30 READ(5,900)NDFFIX(IRESND),(IFPRE(IRESND,
     @JDOFN),JDOFN=1,NDOFN)
      WRITE(6,950)
  950 FORMAT(///,' DEGREES OF FREEDOM AT
     @RESTRAINED NODES',/,'--------------------
     @-------------------------------------',//,
     @'1 = FIXED AND 0 = FREE',//,'NODE NO.',6X,
     @'X-FREEDOM',6X,'Y-FREEDOM',6X,'THETAZ-
     @FREEDOM',/,'--------------------------
     @-----------------------------',//)
      DO 40 IRESND=1,NRESND
   40 WRITE(6,955)NDFFIX(IRESND),(IFPRE(IRESND,
     @JDOFN),JDOFN=1,NDOFN)
  955 FORMAT(I5,3(10X,I5))
C
C***CALL THE SUBROUTINE TO GENERATE THE NODE
C     FREEDOM ARRAY FROM THE DATA ABOUT THE
C     RESTRAINED NODES
C
      CALL NFGEN(NODFRE)
   45 CONTINUE
      WRITE(6,960)
  960 FORMAT(//,' DEGREES OF FREEDOM AT NODES',/,
     @'-------------------------------------
     @----------',//,'NODE',3X,' X-FRDOM ',4X,
     @' Y-FRDOM.',6X,' THETAZ-FRDOM.',//, '---
     @------------------------------------
     @----------------',/)
      DO 50 IPOIN=1,NPOIN
   50 WRITE(6,961)IPOIN,(NODFRE(IPOIN,IDOFN),
     @IDOFN=1,NDOFN)
  961 FORMAT(I3,5X,I5,8X,I5,10X,I5)
C
C***READ AND PRINT THE PROPERTIES IE. AREA,
C     YOUNG'S MODULUS ETC. ABOUT THE DIFFERENT
C     MATERIALS
C
      DO 60 IMATS=1,NMATS
   60 READ(5,970)(PROPS(IMATS,IPROP),IPROP=1,
     @NPROP)
```

```
  970 FORMAT(5E14.6)
      WRITE(6,975)
  975 FORMAT(///,' PROPERTIES OF DIFFERENT
     @MATERIALS' )
      WRITE (6,908)
      WRITE(6,980)
  980 FORMAT(//,' MATERIAL NO.',4X,'YOUNGS MOD.'
     @,7X,'SHEAR MOD.'10X,'INERTIA',14X,'AREA',
     @12X,'AREA SHEAR FACTOR',/,'---------------------
     @------------------------------------------
     @------------------------------------------
     @-----------',//)
      DO 70 IMATS=1,NMATS
   70 WRITE(6,981)IMATS,(PROPS(IMATS,IPROP),
     @IPROP=1,NPROP)
  981 FORMAT(2X,I5,6X,E14.6,4(6X,E14.6))
C
C***READ VALUES OF SUBGRADE MODULUS  AND
C   STIFFNESS OF SEMI-RIGID CONNECTIONS AT THE
C   ENDS OF MEMBERS
C
      WRITE(6,982)
  982 FORMAT(//,' MATERIAL NO.',4X,'SUBGR MOD.'
     @,12X,'SPRNG-1',12X,'SPRNG-2',/,'--------
     @---------------------------------------
     @---------------------------------------
     @',//)
      DO 75 IMATS=1,NMATS
      READ(5,970)(PROPS(IMATS,JPROP),JPROP=
     @NPROP+1,NPROP+3)
   75 WRITE(6,981)IMATS,(PROPS(IMATS,JPROP),JPROP=
     @NPROP+1,NPROP+3)
C
C***READ THE DATA ABOUT NODE NUMBERS AND THE
C   PRESCRIBED VALUE OF DISPLACEMENTS AT NODES
C   WITH PRESCRIBED DISPLACEMENT VALUES.
C
      IF(NVFIX.EQ.0) GOTO 95
      DO 80 IVFIX=1,NVFIX
   80 READ(5,987)NDVFIX(IVFIX),(PRESC(IVFIX,
     @IDOFN),IDOFN=1,NDOFN)
  987 FORMAT(I5,4E14.6)
      WRITE(6,988)
  988 FORMAT(//,' NODES AT WHICH THE VALUE OF
     @DISPLACEMENTS ARE FIXED',//,'NODE',9X,
     @'X-DISP.',12X,'Y-DISP.',8X,'THETAZ-DISP')
      DO 90 IVFIX=1,NVFIX
   90 WRITE(6,989)NDVFIX(IVFIX),(PRESC(IVFIX,
     @IDOFN),IDOFN=1,NDOFN)
  989 FORMAT(2X,I5,3F10.4)
C
C***CALCULATE THE MAXIMUM NUMBER OF EQUATIONS
C
   95 CALL MAXEQNS(NODFRE)
C
C***CALCULATE THE HALF-BANDWIDTH
C
      CALL BANDWIDTH(NODFRE,LNODS)
C
C***FIRST ZERO AND THEN FORM THE STRUCTURAL
```

```
C    STIFFNESS MATRIX
C
      CALL NULLGSTIF
      CALL STIFFRJ(NODFRE,LNODS,MATNO)
C
C***IF NEED BE MODIFY THE DIAGONAL TERM OR TERMS
C   OF THE STRUCTURAL STIFFNESS MATRIX TO
C   INCORPORATE STIFF SPRING APPROACH.
C
      IF(NVFIX.EQ.0) GOTO 105
      CALL STFSPRG(NODFRE)
C
C***REDUCE THE STIFFNESS MATRIX TO UPPER
C   TRIANGULAR FORM.  IF NFAIL =1 THEN THE
C   MATRIX IS SINGULAR INDICATING EITHER AN
C   ERROR IN DATA OR THAT THE STRUCTURE IS NOT
C   PROPERLY BRACED.
C
  105 CALL GAUSRED
      IF(NFAIL.EQ.1) GOTO 140
C
C***START THE LOAD CASE CYCLE
C
      DO 110 ICASE=1,NCASE
      READ(5,'(A60)')TITLE
C
C***FORM LOAD VECTOR DUE TO EXTERNAL LOADING
C
      CALL LOADRJ(NODFRE,LNODS,MATNO)
C
C***DECIDE IF THERMAL STRESSES ARE TO BE
C   INCLUDED.  IF SO MODIFY THE LOAD VECTOR TO
C   INCLUDE THE EFFECTS OF THERMAL RESTRAINTS.
C
      READ(5,900)ITEMP
      WRITE(6,993)ITEMP
  993 FORMAT(//,' ARE THERMAL STRESSES TO BE
     @INCLUDED ? (YES = 1 AND NO = 0) =',I3)
      IF(ITEMP.EQ.1) CALL THERMRJ(NODFRE,LNODS,
     @MATNO)
C
C***IF NECESSARY MODIFY THE LOAD VECTOR TO
C   IMPLEMENT THE STIFF SPRING APPROACH.
C
      IF(NVFIX.EQ.0) GOTO 115
      CALL FIXDISP(NODFRE)
C
C***SOLVE FOR DISPLACEMENTS
C
  115 CALL BACKSUB
      WRITE(6,994)ICASE
  994 FORMAT(//,' DISPLACEMENTS OF JOINTS FOR
     @LOAD CASE = ',I5,/,'---------------------
     @------------------------',//)
      WRITE(6,'(A60)')TITLE
      WRITE(6,997)
  997 FORMAT(//,' NODE',9X,'X-DISP.',12X,'Y-
     @DISP.',8X,'THETAZ-DISP.')
      DO 120 IPOIN=1,NPOIN
      DO 130 IDOFN=1,NDOFN
```

```
      DISP(IDOFN)=0.0
      IF(NODFRE(IPOIN,IDOFN).NE.0)DISP(IDOFN)=
     @ASDIS(NODFRE(IPOIN,IDOFN))
  130 CONTINUE
      WRITE(6,995)IPOIN,(DISP(IDOFN),IDOFN=1,
     @NDOFN)
  120 CONTINUE
  995 FORMAT(I5,6X,3(E14.6,4X))
      WRITE(6,996)ICASE
  996 FORMAT(//,' FORCES IN THE MEMBERS FOR LOAD
     @CASE =',I5,/,'----------------------------
     @-----------------',//,' ELEMENT',3X,'
     @END 1',3X,'END 2',3X,'AXL. FORCE',6X,'
     @SHR. FORCE',8X,'BM',12X,'AXL. FORCE',8X,
     @'SHR. FORCE',10X,'BM')
C
C***CALCULATE THE FORCES IN THE ELEMENTS AT THE
C    ENDS.
C
      CALL STRESSRJ(NODFRE,LNODS,MATNO)
C
C***IF THERE ARE BEAMS ON ELASTIC FOUNDATION,
C    THEN CALCULATE THE PRESSURE, BM AND SF ALONG
C    THE ELEMENT.
C
      CALL PRESURRJ(NODFRE,LNODS,MATNO)
  110 CONTINUE
    1 CONTINUE
      STOP
  140 END
```

```
                  PROGRAM GRID


C
C***THIS PROGRAM CALLED GRID IS USED FOR THE
C    ANALYSIS OF PLANE GRIDS WITH OR WITHOUT
C    WARPING RESTRAINT.  IT CAN ANALYSE GRIDS
C    SUBJECTED TO EXTERNAL LOADING, THERMAL
C    LOADING AND SETTLEMENT OF SUPPORTS.  IT HAS
C    A CHOICE OF ELEMENTS INCLUDING BEF AND
C    SEMI-RIGID ELEMENTS.
C
      COMMON/CONTRO/NPOIN,NELEM,NNODE,NDOFN,
     @NDIME,NPROP,NMATS,NRESND,NVFIX,NEVAB,NFAIL,
     @NLODEL,NEQNS,NBAND,ITEMP,IWARP
      COMMON/LGDATA/COORD(100,2),PROPS(10,12),
     @PRESC(80,4),TEMP(100,2),NDFFIX(80),NDVFIX(80),
     @IFPRE(80,4),FORSEL(80,2),LODEL(80),ASDIS(800),
     @GSTIF(100,30),GLOAD(100)
      DIMENSION DISP(4),NODFRE(100,4),
     @LNODS(100,2),MATNO(100,2)
      READ(5,900)NPROB
  900 FORMAT(8I5)
      WRITE(6,905)NPROB
  905 FORMAT(//,5X,' TOTAL NUMBER OF PROBLEMS =
     @ ',I3)
  908 FORMAT('--------------------------------
     @------------------------',//)
      WRITE(6,908)
```

```
      WRITE(6,908)
C
C***START THE CYCLE ON THE NO. OF STRUCTURES.
C
      DO 1 IPROB=1,NPROB
      READ(5,'(A60)')TITLE
      WRITE(6,910)IPROB
  910 FORMAT(/////,6X,' PROBLEM NO ',I3)
      WRITE(6,'(A60)')TITLE
      WRITE(6,908)
C
C***DECIDE WHETHER THE EFFECTS OF WARPING ARE
C   TO BE INCLUDED OR NOT.
C
      READ(5,900)IWARP
      WRITE(6,911)IWARP
  911 FORMAT(//,'IS WARPING INCLUDED (YES = 1
     @AND NO = 0)',2X,I5)
C
C***SET THE BASIC PARAMETERS OF THE GRID
C   STRUCTURE
C
      NNODE=2
      NDOFN=3
      NDIME=2
      NPROP=6
      IF(IWARP.EQ.1)NDOFN=4
      IF(IWARP.EQ.1)NPROP=7
      NEVAB=NNODE*NDOFN
C
C***READ AND PRINT THE BASIC PARAMETERS OF THE
C   PARTICULAR PROBLEM
C
      READ(5,900)NPOIN,NELEM,NRESND,NVFIX,NMATS,
     @NCASE
      WRITE(6,915)NPOIN,NELEM,NRESND,NVFIX,
     @NMATS,NCASE
  915 FORMAT(//,' NUMBER OF NODES =
     @                ',I3,/,' NUMBER OF MEMBERS
     @=                      ',I3,/,' NUMBER
     @OF RESTRAINED NODES =              ',I3,
     @/,' NUMBER OF NODES WITH FIXED DISPLACEME
     @NTS = ',I3,/,' NUMBER OF DIFFERENT TYPES
     @OF MATERIALS =    ',I3,/,' NUMBER OF LOAD
     @CASES =                 ',I3)
C
C***READ THE COORDINATES OF SOME OF THE JOINTS,
C   GENERATE THE OMITTED DATA AND PRINT THE
C   COORDINATES OF ALL THE NODES.  IF NFAIL = 1
C   IT MEANS THERE IS AN ERROR IN DATA.
C   THEREFORE ALL CALCULATIONS ARE HALTED.
C
      CALL CORDGEN
      IF(NFAIL.EQ.1) GOTO 140
      WRITE(6,920)
  920 FORMAT(//,' COORDINATES OF NODES')
      WRITE(6,908)
      WRITE(6,925)
  925 FORMAT(//,'    NODE',7X,'X-CORD',9X,'
     @Y-CORD')
```

```fortran
      WRITE(6,908)
      DO 10 IPOIN=1,NPOIN
   10 WRITE(6,935)IPOIN,(COORD(IPOIN,JDIME),
     @JDIME=1,NDIME)
  935 FORMAT(2X,I5,3(5X,F10.3))
C
C***READ THE NODE NUMBERS DEFINING THE ENDS OF
C    SOME OF THE ELEMENTS AND GENERATE THE
C    OMITTED DATA.  IF NFAIL = 1 IT MEANS AN
C    ERROR IN DATA.  THEREFORE ALL CALCULATIONS
C    STOP.
C
      CALL INTGEN(LNODS,NNODE,NELEM)
      IF(NFAIL.EQ.1) GOTO 140
C
C***READ THE MATERIAL NUMBERS OF SOME OF THE
C    ELEMENTS AND GENERATE THE OMITTED DATA.
C    NFAIL = 1 INDICATES AN ERROR IN DATA AND
C    ALL CALCULATIONS ARE STOPPED.
C
      CALL INTGEN(MATNO,2,NELEM)
      IF(NFAIL.EQ.1) GOTO 140
C
C***PRINT THE MATERIAL NUMBERS AND THE NODE
C    NUMBERS AT THE ENDS FOR ALL THE ELEMENTS
C
      WRITE(6,940)
  940 FORMAT(///,' ELEMENT NO.',3X,' MATERIAL
     @NO.',6X,'TYPE',6X,'NODE-1',6X,' NODE-2',/,
     @'----------------------------------------
     @-------------------------',//)
      DO 20 IELEM = 1, NELEM
   20 WRITE(6,945)IELEM,MATNO(IELEM,1),
     @MATNO(IELEM,2),(LNODS(IELEM,JNODE),
     @JNODE=1,NNODE)
  945 FORMAT(1X,I5,12X,I5,8X,I5,6X,I5,9X,I5)
C
C***READ NOPTN.  THIS TELLS WHETHER THE NODE
C    FREEDOM ARRAY IS EITHER INPUT AS DATA
C    (FOR NOPTN =1 ) OR TO BE GENERATED FROM
C    THE DATA ABOUT THE NODES WHERE ONE MORE
C    DISPLACEMENTS ARE ZERO (NOPTN = 0)
C
      READ(5,900)NOPTN
C
C***READ NODE FREEDOMS OF SOME OF THE NODES AND
C    GENERATE THE OMITTED DATA.  NFAIL = 1 MEANS
C    THERE IS AN ERROR IN DATA AND THIS HALTS ALL
C    CALCULATIONS.
C
      IF(NOPTN.EQ.1) CALL INTGEN(NODFRE,NDOFN,
     @NPOIN)
      IF(NFAIL.EQ.1) GOTO 140
      IF(NOPTN.EQ.1) GOTO 45
C
C***READ DATA ABOUT THE NODE NUMBERS AND THE
C    FIXITY STATE AT NODES WHERE ONE OR MORE
C    FREEDOMS ARE EQUAL TO ZERO
C
      DO 30 IRESND=1,NRESND
```

```
   30 READ(5,900)NDFFIX(IRESND),(IFPRE(IRESND,
      @JDOFN),JDOFN=1,NDOFN)
        WRITE(6,950)
  950 FORMAT(///,' DEGREES OF FREEDOM AT RESTR
      @AINED NODES',/,'--------------------------
@-------------------------------',//,'1 = FIXED
@AND 0 = FREE',//,'NODE NO.',6X,'Z-FREEDOM',6X,
      @'THETAX-FREEDOM',6X,'THETAY-FREEDOM',6X,
      @'THETAZ-FREEDOM',/,'-------------------------
      @------------------------------------------
      @-'//)
        DO 40 IRESND=1,NRESND
   40 WRITE(6,955)NDFFIX(IRESND),(IFPRE(IRESND,
      @JDOFN),JDOFN=1,NDOFN)
  955 FORMAT(I5,7X,I5,13X,I5,16X,I5,16X,I5)
C
C***CALL THE SUBROUTINE TO GENERATE THE NODE
C   FREEDOM ARRAY FROM THE DATA ABOUT THE
C   RESTRAINED NODES
C
        CALL NFGEN(NODFRE)
   45 CONTINUE
        WRITE(6,960)
  960 FORMAT(//,' DEGREES OF FREEDOM AT NODES',/,
      @'---------------------------------------
      @--------------------',//,'NODE',10X,'Z-
      @FRDOM',6X,' THETAX-FRDOM.',7X,' THETAY-
      @FRDOM.',7X,'THETAZ-FRDOM.',/)
        WRITE(6,908)
        DO 50 IPOIN=1,NPOIN
   50 WRITE(6,955)IPOIN,(NODFRE(IPOIN,IDOFN),
      @IDOFN=1,NDOFN)
C
C***READ AND PRINT THE PROPERTIES IE. AREA,
C   YOUNG'S MODULUS ETC. ABOUT THE DIFFERENT
C   MATERIALS
C
        DO 60 IMATS=1,NMATS
   60 READ(5,970)(PROPS(IMATS,IPROP),IPROP=1,
      @NPROP)
  970 FORMAT(7E14.6)
        WRITE(6,975)
  975 FORMAT(///,' PROPERTIES OF DIFFERENT
      @MATERIALS' )
        WRITE(6,908)
        WRITE(6,980)
  980 FORMAT(//,' MATERIAL NO.',4X,'YOUNGS MOD.
      @',5X,'SHEAR MOD.',4X,'INERTIA',6X,'TOR.
      @INERTIA',6X,'AREA',6X,'AREA SHEAR FACTOR',4X,'
      @WARP.INERTIA',/,'---------------------
      @------------------------------------
      @------------------------------------
      @-----',//)
        DO 70 IMATS=1,NMATS
   70 WRITE(6,981)IMATS,(PROPS(IMATS,IPROP),
      @IPROP=1,NPROP)
  981 FORMAT(2X,I5,10X,E10.4,5X,E10.4,5X,E10.4,
      @2X,E10.4,4(8X,E10.4))
C
C***READ VALUES OF SUBGRADE MODULUS  AND
```

```
C     STIFFNESS OF SEMI-RIGID CONNECTIONS AT THE
C     ENDS OF MEMBERS
C
      WRITE(6,982)
  982 FORMAT(//,' MATERIAL NO.',4X,'SUBGR MOD.'
     @,8X,'SPRNG-1',6X,'SPRNG-2')
      WRITE(6,908)
      DO 75 IMATS=1,NMATS
      READ(5,970)(PROPS(IMATS,JPROP),JPROP=
     @NPROP+1,NPROP+3)
   75 WRITE(6,981)IMATS,(PROPS(IMATS,JPROP),
     @JPROP=NPROP+1,NPROP+3)
C
C***READ THE DATA ABOUT NODE NUMBERS AND THE
C   FIXED VALUE OF DISPLACEMENTS AT NODES WHERE
C   CERTAIN DISPALCEMENTS ARE PRESCRIBED
C
      IF(NVFIX.EQ.0) GOTO 95
      DO 80 IVFIX=1,NVFIX
   80 READ(5,987)NDVFIX(IVFIX),(PRESC(IVFIX,
     @IDOFN),IDOFN=1,NDOFN)
  987 FORMAT(I5,4E14.6)
      WRITE(6,988)
  988 FORMAT(//,' NODES AT WHICH THE VALUE OF
     @DISPLACEMENTS ARE FIXED',//,'NODE',20X,'Z-
     @DISP.',13X,'THETAX-DISP.',9X,'THETAY-
     @DISP',9X,'THETAZ-DISP.',/,'------------
     @----------------------------------------
     @-----------------------',//)
      DO 90 IVFIX=1,NVFIX
   90 WRITE(6,989)NDVFIX(IVFIX),(PRESC(IVFIX,
     @IDOFN),IDOFN=1,NDOFN)
  989 FORMAT(I5,4F10.4)
C
C***CALCULATE THE MAXIMUM NUMBER OF EQUATIONS
C
   95 CALL MAXEQNS(NODFRE)
C
C***CALCULATE THE HALF-BANDWIDTH
C
      CALL BANDWIDTH(NODFRE,LNODS)
C
C***FIRST ZERO AND THEN FORM THE STRUCTURAL
C   STIFFNESS MATRIX
C
      CALL NULLGSTIF
      CALL STIFFGRID(NODFRE,LNODS,MATNO)
C
C***IF NEED BE MODIFY THE DIAGONAL ELEMENTS OF
C   THE STIFFNESS MATRIX TO INCORPORATE STIFF
C   SPRING APPROACH.
C
      IF(NVFIX.EQ.0) GOTO 105
      CALL STFSPRG(NODFRE)
C
C***REDUCE THE STIFFNESS MATRIX TO UPPER
C   TRIANGULAR FORM.  NFAIL =1 MEANS THAT THE
C   STIFFNESS MATRIX IS SINGULAR INDICATING
C   EITHER AN ERROR IN DATA OR THAT THE
C   STRUCTURE OR PARTS OF IT CAN UNDERGO RIGID
```

```
C    BODY DISPLACEMENTS DUE TO LACK OF SUFFICIENT
C    BRACING.
C
  105 CALL GAUSRED
      IF(NFAIL.EQ.1) GOTO 140
C
C***START THE LOAD CASE CYCLE
C
      DO 110 ICASE=1,NCASE
      READ(5,'(A60)')TITLE
C
C***FORM THE LOAD VECTOR DUE TO EXTERNAL LOADS.
C
      CALL LOADGRID(NODFRE,LNODS,MATNO)
C
C***DECIDE IF THERMAL LOADS ARE TO BE INCLUDED.
C   IF SO MODIFY THE LOAD VECTOR BY INCLUDING
C   THE EFFECTS OF THERMAL RESTRAINT.
C
      READ(5,900)ITEMP
      WRITE(6,991)ITEMP
  991 FORMAT(//,' ARE THERMAL STRESSES TO BE
     @ INCLUDED ? (YES = 1 NO = 0)',I5,//)
      IF(ITEMP.EQ.1) CALL THERMGRID(NODFRE,LNODS
     @,MATNO)
C
C***IF NECESSARY MODIFY THE LOAD VECTOR TO
C   INCORPORATE THE STIFF SPRING APPROACH.
C
      IF(NVFIX.EQ.0) GOTO 115
      CALL FIXDISP(NODFRE)
C
C***CALCULATE THE DISPLACEMENTS.
C
  115 CALL BACKSUB
      WRITE(6,992)ICASE
  992 FORMAT(//,' DISPLACEMENTS OF JOINTS FOR
     @LOAD CASE = ',I5,/,'--------------------------
     @------------------------------',//)
      WRITE(6,'(A60)')TITLE
      WRITE(6,903)
  903 FORMAT(//,' NODE',15X,'Z-DISP.',11X,'THET
     @AX-DISP.',6X,'THETAY-DISP.',6X,'THETAZ-
     @DISP.')
      WRITE(6,908)
      DO 120 IPOIN=1,NPOIN
      DO 130 IDOFN=1,NDOFN
      DISP(IDOFN)=0.0
      IF(NODFRE(IPOIN,IDOFN).NE.0)DISP(IDOFN)=
     @ASDIS(NODFRE(IPOIN,IDOFN))
  130 CONTINUE
      WRITE(6,994)IPOIN,(DISP(IDOFN),IDOFN=1,
     @NDOFN)
  120 CONTINUE
  994 FORMAT(I5,11X,4(E14.6,4X))
      IF(IWARP.EQ.0)WRITE(6,995)ICASE
  995 FORMAT(//,' FORCES IN THE MEMBERS FOR LOAD
     @ CASE =',I3,/,'-----------------------
     @----------------------------------',//,'
     @ELEMENT',3X,'END 1',3X,'END 2',3X,'SHR.
```

```
      @FORCE',7X,'BM.',6X,'TORQUE',4X,'SHR.
      @FORCE',3X,'BM',8X,'TORQUE',/,'---------
      @----------------------------------------
      @----------------------------------------
      @-----',//)
       IF(IWARP.EQ.1)WRITE(6,997)ICASE
  997 FORMAT(//,' FORCES IN THE MEMBERS FOR
      @LOAD CASE =',I3,/,'--------------------
      @----------------------------------------
      @',//,' ELEMENT',3X,'END 1',3X,'END 2',3X
      @,'SHR. FORCE',3X,'BM.',10X,'TORQUE',6X,
      @'BI-MOM',4X,'SHR.FORCE',3X,'BM',6X,'
      @TORQUE',5X,'BI-MOM',/,'----------------
      @----------------------------------------
      @---------------------------------------',
      @//)
C
C***CALCULATE THE END FORCES IN THE ELEMENTS.
C
      CALL STRESSGRID(NODFRE,LNODS,MATNO)
C
C***IF THERE ARE BEAMS ON ELASTIC FOUNDATION,
C   THEN CALCULATE THE PRESSURE, BM AND SF FOR
C   THEM.
C
      CALL PRESURGRID(NODFRE,LNODS,MATNO)
  110 CONTINUE
    1 CONTINUE
      STOP
  140 END
```

```
                   PROGRAM STABIL
C
C***THIS PROGRAM CALLED STABIL IS DESIGNED TO
C   CARRY OUT THE ELASTIC STABILITY ANALYSIS OF
C   2-D RIGID-JOINTED STRUCTURES.  IT HAS THE
C   OPTION OF EITHER DO A FULL NONLINEAR
C   ANALYSIS OR JUST CALCULATE THE SMALLEST LOAD
C   FACTOR AT WHICH THE STRUCTURES BECOMES
C   UNSTABLE.
C
      COMMON/CONTRO/NPOIN,NELEM,NNODE,NDOFN,
      @NDIME,NPROP,NMATS,NRESND,NEVAB,NFAIL,
      @NLODEL,NEQNS,NBAND,NLODND,NSINAG,ICONV,
      @DETNEW,NNEW,FACTLD,CONFAC
      COMMON/LGDATA/COORD(100,3),PROPS(10,3),
      @PNEW(100),POLD(100),NDFFIX(80),IFPRE(80,3),
      @FORSEL(80,4),LODEL(80),FORSND(80,3),
      @LODPT(80),ASDIS(800),GSTIF(100,30),
      @GLOAD(100)
      DIMENSION DISP(3),NODFRE(100,3),
      @LNODS(100,2),MATNO(100,2)
      READ(5,900)NPROB
  900 FORMAT(8I5)
      WRITE(6,905)NPROB
  905 FORMAT(//,5X,' TOTAL NUMBER OF PROBLEMS
      @= ',I3)
  911 FORMAT('--------------------------------
      @----------------------------------------
```

```
      @-----------',//)
        WRITE(6,911)
C
C***START THE CYCLE ON NUMBER OF STRUCTURES
C
        DO 1 IPROB=1,NPROB
        READ(5,'(A60)')TITLE
        WRITE(6,910)IPROB
   910 FORMAT(/////,6X,' PROBLEM NO ',I3)
        WRITE(6,'(A60)')TITLE
        WRITE(6,911)
C
C***SET THE BASIC PARAMETERS OF THE 2-D
C   RIGID-JOINTED STRUCTURE
C
        NFAIL=0
        NNODE=2
        NDOFN=3
        NDIME=2
        NPROP=3
        NEVAB=NNODE*NDOFN
C
C***READ AND PRINT THE BASIC PARAMETERS OF THE
C   PARTICULAR PROBLEM
C
        READ(5,900)NPOIN,NELEM,NRESND,NMATS,ISTRES
        WRITE(6,915)NPOIN,NELEM,NRESND,NMATS,
       @ISTRES
   915 FORMAT(//,' NUMBER OF NODES =
       @                     ',I5,/,' NUMBER OF MEMBERS
       @=                            ',I3,/,
       @' NUMBER OF RESTRAINED NODES =
       @            ',I3,/,' NUMBER OF DIFFERENT TYPES OF
       @MATERIALS =         ',I3,/,' IS STRESS AND
       @DISPLACEMENT TO BE OUTPUT? (YES =1, NO = 0) =',
       @I3,//)
C
C***READ THE COORDINATES OF SOME OF THE JOINTS,
C   GENERATE THE OMITTED DATA AND PRINT THE
C   COORDINATES OF ALL THE NODES.  IF NFAIL = 1
C   IT MEANS ERROR IN DATA AND ALL CALCULATIONS
C   STOP.
C
        CALL CORDGEN
        IF(NFAIL.EQ.1) GOTO 150
        WRITE(6,920)
   920 FORMAT(//,' COORDINATES OF NODES')
        WRITE(6,911)
        WRITE(6,925)
   925 FORMAT(//,'    NODE',7X,'X-CORD',5X,'
       @Y-CORD')
        WRITE (6,911)
        DO 10 IPOIN=1,NPOIN
    10 WRITE(6,935)IPOIN,(COORD(IPOIN,JDIME),
       @JDIME=1,NDIME)
   935 FORMAT(1X,I5,3(5X,F10.3))
C
C***READ THE NODE NUMBERS DEFINING THE ENDS OF
C   SOME OF THE ELEMENTS AND GENERATE THE
C   OMITTED DATA.  NFAIL = 1 MEANS ERROR IN DATA.
```

```
C     THEREFORE ALL COMPUTATIONS STOP.
C
      CALL INTGEN(LNODS,NNODE,NELEM)
      IF(NFAIL.EQ.1) GOTO 150
C
C***READ THE MATERIAL NUMBERS OF SOME OF THE
C   ELEMENTS AND GENERATE THE OMITTED DATA.  IF
C   NFAIL =1 IT MEANS DATA ERROR AND ALL
C   CALCULATIONS STOP.
C
      CALL INTGEN(MATNO,1,NELEM)
      IF(NFAIL.EQ.1) GOTO 150
C
C***PRINT THE MATERIAL NUMBERS AND THE NODE
C   NUMBERS AT THE ENDS FOR ALL THE ELEMENTS
C
      WRITE(6,940)
  940 FORMAT(///,' ELEMENT NO.',3X,' MATERIAL
     @NO.',6X,'NODE-1',4X,' NODE-2')
      WRITE(6,911)
      DO 20 IELEM = 1, NELEM
   20 WRITE(6,945)IELEM,MATNO(IELEM,1),(LNODS
     @(IELEM,JNODE),JNODE=1,NNODE)
  945 FORMAT(1X,I5,11X,I5,9X,I5,5X,I5,6X,I5)
C
C***READ NOPTN. THIS TELLS WHETHER THE NODE
C   FREEDOM ARRAY IS EITHER INPUT AS DATA
C   (FOR NOPTN = 1) OR TO BE GENERATED FROM
C   THE DATA ABOUT THE NODES WHERE ONE MORE
C   DISPLACEMENTS ARE ZERO (NOPTN = 0)
C
      READ(5,900)NOPTN
C
C***READ NODE FREEDOMS OF SOME OF THE NODES AND
C   GENERATE THE OMITTED DATA
C
      IF(NOPTN.EQ.1) CALL INTGEN(NODFRE,NDOFN,
     @NPOIN)
      IF(NFAIL.EQ.1) GOTO 150
      IF(NOPTN.EQ.1) GOTO 45
C
C***READ DATA ABOUT THE NODE NUMBERS AND THE
C   FIXITY CONDITION AT  NODES WHERE ONE OR MORE
C   FREEDOMS ARE EQUAL TO ZERO
C
      DO 30 IRESND=1,NRESND
   30 READ(5,900)NDFFIX(IRESND),(IFPRE(IRESND,
     @JDOFN),JDOFN=1,NDOFN)
      WRITE(6,950)
  950 FORMAT(///,' DEGREES OF FREEDOM AT RESTRA
     @INED NODES',//,'1 = FIXED AND 0 = FREE',
     @//,'NODE NO.',6X,'X-FREEDOM',6X,'Y-FREE
     @DOM',6X,'THETAZ-FREEDOM')
      WRITE(6,911)
      DO 40 IRESND=1,NRESND
   40 WRITE(6,955)NDFFIX(IRESND),(IFPRE(IRESND,
     @JDOFN),JDOFN=1,NDOFN)
  955 FORMAT(I5,10X,I5,8X,I5,16X,I5)
C
C***CALL THE SUBROUTINE TO GENERATE THE NODE
```

```
C    FREEDOM ARRAY FROM THE DATA ABOUT THE
C    RESTRAINED NODES
C
     CALL NFGEN(NODFRE)
  45 CONTINUE
     WRITE(6,960)
 960 FORMAT(//,' DEGREES OF FREEDOM AT NODES',
    @/,'------------------------------------
    @----------------------------',//,'NODE
    @',6X,' X-FRDOM.',6X,' Y-FRDOM.',6X,'
    @THETAZ-FRDOM.')
     WRITE(6,911)
     DO 50 IPOIN=1,NPOIN
  50 WRITE(6,903)IPOIN,(NODFRE(IPOIN,IDOFN),
    @IDOFN=1,NDOFN)
 903 FORMAT(I3,5X,I5,10X,I5,9X,I5)
C
C***READ AND PRINT THE PROPERTIES IE. AREA,
C    YOUNG'S MODULUS ETC. ABOUT THE DIFFERENT
C    MATERIALS
C
     DO 60 IMATS=1,NMATS
  60 READ(5,970)(PROPS(IMATS,IPROP),IPROP=1,
    @NPROP)
 970 FORMAT(5E14.6)
     WRITE(6,975)
 975 FORMAT(///,' PROPERTIES OF DIFFERENT
    @MATERIALS' )
     WRITE(6,911)
     WRITE(6,980)
 980 FORMAT(//,' MATERIAL NO.',6X,'YOUNGS MOD
    @.',6X,'INERTIA',10X,'AREA')
     WRITE(6,911)
     DO 70 IMATS=1,NMATS
  70 WRITE(6,981)IMATS,(PROPS(IMATS,IPROP),
    @IPROP=1,NPROP)
 981 FORMAT(2X,I5,8X,E14.6,2X,E14.6,2X,E14.6,
    @2X,E14.6)
C
C***ZERO AXIAL FORCE IN ALL THE ELEMENTS.
C
     DO 80 IELEM=1,NELEM
  80 POLD(IELEM)=0.0
C
C***CALCULATE THE MAXIMUM NUMBER OF EQUATIONS
C
     CALL MAXEQNS(NODFRE)
C
C***CALCULATE THE HALF-BANDWIDTH
C
     CALL BANDWIDTH(NODFRE,LNODS)
C
C***READ THE LOADS AT THE NODES AND ON THE
C    ELEMENTS AND STORE IT IN THE ARRAY LODPT
C    (NLODND),FORSND(NLODND,NDOFN) AND
C    LODEL(NLODEL),FORSEL(NLODEL,4).
C
     CALL READLOD
C
C***INPUT THE STARTING VALUE OF LOAD FACTOR,
```

```
C    MAXIMUM NUMBER OF ITERATIONS AND THE
C    INCREMENT IN THE LOAD FACTOR PER CYCLE AS A
C    PERCENTAGE OF THE INITIAL LOAD FACTOR.
C
      READ(5,982)STRFAC,NITER,INCREM
  982 FORMAT(F10.4,2I5)
      WRITE(6,983)STRFAC,NITER,INCREM
  983 FORMAT(//,'LOAD FACTOR AT START = ',F10.4
     @,/,'NUMBER OF ITERATIONS = ',I5,/,'INCRE
     @MENT OF LOAD FACTOR PER CYCLE AS A
     @PERCENTAGE = ',I5,///)
C
***START THE 'OUTERMOST' ITERATIONS ON THE LOAD
C   FACTOR.
C
      DO 110 IITER = 1, NITER
C
C***CALCULATE THE LOAD FACTOR FOR THE ITERATION
C
      FACTLD=STRFAC*(1.0+(IITER-1)*INCREM*0.01)
C
C***KEEP COUNT OF THE NUMBER OF INNER CYCLES TO
C   ACHIEVE CONVERGENCE OF THE AXIAL LOAD IN THE
C   ELEMENTS.
C
      KOUNT=0
C
C***FORM THE LOAD VECTOR FROM APPLIED EXTERNAL
C   LOADS
C
  115 CALL LOADRJS(NODFRE,LNODS,MATNO)
C
C***FIRST ZERO AND THEN FORM THE STRUCTURAL
C   STIFFNESS MATRIX
C
      CALL NULLGSTIF
      CALL STIFFRJS(NODFRE,LNODS,MATNO)
C
C***REDUCE THE STIFFNESS MATRIX TO UPPER
C   TRIANGULAR FORM.  IF NFAIL =0 THIS MEANS
C   THAT THE STRUCTUAL STIFFNESS MATRIX IS
C   SINGULAR.  THIS CAN BE EITHER DUE TO AN
C   ERROR IN DATA OR LUCKILY (OR UNLUCKILY) THE
C   EXACT LOAD FACTOR AT WHICH THE STRUCTURE
C   BECOMES UNSTABLE HAS BEEN REACHED.
C
      CALL GAUSRED
      IF(NFAIL.EQ.1) GOTO 150
C
C***SOLVE FOR DISPLACEMENTS.
C
      CALL BACKSUB
C
C***CHECK IF THE ASSUMED AND CALCULATED LOADS
C   ARE REASONABLY EQUAL.
C
      CALL CONVER(NODFRE,LNODS,MATNO)
C
C***CHECK TO SEE HOW MANY TIMES THE AXIAL LOAD
C   CYCLE HAS BEEN DONE.
```

```fortran
C
      KOUNT=KOUNT+1
      IF(KOUNT.GT.30) WRITE(6,994)KOUNT,CONFAC
      IF(KOUNT.GT.30) GOTO 125
      IF(ICONV.NE.1) GOTO 115
C
C***CALL THE SUBROUTINE TO CALCULATE THE
C   DETERMINANT AND THE NUMBER OF AGREEMENTS IN
C   THE SIGN OF THE SUBMATRICES OF THE
C   STRUCTURAL STIFFNESS MATRIX.
C
  125 CALL STURM
      IF(ISTRES.EQ.0) GOTO 145
      WRITE(6,990)FACTLD
  990 FORMAT(//,' DISPLACEMENTS OF JOINTS AT A
      @LOAD FACTOR =',F10.4,/,'---------------
      @----------------------------------------
      @--------',//,' NODE',11X,'X-DISP.',10X,'
      @Y-DISP.',11X,'THETAZ-DISP.',/,'--------
      @----------------------------------------
      @-------------------',//)
      DO 130 IPOIN=1,NPOIN
      DO 140 IDOFN=1,NDOFN
      DISP(IDOFN)=0.0
      IF(NODFRE(IPOIN,IDOFN).NE.0)DISP(IDOFN)=
      @ASDIS(NODFRE(IPOIN,IDOFN))
  140 CONTINUE
      WRITE(6,991)IPOIN,(DISP(IDOFN),IDOFN=1,
      @NDOFN)
  130 CONTINUE
  991 FORMAT(I5,6X,3(E14.6,4X))
      WRITE(6,992)FACTLD
  992 FORMAT(//,' FORCES IN THE MEMBERS AT LOAD
      @ FACTOR =',F10.4,/,'---------------
      @----------------------------------------
      @--------'//,' ELEMENT',2X,'END 1',2X,'
      @END2',3X,'AXL. FORCE',3X,'SHR. FORCE'
      @,6X,'BM',11X,'AXL. FORCE',3X,'SHR. FORCE'
      @,6X,'BM',/,'----------------------------
      @----------------------------------------
      @--------',//)
C
C***IF NEED BE CALL THE SUBROUTINE TO CALCULATE
C   THE STRESSES
C
      CALL STRESSRJS(NODFRE,LNODS,MATNO)
  145 IF(DETNEW.LT.0) GOTO 135
      DETOLD=DETNEW
      NOLD=NNEW
  110 CONTINUE
C
C***CALCULATE THE CRITICAL LOAD FACTOR AT WHICH
C   THE DETERMINANT BECOMES ZERO BY LINEAR
C   INTERPOLATION.
C
  135 CRITLD=FACTLD+DETNEW*INCREM*0.01/(DETOLD*
      @(10**(NOLD-NNEW))-DETNEW)
      WRITE(6,993)CRITLD,NSINAG
    1 CONTINUE
  993 FORMAT(/,' CRITICAL LOAD FACTOR =',F10.4,
```

```
      @/,'---------------------',//,' NUMBER
      @ OF AGREEMENTS IN SIGN = ',I5,/,'------
      @---------------------',//)
  994 FORMAT(/,' NO CONVERGENCE IN THE AXIAL
      @LOAD AFTER',I5,'CYCLES',' ---------
      @-------------------------------------
      @------------',//,' AXIAL LOAD CONVERGENCE
      @FACTOR =',E14.6,/,'---------------
      @-------------------------------------
      @-------',/)
      STOP
  150 END
```

<u>PROGRAM VIBES</u>

```
C
C***THIS PROGRAM CALLED VIBES CALCULATES ALL THE
C   NATURAL FREQUENCIES IN A GIVEN BANDWIDTH.
C   THIS PROGRAM CAN HANDLE THE FREQUENCY
C   ANALYSIS OF BOTH 2-D RIGID-JOINTED
C   STRUCTURES AND PLANE GRID STRUCTURES.
C
      COMMON/CONTRO/NPOIN,NELEM,NNODE,NDOFN,
      @NDIME,NPROP,NMATS,NRESND,NEVAB,NFAIL,
      @NLODEL,NEQNS,NBAND,NLODND,NSINAG,OMEGA,
      @DETNEW,NNEW
      COMMON/LGDATA/COORD(100,2),PROPS(10,6),
      @NDFFIX(80),IFPRE(80,3),FORSEL(80),
      @LODEL(80),FORSND(80),LODPT(80),
      @GSTIF(100,30)
      DIMENSION NODFRE(100,3),LNODS(100,2),
      @MATNO(100,1),FRQTAB(30,6)
      READ(5,900)NPROB
  900 FORMAT(8I5)
      WRITE(6,905)NPROB
  905 FORMAT(//,5X,' TOTAL NUMBER OF PROBLEMS
      @= ',I3)
  911 FORMAT('---------------------------------
      @----------------------------------------@
      @---------',//)
      WRITE(6,911)
C
C***START THE CYCLE ON THE NUMBER OF PROBLEMS.
C
      DO 1 IPROB=1,NPROB
      READ(5,'(A60)')TITLE
      WRITE(6,910)IPROB
  910 FORMAT(/////,6X,' PROBLEM NO ',I3)
      WRITE(6,'(A60)')TITLE
      WRITE(6,911)
C
C***SET THE BASIC PARAMETERS OF THE 2-D RIGID-
C   JOINTED STRUCTURE
C
      NFAIL=0
      NNODE=2
      NDOFN=3
      NDIME=2
      NEVAB=NNODE*NDOFN
```

```
      PI=3.1415926534
      READ(5,900)ISTRUCT
      IF(ISTRUCT.EQ.1)WRITE(6,912)
      IF(ISTRUCT.EQ.2)WRITE(6,913)
      WRITE(6,911)
  912 FORMAT(//,' THIS IS A 2-D RIGID-JOINTED
     @STRUCTURE',//)
  913 FORMAT(//,' THIS A PLANE GRID STRUCTURE',
     @//)
      IF(ISTRUCT.EQ.1) NPROP=3
      IF(ISTRUCT.EQ.2) NPROP=6
C
C***READ AND PRINT THE BASIC PARAMETERS OF THE
C   PARTICULAR PROBLEM
C
      READ(5,900)NPOIN,NELEM,NRESND,NMATS
      WRITE(6,915)NPOIN,NELEM,NRESND,NMATS
  915 FORMAT(//,' NUMBER OF NODES =
     @                    ',I5,/,' NUMBER OF MEMB
     @ERS =                         ',I3,/,
     @' NUMBER OF RESTRAINED NODES =
     @            ',I3,/,' NUMBER OF DIFFERENT
     @TYPES OF MATERIALS =         ',I3,//)
C
C***READ THE COORDINATES OF SOME OF THE JOINTS,
C   GENERATE THE OMITTED DATA AND PRINT THE
C   COORDINATES OF ALL THE NODES.  NFAIL = 1
C   THAT ERROR IN DATA AND THIS HALTS ALL
C   CALCULATIONS.
C
      CALL CORDGEN
      IF(NFAIL.EQ.1) GOTO 180
      WRITE(6,920)
  920 FORMAT(//,' COORDINATES OF NODES')
      WRITE(6,911)
      WRITE(6,925)
  925 FORMAT(//,'    NODE',7X,'X-CORD',10X,'
     @Y-CORD')
      WRITE (6,911)
      DO 10 IPOIN=1,NPOIN
   10 WRITE(6,935)IPOIN,(COORD(IPOIN,JDIME),
     @JDIME=1,NDIME)
  935 FORMAT(1X,I5,3(5X,F10.3))
C
C***READ THE NODE NUMBERS DEFINING THE ENDS OF
C   SOME OF THE ELEMENTS AND GENERATE THE
C   OMITTED DATA.  NFAIL = 1 MEANS ERROR IN
C   DATA AND THIS HALTS ALL CALCULATIONS.
C
      CALL INTGEN(LNODS,NNODE,NELEM)
      IF(NFAIL.EQ.1) GOTO 180
C
C***READ THE MATERIAL NUMBERS OF SOME OF THE
C   ELEMENTS AND GENERATE THE OMITTED DATA.
C   NFAIL = 1 MEANS THAT THERE IS ERROR IN
C   DATA AND THIS HALTS ALL CALCULATIONS.
C
      CALL INTGEN(MATNO,1,NELEM)
      IF(NFAIL.EQ.1) GOTO 180
C
```

```fortran
C***PRINT THE MATERIAL NUMBERS AND THE NODE
C   NUMBERS AT THE ENDS FOR ALL THE ELEMENTS
C
      WRITE(6,940)
  940 FORMAT(///,' ELEMENT NO.',3X,' MATERIAL
     @NO.',6X,'NODE-1',3X,' NODE-2')
      WRITE(6,911)
      DO 20 IELEM=1,NELEM
   20 WRITE(6,945)IELEM,MATNO(IELEM,1),(LNODS
     @(IELEM,JNODE),JNODE=1,NNODE)
  945 FORMAT(1X,I5,11X,I5,9X,I5,5X,I5,6X,I5)
C
C***READ NOPTN.  THIS TELLS WHETHER THE NODE
C   FREEDOM ARRAY IS EITHER INPUT AS DATA
C   (FOR NOPTN = 1) OR TO BE GENERATED FROM
C   THE DATA ABOUT THE NODES WHERE ONE MORE
C   DISPLACEMENTS ARE ZERO (NOPTN = 0)
C
      READ(5,900)NOPTN
C
C***READ NODE FREEDOMS OF SOME OF THE NODES AND
C   GENERATE THE OMITTED DATA.  NFAIL=1 MEANS
C   ERROR IN DATA AND THIS HALTS ALL CALCULATIONS.
C
      IF(NOPTN.EQ.1) CALL INTGEN(NODFRE,NDOFN,
     @NPOIN)
      IF(NFAIL.EQ.1) GOTO 180
      IF(NOPTN.EQ.1) GOTO 45
C
C***READ DATA ABOUT THE NODE NUMBERS AND FIXITY
C   STATE AT  NODES WHERE ONE OR MORE FREEDOMS
C   ARE EQUAL TO ZERO
C
      DO 30 IRESND=1,NRESND
   30 READ(5,900)NDFFIX(IRESND),(IFPRE(IRESND,
     @JDOFN),JDOFN=1,NDOFN)
      IF(ISTRUCT.EQ.1)WRITE(6,950)
      IF(ISTRUCT.EQ.2) WRITE(6,951)
  950 FORMAT(///,' DEGREES OF FREEDOM AT
     @ RESTRAINED NODES',//,'1 = FIXED AND 0 =
     @FREE',//,'NODE NO.',6X,'X-FREEDOM',6X,'
     @Y-FREEDOM',6X,'THETAZ-FREEDOM')
  951 FORMAT(///,' DEGREES OF FREEDOM AT
     @RESTRAINED NODES',//,'1 = FIXED AND 0 =
     @FREE',//,'NODE NO.',6X,'Z-FREEDOM',6X,
     @'THETAX-FREEDOM',6X,'THETAY-FREEDOM')
      WRITE(6,911)
      DO 40 IRESND=1,NRESND
   40 WRITE(6,955)NDFFIX(IRESND),(IFPRE(IRESND,
     @JDOFN),JDOFN=1,NDOFN)
  955 FORMAT(I5,10X,I5,8X,I5,16X,I5)
C
C***CALL THE SUBROUTINE TO GENERATE THE NODE
C   FREEDOM ARRAY FROM THE DATA ABOUT THE
C   RESTRAINED NODES
C
      CALL NFGEN(NODFRE)
   45 CONTINUE
      IF(ISTRUCT.EQ.1) WRITE(6,960)
      IF(ISTRUCT.EQ.2) WRITE(6,961)
```

```
  960 FORMAT(//,' DEGREES OF FREEDOM AT NODES',
     @/,'-----------------------------------
     @----------------------------',//,'NODE',
     @6X,' X-FRDOM.',6X,' Y-FRDOM.',6X,' THETAZ
     @-FRDOM.')
  961 FORMAT(//,' DEGREES OF FREEDOM AT NODES',
     @/,'-----------------------------------
     @---------------------------',//,'NODE
     @',6X,' Z-FRDOM.',4X,' THETAX-FRDOM.',2X,'
     @ THETAY-FRDOM.')
      WRITE(6,911)
      DO 50 IPOIN=1,NPOIN
   50 WRITE(6,903)IPOIN,(NODFRE(IPOIN,IDOFN),
     @IDOFN=1,NDOFN)
  903 FORMAT(I3,5X,I5,10X,I5,9X,I5)
C
C***READ AND PRINT THE PROPERTIES IE. AREA,
C   YOUNG'S MODULUS ETC. ABOUT THE DIFFERENT
C   MATERIALS
C
      DO 60 IMATS=1,NMATS
   60 READ(5,970)(PROPS(IMATS,IPROP),IPROP=1,
     @NPROP)
  970 FORMAT(6E14.6)
      WRITE(6,975)
  975 FORMAT(///,' PROPERTIES OF DIFFERENT
     @MATERIALS' )
      WRITE(6,911)
      IF(ISTRUCT.EQ.1) WRITE(6,977)
      IF(ISTRUCT.EQ.2) WRITE(6,979)
  977 FORMAT(//,' MATERIAL NO.',6X,'YOUNGS MOD
     @.',6X,'INERTIA',10X,'AREA')
  979 FORMAT(//,' MATERIAL NO.',6X,'YOUNGS MOD
     @.',6X,'SHEAR MOD.',6X,'INERTIA',8X,'TOR
     @INERTIA',8X,'AREA',10X,'POLAR INERTIA')
      WRITE(6,911)
      DO 70 IMATS=1,NMATS
   70 WRITE(6,981)IMATS,(PROPS(IMATS,IPROP),
     @IPROP=1,NPROP)
  981 FORMAT(2X,I5,8X,6(E14.6,2X))
C
C***CALCULATE THE MAXIMUM NUMBER OF EQUATIONS
C
      CALL MAXEQNS(NODFRE)
C
C***CALCULATE THE HALF-BANDWIDTH
C
      CALL BANDWIDTH(NODFRE,LNODS)
C
C***READ THE CONCENTRATED MASSES AT THE NODES
C   AND UD MASS ON THE ELEMENTS AND STORE IN
C   THE ARRAY LODPT(NLODND),FORSND(NLODND) AND
C   LODEL(NLODEL),FORSEL(NLODEL).
C
      CALL READMAS
C
C***THIS SECTION CALCULATES THE NATURAL
C   FREQUENCIES IN A GIVEN RANGE BY ROOT
C   BISECTION IN COMBINATION WITH STURM SEQUENCE.
C
```

```
C
C***READ THE VALUES OF THE LOWER LIMIT AND THE
C   UPPER LIMIT OF THE BAND IN WHICH THE
C   NATURAL FREQUENCIES ARE DESIRED.  ALSO INPUT
C   THE MAXIMUM ACCEPTABLE DIFFERENCE BETWEEN
C   THE UPPER AND LOWER LIMITS WITHIN WHICH A
C   NATURAL FREQUENCY LIES.
C
      READ(5,982)STRFRQ,ENDFRQ,ACURCY
  982 FORMAT(3F10.4)
      WRITE(6,983)STRFRQ,ENDFRQ,ACURCY
  983 FORMAT(//,'LOWER LIMIT OF FREQUENCY = ',
     @F10.4,/,'UPPER LIMIT OF FREQUENCY = ',
     @F10.4,/,'ACCEPTABLE ACCURACY IN LIMITS
     @FOR A FREQUENCY = ',F10.4,///)
C
C***CALCULATE THE DETERMINANT AND THE NUMBER OF
C   THE SIGN AGREEMENTS FOR THE UPPER AND LOWER
C   LIMITS OF THE BAND IN WHICH THE NATURAL
C   FREQUENCIES ARE DESIRED.  THIS WILL ENABLE
C   THE CALCULATION OF HOW MANY NATURAL
C   FREQUENCIES ARE IN THE GIVEN BAND.
C
      DO 110 IITER=1,2
      FREQ=STRFRQ
      IF(IITER.EQ.2)FREQ=ENDFRQ
      OMEGA=2.0*PI*FREQ
C
C***FIRST ZERO AND THEN FORM THE STRUCTURAL
C   STIFFNESS MATRIX
C
      CALL NULLGSTIF
      IF(ISTRUCT.EQ.1)CALL STIFFRJV(NODFRE,
     @LNODS,MATNO)
      IF(ISTRUCT.EQ.2)CALL STIFFGRV(NODFRE,
     @LNODS,MATNO)
C
C***REDUCE THE STIFFNESS MATRIX TO UPPER
C   TRIANGULAR FORM. NFAIL = 1 MEANS THAT THE
C   MATRIX IS SINGULAR EITHER DUE TO ERROR IN
C   DATA OR DUE TO THE FACT THAT THE FREQUENCY
C   AT WHICH THE CALCULATIONS ARE MADE IS VERY
C   NEAR TO A NATURAL FREQUENCY.
C
      CALL GAUSRED
C
C***CALCULATE THE VALUE OF THE DETERMINANT AND
C   THE NUMBER OF SIGN AGREEMENTS.
C
      CALL STURM
      IF(IITER.EQ.2) GOTO 115
C
C***CALCULATE THE DETERMINANT AND THE NUMBER OF
C   SIGN AGREEMENTS AT THE LOWER LIMIT OF
C   FREQUENCY BAND.
C
      STRDET=DETNEW
      NSTR=NNEW
      NSNSTR=NSINAG
      GOTO 110
```

```
C
C***CALCULATE THE DETERMINANT AND THE NUMBER OF
C   SIGN AGREEMENTS AT THE UPPER LIMIT OF
C   FREQUENCY BAND.
C
   115 ENDDET=DETNEW
       NEND=NNEW
       NSNEND=NSINAG
   110 CONTINUE
C
C***CALCULATE THE NUMBER OF NATURAL FREQUENCIES
C   TO BE DETERMINED.
C
       NFREQ=NSNEND-NSNSTR
C
C***FOR ALL THE NATURAL FREQUENCIES SET THE
C   UPPER AND LOWER LIMITS IN WHICH THEY LIE AND
C   THE CORRESPONDING DETERMINANTS.
C
       DO 120 IFREQ=1,NFREQ
       FRQTAB(IFREQ,1)=STRFRQ
       FRQTAB(IFREQ,2)=ENDFRQ
       FRQTAB(IFREQ,3)=STRDET
       FRQTAB(IFREQ,4)=NSTR
       FRQTAB(IFREQ,5)=ENDDET
       FRQTAB(IFREQ,6)=NEND
   120 CONTINUE
C
C***START THE ROOT BISECTION CYCLE TO DETERMINE
C   THE CLOSE UPPER AND LOWER BOUND LIMITS FOR
C   EACH NATURAL FREQUENCY IN THE GIVEN BAND.
C
       DO 130 IFREQ=1,NFREQ
   131 FREQ=0.5*(FRQTAB(IFREQ,1)+FRQTAB(IFREQ,
      @2))
       OMEGA=2.0*PI*FREQ
C
C***FIRST ZERO AND THEN FORM THE STRUCTURAL
C   STIFFNESS MATRIX
C
       CALL NULLGSTIF
       IF(ISTRUCT.EQ.1) CALL STIFFRJV(NODFRE,
      @LNODS,MATNO)
       IF(ISTRUCT.EQ.2) CALL STIFFGRV(NODFRE,
      @LNODS,MATNO)
C
C***REDUCE THE STIFFNESS MATRIX TO UPPER
C   TRIANGULAR FORM
C
       CALL GAUSRED
C
C***CALCULATE THE VALUE OF THE DETERMINANT AND
C   THE NUMBER OF SIGN AGREEMENTS.
C
       CALL STURM
C
C***CALCULATE HOW MANY NATURAL FREQUENCIES ARE
C   BELOW AND HOW MANY ABOVE THE FREQUENCY AT
C   WHICH THE CALCULATION HAS BEEN DONE.
C
```

```fortran
      IBELOW=NSINAG-NSNSTR
      IABOVE=NSNEND-NSINAG
      IF(IBELOW.EQ.0) GOTO 140
C
C***RESET IF NEED BE THE UPPER LIMITS OF ALL
C   THE NATURAL FREQUENCIES BELOW THE FREQUENCY
C   AT WHICH THE CALCULATION HAS BEEN DONE.
C
      DO 135 JFREQ=1,IBELOW
      IF(FRQTAB(JFREQ,2).LT.FREQ)GOTO 135
      FRQTAB(JFREQ,2)=FREQ
      FRQTAB(JFREQ,5)=DETNEW
      FRQTAB(JFREQ,6)=NNEW
  135 CONTINUE
  140 IF(IABOVE.EQ.0) GOTO 160
C
C***RESET IF NEED BE THE LOWER LIMITS OF THE
C   NATURAL FREQUENCIES ABOVE THE FREQUENCY AT
C   WHICH THE CALCULATION HAS BEEN DONE.
C
      DO 150 JFREQ=(IBELOW+1),NFREQ
      IF(FRQTAB(JFREQ,1).GT.FREQ) GOTO 150
      FRQTAB(JFREQ,1)=FREQ
      FRQTAB(JFREQ,3)=DETNEW
      FRQTAB(JFREQ,4)=NNEW
  150 CONTINUE
  160 CONTINUE
      IF((FRQTAB(IFREQ,2) - FRQTAB(IFREQ,1)).
     @GT.ACURCY) GOTO 131
  130 CONTINUE
C
C***CALCULATE BY LINEAR INTERPOLATION FROM THE
C   UPPER AND LOWER BOUNDS OF EACH NATURAL
C   FREQUENCY THE 'CORRECT' NATURAL FREQUENCY.
C
      DO 165 IFREQ=1,NFREQ
      FREQCY=FRQTAB(IFREQ,1)+(FRQTAB(IFREQ,2)-
     @FRQTAB(IFREQ,1))/(1.0-FRQTAB(IFREQ,5)/
     @FRQTAB(IFREQ,3)*(10**(FRQTAB(IFREQ,6)-
     @FRQTAB(IFREQ,4))))
      FRQTAB(IFREQ,1)=FREQCY
  165 CONTINUE
      WRITE(6,994)
  994 FORMAT(//,' FREQUENCY NO.',5X,'VALUE OF
     @ FREQUENCY IN HERTZ')
      DO 170 IFREQ=1, NFREQ
      WRITE(6,995)IFREQ,FRQTAB(IFREQ,1)
  170 CONTINUE
  995 FORMAT(I5,12X,F10.4)
    1 CONTINUE
      STOP
  180 END
```

APPENDIX 1 Glossary of Names of Variables

This glossary contains an explanation of the names of the most important variables used in the programs presented in this book.

AREA	Area of cross section
ALPHA1 and ALPHA2	Parameters of a semi-rigid beam element.
ALPHAT	α_t, the coefficient of linear thermal expansion.
AMDAL	λL, a parameter of a beam on an elastic foundation.
ASDIS(NEQNS)	Array storing the nonzero displacements of the nodes.
BETA	β, a parameter to account for the effect of shearing deformations in beam elements.
BETAL	βL, a parameter in torsion with warping restraint.
BOTTEM	T_b, temperature change at the bottom face of a beam element.
CN	Total mid-span concentrated normal load component.
CONLX, CONLY, CONLZ	Mid-span concentrated load in x, y and z directions.
COORD(NPOIN,NDIME)	Array storing the coordinates of all the nodal points.
DEPTH	d, the depth of a beam element.
DETNEW, DETOLD	Determinant of the structural stiffness matrix at the current and previous load increments. See section 7.5.
DISFOR(NEVAB)	Array storing the stresses at the ends of the element due to nodal displacements.

ELOAD(NEVAB)	Temporary array which stores either the reactions or the stresses at the ends of an element due to lateral loads on a beam element.
ENDFRQ	Upper limit of the frequency band.
ESTIF(NEVAB,NEVAB)	Array storing the element stiffness matrix.
FACTLD	Multiplication factor on loads.
FORSEL(NLODEL,4)	Array which stores the lateral loads on the elements.
FORSND(NLODND,NDOFN)	In stability analysis, this array stores the applied concentrated loads at the nodes.
FORSTM(NEVAB)	Similar to ELOAD array except the effects due to the prevention of free thermal expansion are stored.
GLOAD(NEQNS)	Global or assembled load vector.
GSTIF(NEQNS,NEQNS)	Global or assembled structural stiffness matrix.
ICONV	In stability analysis, if ICONV = 1, it means that the assumed and calculated axial loads in elements have converged.
IEND	Node number at end 1 of an element.
IFPRE(NRESND,NDOFN)	Array which stores the fixity condition in code form (1 = fixed and 0 = free) of the NRESND restrained nodes.
INCREM	In stability analysis, the percentage increment in the initial load factor for iterations on the load cycle.
ISTRES	In stability, ISTRES = 0 gives only the load factor to cause instability, and ISTRES = 1 gives the output of stresses and displacements.
IWARP	In grid analysis, IWARP = 0 ignores the effects of warping restraint effects, and IWARP = 1 includes the effects of warping restraint.
JEND	Node number at end 2 of an element.
LNODS(NELEM,NNODE)	Array which stores the node numbers at the ends of elements.

LODEL(NLODEL)	Array which stores the element number of the elements with lateral loads on them.

LODPT	Node number of a node where nodal loads are acting.

LPROP	Group number of an element.

MATNO(NELEM,2)	Array which stores the group number and the type number of the beam element for all NELEM elements.

MEMDIS(NEVAB)	Array which stores the node freedom numbers of the nodes at the ends of an element.

NBAND	Half-bandwidth of GSTIF matrix is NBAND + 1.

NCASE	Number of load cases to be analysed for a structure.

NDFIX(NRESND)	Array which stores the node numbers of all NRESND restrained nodes.

NDIME	Number of dimensions, i.e. 2-D or 3-D.

NDOFN	Number of degrees of freedom at a node of an element.

NDVFIX(NVFIX)	Array storing the node number of nodes with one or more degrees of freedom required to have a fixed nonzero value.

NEQNS	Number of simultaneous equations to be solved.

NEVAB	Equals NNODE∗NDOFN, the number of element variables.

NFREQ	Number of natural frequencies of the structure in a given frequency band.

NITER	Maximum number of iterations for the axial forces in elements to converge in stability analysis (Chapter 7).

NLODEL	Number of elements with lateral loads on them.

NLODPT	Number of nodes where concentrated nodal loads are applied.

NMATS	Number of different groups of elements in a structure.

NNEW, NOLD	Scale factors on determinants, See section 7.5.

NNODE	Equals 2 for skeletal structures.
NODFRE(NPOIN,NDOFN)	Array which stores the unknown displacement numbers for all the NDOFN degrees of freedom for NPOIN nodes. Restrained displacements are given a number zero.
NOPTN	Number of option. This decides whether the unknown displacement numbers are input as data or whether they are generated from information about restrained nodes.
NPOIN	Number of nodes in a structure.
NPROB	Number of problems to be analysed.
NPROP	Number of basic material and geometrical properties required to calculate the element stiffness matrix of an element.
NRESND	Numbers of nodes where one or more displacements are fully restrained.
NSINAG	Number of agreements in sign of consecutive members of sequence S_m and S_{m+1} as explained in section 7.4.
NTYPE	Type number of a beam element: 1, an orthodox beam; 2, a beam on an elastic foundation; 3, a semi-rigid beam; 4, a nonstandard beam.
NVFIX	Number of nodes where one or more displacements are required to have a fixed nonzero value as for example the settlement of a support.
OMEGA	ω, the circular frequency of vibration.
PNEW(NELEM), POLD(NELEM)	Arrays storing the calculated and 'assumed' value of axial loads in elements in stability analysis.
PRESC(NVFIX,NDOFN)	Array which stores the fixed value of displacements of NVFIX nodes. The code used is that, if a displacement is zero, it means that it is free!

PROPS(NMATS,NPROP + ?)	Array which stores the 'properties' of all NMATS group of elements.
Q11, Q12, Q21, Q22	Cross-rotation-translation stiffness coefficients of a beam element. See Fig. 3.2.
S11, S12, S22	Pure rotational stiffness coefficients of a beam element. See Fig. 3.2.
SHEARM	G, the shear modulus.
SPAN	L, the length of an element.
STRFAC	Starting value of the load factor in stability analysis.
STRFRQ	Lower limit of frequency band.
SUBGRM	k, the subgrade modulus.
T11, T12, T22	Pure translational stiffness coefficients of a beam element. See Fig. 3.2.
TEMP(NELEM,2)	Array which stores the temperature change in the elements.
TINERA	J, Saint Venant's torsion inertia.
TOPTEM	T_u, the temperature change in the top face of a beam element.
TOTFOR(NEVAB)	The final stresses at the ends of an element are stored in this array before printing.
UDLN, UDLXT, UDLYT, UDLZT	Total lateral loads on a beam element in the normal, x, y and z directions respectively.
WINERA	I_w, the warping inertia.
XL	l, a direction cosine.
XPROJ	Projection of an element on the x axis.
YDOWN	Y_b, the distance from the neutral axis to the bottom face of beam element.
YINERA	I, the second moment of area about the y_m axis.
YM	m, a direction cosine.
YOUNG	E, Young's modulus.
YPROJ	Projection of an element on the y axis.
YUP	Y_u, the distance from the neutral axis to the top face of a beam element.
ZSAF	Area shear factor for shear force along the z_m axis. Equals AREA/A_s.
ZN	n, a direction cosine.

APPENDIX 2 A beam on an Elastic Foundation

A2.1 DERIVATION OF THE EQUATION FOR A BEAM ON AN ELASTIC FOUNDATION

Consider the prismatic beam element shown in Fig. A2.1(a). The beam rests on a continuously distributed subgrade whose reaction is directly proportional to the normal displacement of the beam at that particular section. As shown in Fig. A2.1(b), one can consider the beam to be subjected to a vertically upward lateral load of $q - kv$, where k is the modulus of subgrade reaction, v the normal displacement and q the distributed external normal load. The differential equation for the beam on an elastic foundation is, using equation (1.14), given by

$$EIv'''' = q - kv$$

$$EIv'''' + kv = q$$

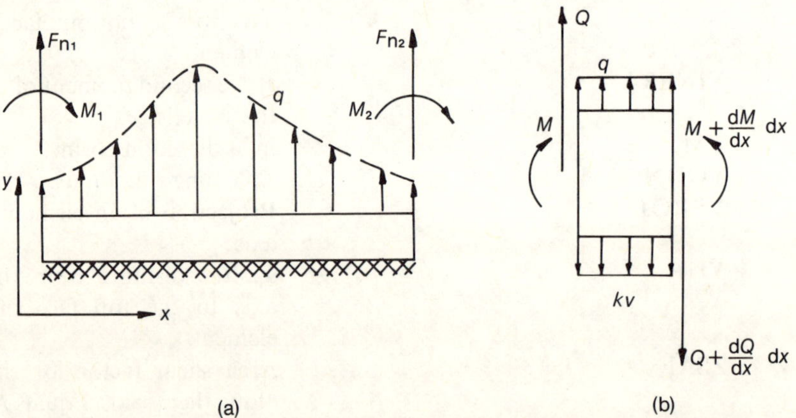

Fig. A2.1   A general beam on an elastic foundation: (a) forces on the element; (b) a segment of the beam.

A2.2 DERIVATION OF THE ELEMENT STIFFNESS RELATIONSHIP

The solution to the equation is obtained in two parts as explained in Chapter 1.

(i) *Fixed end solution.*

$$v = C_1 \cosh \lambda x \cos \lambda x + C_2 \sinh \lambda x \sin \lambda x + C_3 \cosh \lambda x \sin \lambda x$$

$$+ C_4 \sinh \lambda x \cos \lambda x + \frac{q}{k}$$

where $\lambda^4 = k/4EI$. The boundary conditions are given by

$$\text{at end 1, } v = \frac{dv}{dx} = 0, \quad x = 0$$

$$\text{at end 2, } v = \frac{dv}{dx} = 0, \quad x = L$$

The constants of integration are given by

$$C_1 = -\frac{q}{k}, \quad C_2 = \frac{q(S - s)}{k(S + s)}, \quad C_3 = \frac{q(C - c)}{k(S + s)}, \quad C_4 = -C_3$$

where $S = \sinh \lambda L$, $C = \cosh \lambda L$, $s = \sin \lambda L$ and $c = \cos \lambda L$.

Substituting for the constants of integration, the displacement v can be expressed as

$$v = \frac{q}{k}\left[1 - \cosh \lambda x \cos \lambda x + \frac{S - s}{s + s} \sinh \lambda x \sin \lambda x \right.$$

$$\left. - \frac{C - c}{S + s}(\cosh \lambda x \sin \lambda x - \sinh \lambda x \cos \lambda x) \right]$$

The bending moment M and the shear force Q are given by

$$M = EIv'', \quad Q = -EIv'''$$

$$M = 0.5q\lambda^{-2}\left[\sinh \lambda x \sin \lambda x + \frac{S - s}{S + s} \cosh \lambda x \cos \lambda x \right.$$

$$\left. - \frac{C - c}{S + s}(\cosh \lambda x \sin \lambda x + \sinh \lambda x \cos \lambda x) \right]$$

$$Q = -0.5q\lambda^{-1}\left[\cosh \lambda x \sin \lambda x + \sinh \lambda x \cos \lambda x \right.$$

$$+ \frac{S - s}{S + s}(\sinh \lambda x \cos \lambda x - \cosh \lambda x \sin \lambda x)$$

$$\left. - \frac{C - c}{S + s} 2 \cosh \lambda x \cos \lambda x \right]$$

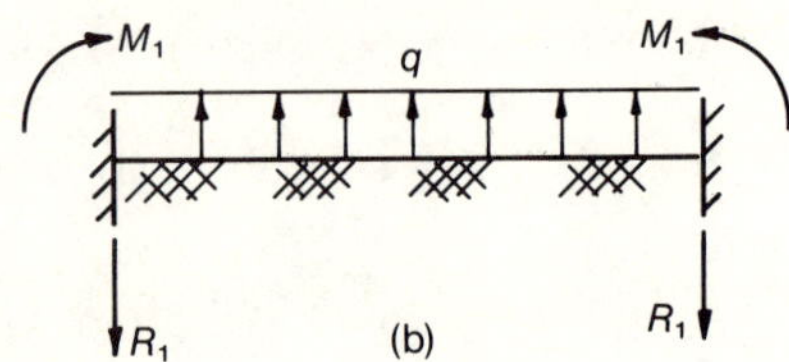

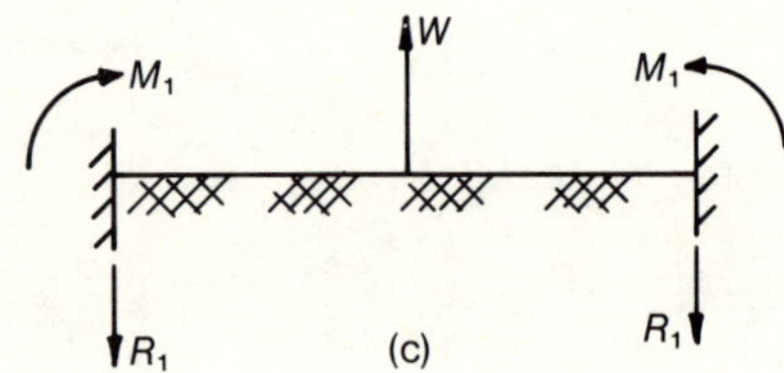

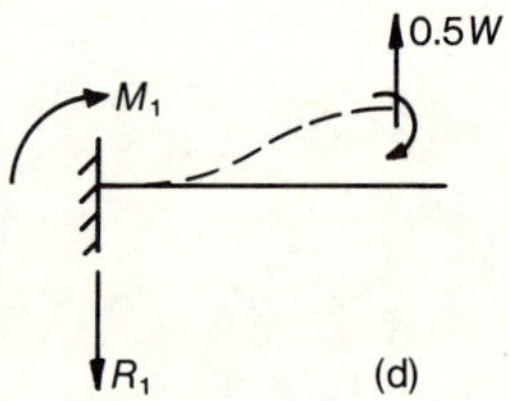

Fig. A2.2 Forces on a beam on an elastic foundation: (a) displacements and corresponding forces at the ends; (b) fixed end forces due to a uniformly distributed load; (c) fixed end forces due to a mid-span concentrated load; (d) a symmetrical half of the beam shown in (c).

The fixed end reactions, as shown in Fig. A2.2(b), are given by

$$M_1(= M \text{ at } x = 0) = \frac{qL^2}{12}\left[\frac{6(S - s)}{(\lambda L)^2(S + s)}\right]$$

$$R_1(= Q \text{ at } x = 0) = 0.5qL\left[\frac{2(C - c)}{\lambda L(S + s)}\right]$$

It is useful to interpret the terms in square brackets in the above equations as the lateral elastic support modification factors of the usual fixed end reactions.

(ii) *Fixed displacement case.*   In this case, since $q = 0$,

$$v = C_1 \cosh \lambda x \cos \lambda x + C_2 \sinh \lambda x \sin \lambda x$$
$$+ C_3 \cosh \lambda x \sin \lambda x + C_4 \sinh \lambda x \cos \lambda x$$

The boundary conditions are given by

$$\text{at end 1,} \quad v = V_{n1}, \quad \frac{dv}{dx} = -\theta_1, \quad x = 0$$

$$\text{at end 2,} \quad v = V_{n2}, \quad \frac{dv}{dx} = -\theta_1, \quad x = L$$

The constants of integration are given by

$$C_1 = V_{n1}$$

$$C_2 = \frac{-V_{n1}(C^2 - c^2) + V_{n2}(2Ss) + \theta_1\lambda^{-1}(SC - sc) + \theta_2\lambda^{-1}(Cs - Sc)}{D}$$

$$C_3 = \frac{V_{n1}(SC + sc) - V_{n2}(Cs + Sc) - \theta_1\lambda^{-1}S^2 - \theta_2\lambda^{-1}Ss}{D}$$

$$C_4 = \frac{-V_{n1}(SC + sc) + V_{n2}(Cs + Sc) + \theta_1\lambda^{-1}s^2 + \theta_2\lambda^{-1}Ss}{D}$$

where $D = S^2 - s^2$, $S = \sinh \lambda L$, $C = \cosh \lambda L$, $s = \sin \lambda L$ and $c = \cos \lambda L$.

Using the above constants of integration, the displacement v can be expressed in terms of shape function N_1 to N_4 as follows:

$$V = N_1 V_{n1} + N_2 \theta_1 + N_3 V_{n2} + N_4 \theta_2$$

where

$$N_1 = \frac{(S^2 - s^2)\cosh \lambda x \cos \lambda x - (C^2 - c^2)\sinh \lambda x \sin \lambda x + (SC + sc)(\cosh \sin \lambda x - \sinh \lambda x \cos \lambda x)}{D}$$

$$N_2 = \frac{(SC - sc)\sinh \lambda x \sin \lambda x - S^2 \cosh \lambda x \sin \lambda x + s^2 \sinh \lambda x \cos \lambda x}{D\lambda}$$

$$N_3 = \frac{2Ss \sinh \lambda x \sin \lambda x - (Cs + Sc)(\cosh \lambda x \sin \lambda x - \sinh \lambda x \cos \lambda x)}{D}$$

$$N_4 = \frac{(Cs - Sc)\sinh \lambda x \sin \lambda x - Ss(\cosh \lambda x \sin \lambda x - \sinh \lambda x \cos \lambda x)}{D\lambda}$$

The bending moment M and shear force Q at a section are given by

$$M = EIv'', \quad Q = -EIv'''$$

The forces at the ends are given by

$$\text{at end 1,} \quad x = 0, \quad M = M_1, \quad Q = -F_{n1}$$

$$\text{at end 2,} \quad x = L, \quad M = -M_2, \quad Q = F_{n2}$$

In order to assist in the calculation of these forces, it is necessary to have the derivatives of shape functions as given below. They will also be needed for calculating the forces at intermediate sections of the beam element. The second derivatives of the shape functions are as follows:

$$N_1'' = \frac{2\lambda^2[(-S^2 - s^2)\sinh \lambda x \sin \lambda x - (C^2 - c^2)\cosh \lambda x \cos \lambda x + (Sc + sc)(\sinh \lambda x \cos \lambda x + \cosh \lambda x \sin \lambda x)]}{D}$$

$$N_2'' = \frac{2\lambda[(SC - sc)\cosh \lambda x \cos \lambda x - S^2 \sinh \lambda x \cos \lambda x - s^2 \cosh \lambda x \sin \lambda x]}{D}$$

$$N_3'' = \frac{2\lambda^2[2Ss \cosh \lambda x \cos \lambda x - (Cs + Sc)(\sinh \lambda x \cos \lambda x + \cosh \lambda x \sin \lambda x)]}{D}$$

$$N_4'' = \frac{2\lambda[(Cs - Sc)\cosh \lambda x \cos \lambda x - Ss(\sinh \lambda x \cos \lambda x + \cosh \lambda x \sin \lambda x)]}{D}$$

The third derivatives of the shape functions are as follows:

$$N_1''' = \frac{2\lambda^3[2(SC + sc)\cosh \lambda x \cos \lambda x + (1 - c^2 + s^2)\cosh \lambda x \sin \lambda x + (1 - C^2 - S^2)\sinh \lambda x \cos \lambda x]}{D}$$

$$N_2''' = \frac{2\lambda^2[-(S^2 + s^2)\cosh \lambda x \cos \lambda x + (S^2 - s^2)\sinh \lambda x \sin \lambda x + (SC - sc)(\sinh \lambda x \cos \lambda x - \cosh \lambda x \sin \lambda x)]}{D}$$

$$N_3''' = \frac{2\lambda^3[-2(Cs + Sc)\cosh \lambda x \cos \lambda x + 2Ss\{\sinh \lambda x \cos \lambda x - \cosh \lambda x \sin \lambda x)]}{D}$$

$$N_4''' = \frac{2\lambda^2[-2Ss \cosh \lambda x \cos \lambda x + (Cs - Sc)(\sinh \lambda x \cos \lambda x - \cosh \lambda x \sin \lambda x)]}{D}$$

The above calculations lead to the following stiffness relationship:

$$\begin{bmatrix} F_{n1} \\ M_1 \\ F_{n2} \\ M_2 \end{bmatrix} = \frac{EI}{L} \begin{bmatrix} \dfrac{T_{11}}{L^2} & \dfrac{-Q_{11}}{L} & \dfrac{-T_{12}}{L^2} & \dfrac{-Q_{12}}{L} \\[2mm] & S_{11} & \dfrac{Q_{12}}{L} & S_{12} \\[2mm] & & \dfrac{T_{11}}{L^2} & \dfrac{Q_{11}}{L} \\[2mm] \text{symmetric} & & & S_{11} \end{bmatrix} \begin{bmatrix} V_{n1} \\ \theta_1 \\ V_{n2} \\ \theta_2 \end{bmatrix}$$

$$S_{11} = (SC - sc)D_1, \quad S_{12} = (Cs - Sc)D_1$$

$$Q_{11} = \lambda L(C^2 - c^2)D_1, \quad Q_{12} = \lambda L(2Ss)D_1$$

$$T_{11} = 2(\lambda L)^2(SC + sc)D_1, \quad T_{12} = 2(\lambda L)^2(Cs + Sc)D_1$$

where $D_1 = 2\lambda L/(S^2 - s^2)$, $S = \sinh \lambda L$, $C = \cosh \lambda L$, $s = \sin \lambda L$, $c = \cos \lambda L$ and $\lambda^4 = k/4EI$.

A2.3 FIXED END FORCES FOR A MID-SPAN CONCENTRATED LOAD

Considering only a symmetrical half of the beam as shown in Fig. A2.2(d), the lateral load on the element is zero. Therefore the solution to the problem is the same as the fixed displacement case, i.e.

$$v = C_1 \cosh \lambda x \cos \lambda x + C_2 \sinh \lambda x \sin \lambda x$$
$$+ C_3 \cosh \lambda x \sin \lambda x + C_4 \sinh \lambda x \cos \lambda x$$

The boundary conditions are $x = 0$, $v = dv/dx = 0$ and $x = 0.5L$, $dv/dx = 0$, $EIv''' = 0.5W$. Using the boundary conditions, the integration constants are given by

$$C_1 = 0, \quad C_2 = \frac{W\lambda(2\,\text{SH sh})}{k(S + s)}, \quad C_3 = \frac{-W\lambda(\text{SH ch} + \text{CH sh})}{k(S + s)}, \quad C_4 = -C_3$$

where $\text{SH} = \sinh 0.5\lambda L$, $\text{CH} = \cosh 0.5\lambda L$, $\text{sh} = \sin 0.5\lambda L$, $\text{ch} = \cos 0.5\lambda L$, $S = \sinh \lambda L$, $C = \cosh \lambda L$, $c = \cos \lambda L$ and $s = \sin \lambda L$.

The displacement and its derivatives can be expressed as

$$v = C_2 \sinh \lambda x \sin \lambda x + C_3(\cosh \lambda x \sin \lambda x - \sinh \lambda x \cos \lambda x)$$

$$v'' = 2\lambda^2[C_2 \cosh \lambda x \cos \lambda x + C_3(\cosh \lambda x \sin \lambda x + \sinh \lambda x \cos \lambda x)]$$

$$v''' = 2\lambda^3[C_2(\sinh \lambda x \cos \lambda x - \cosh \lambda x \sin \lambda x) + C_3(2 \cosh \lambda x \cos \lambda x)]$$

The fixed end reactions are, as shown in Fig. A2.2(c), given by

$$M_1(= EIv'' \text{ at } x = 0) = \frac{WL}{8}\left[\frac{8\,\text{SH sh}}{\lambda L(S + s)}\right]$$

$$R_1(= EIv''' \text{ at } x = 0) = 0.5W\left[\frac{2(\text{SH ch} + \text{CH sh})}{S + s}\right]$$

A2.4 REFERENCES

[1] S. P. Timoshenko, *Strength of Materials*, Vol. 2, Van Nostrand, 1956, Chapter 1.

[2] K. Terzaghi, Evaluation of coefficients of subgrade reaction, *Geotechnique*, December 1955, pp. 297–326.

[3] J. E. Bowles, *Foundation Analysis and Design*, McGraw-Hill, 1968, pp. 229–256.

APPENDIX 3   A Semi-rigid Beam Element

Consider the beam element shown in Fig. A3.1(a). When moments are applied at the two ends, the joints 1 and 2 rotate by θ_1 and θ_2 respectively. However, because of the semi-rigid connection between the beam and the joint, the ends of the beam rotate by ϕ_1 and ϕ_2 at ends 1 and 2 respectively.

A3.1   STIFFNESS RELATIONSHIP

The stiffness relationship for the beam element and the two springs can be written as follows.

(i) *A beam element.*   Using the rotation stiffness factors derived in Chapter 1,

$$M_1 = \frac{EI}{L}(S_{11}\phi_1 + S_{12}\phi_2), \quad M_2 = \frac{EI}{L}(S_{12}\phi_1 + S_{22}\phi_2) \qquad \text{(A3.1)}$$

where, when shear deformation of the beam is included,

$$S_{11} = S_{22} = \frac{4+\beta}{1+\beta}, \quad S_{12} = \frac{2-\beta}{1+\beta}$$

(ii) *Springs.*   If the rotational stiffness of the springs at ends 1 and 2 are respectively k_1 and k_2, then

$$M_1 = k_1(\theta_1 - \phi_1), \quad M_2 = k_2(\theta_2 - \phi_2) \qquad \text{(A3.2)}$$

The stiffness relationship is established between the moments M_1 and M_2 and the joint rotations θ_1 and θ_2 by eliminating ϕ_1 and ϕ_2 from equations (A3.1) and (A3.2).

$$M_1 = \frac{EI}{L}(\overline{S_{11}}\theta_1 + \overline{S_{12}}\theta_2), \quad M_2 = \frac{EI}{L}(\overline{S_{12}}\theta_1 + \overline{S_{22}}\theta_2)$$

where $\overline{S_{11}} = (4 + \beta + 12\alpha_2)/D$, $\overline{S_{22}} = 4 + \beta + 12\alpha_1$, $\overline{S_{12}} = (2 - \beta)/D$, $D = 1 + \beta + (4 + \beta)(\alpha_1 + \alpha_2) + 12\alpha_1\alpha_2$, $\alpha_1 = EI/k_1L$ and $\alpha_2 = EI/k_2L$.

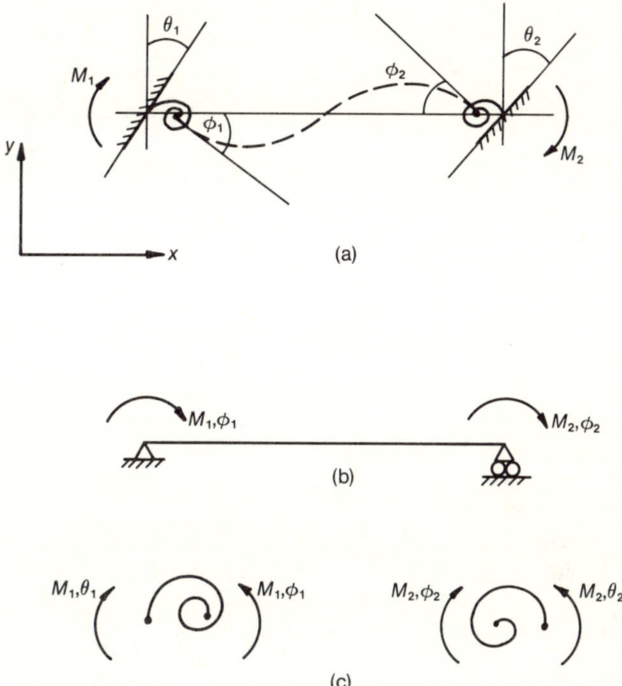

Fig. A3.1   Deformations of a semi-rigid beam element: (a) a semi-rigid beam element; (b) deformations and forces at the ends of the beam; (c) rotations and moments on the 'springs'.

The above rotational stiffness coefficients can be used to calculate the translational and cross-translational-rotational stiffness coefficients as explained in section 3.3.

A3.2   FIXED END REACTIONS FOR A SEMI-RIGID BEAM ELEMENT

The fixed end reactions will be derived for the following two types of loads on the beam.

(i) *Uniformly distributed load.*   Consider the beam element shown in Fig. A3.2(a). Because of the action of the distributed load q (Fig. A3.2(b)) and the end moments (Fig. A3.2(c)), the net rotations at the ends of the beam are ϕ_1 and ϕ_2 at ends 1 and 2 respectively. Using the well-known results from elementary beam theory, the following expressions can be derived:

$$\phi_1 = \frac{qL^3}{24EI} - \frac{M_1 L}{3EI} - \frac{M_2 L}{6EI}$$

$$\phi_2 = \frac{qL^3}{24EI} - \frac{M_1 L}{6EI} - \frac{M_2 L}{3EI}$$

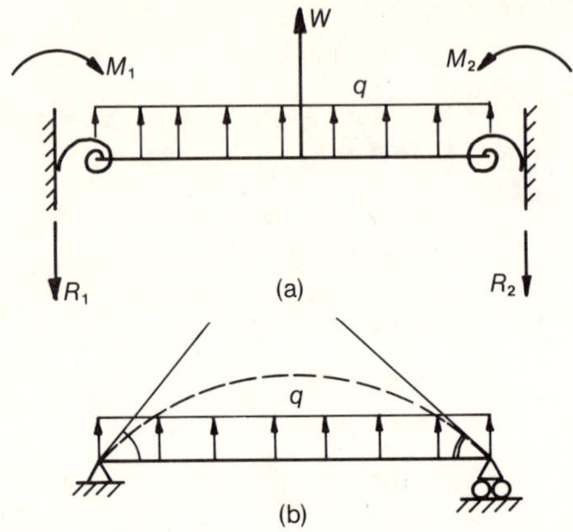

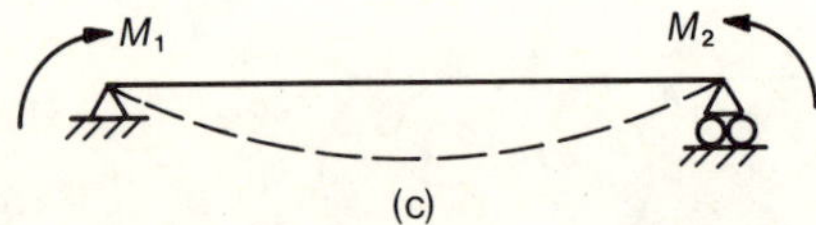

Fig. A3.2   Fixed end forces due to a uniformly distributed load on a semi-rigid beam element: (a) loads on the semi-rigid beam element; (b) deformations and forces at the ends of the beam due to the uniformly distributed load; (c) deformation of the beam due to end moments.

For the two springs, the moment–rotation relationship is given by $M_1 = k_1\phi_1$ and $M_2 = k_2\phi_2$.

Eliminating the rotations ϕ_1 and ϕ_2 from the above equations, the expressions for M_1 and M_2 are given by

$$M_1 = \frac{qL^2}{12}\left[\frac{3(1 + 6\alpha_2)}{D_1}\right], \quad M_2 = \frac{qL^2}{12}\left[\frac{3(1 + 6\alpha_1)}{D_1}\right]$$

where $D_1 = (2 + 6\alpha_1)(2 + 6\alpha_2) - 1$.

The normal reactions can be obtained from statics as

$$R_1 = 0.5qL + \frac{M_1 - M_2}{L}, \quad R_2 = 0.5qL - \frac{M_1 - M_2}{L}$$

(ii)   *Midspan concentrated load.*   Following the procedure outlined for a uniformly distributed load, we have

$$\phi_1 = \frac{WL^2}{8EI} - \frac{M_1 L}{3EI} - \frac{M_2 L}{6EI}$$

$$\phi_2 = \frac{WL^2}{8EI} - \frac{M_1 L}{6EI} - \frac{M_2 L}{3EI}$$

and $M_1 = k_1\phi_1$ and $M_2 = k_2\phi_2$.

Eliminating ϕ_1 and ϕ_2 from the above equations, the fixed end reactions are given by

$$M_1 = \frac{WL}{8}\left[\frac{6(1 + 6\alpha_2)}{D_1}\right], \quad M_2 = \frac{WL}{8}\left[\frac{6(1 + 6\alpha_1)}{D_1}\right]$$

$$R_1 = 0.5W + \frac{M_1 - M_2}{L}, \quad R_2 = 0.5W - \frac{M_1 - M_2}{L}$$

A3.3 REFERENCES

[1] J. F. Baker, *Steel Skeleton*, Vol. 1, Cambridge University Press, 1960, Chapters 8 and 9.

[2] D. M. Vanderbilt and W. G. Corley, Frame analysis of concrete buildings, *Concrete International*, December 1983, pp. 33–43.

[3] D. A. Nethercot, Steel beam to column connections—A review of test data and its applicability to the evaluation of joint behaviour in the performance of steel frames, CIRIA project record 338, CIRIA, 1985.

[4] Wai-Fah Chen (Ed.), Connection flexibility and steel frames, American Society of Civil Engineers, 1985.

APPENDIX 4 Nonuniform Torsion

A4.1 DERIVATION OF THE NONUNIFORM TORSION EQUATION

In pure torsion (also called free warping or Saint Venant's torsion), it is assumed that the torque is constant throughout the length of the member and also that the warping displacements (which are axial displacements) are unrestrained. In such a case the relationship between the torque and twist is given by

$$T = GJ \frac{d\psi}{dx} \tag{A4.1}$$

where T is the torque, ψ is the twist, J is Saint Venant's torsion constant and x is the distance along the axis of the element.

However, if the torque is not constant along the axis or if the warping of the cross section is restrained, then part of the torque is resisted by bending action and the rest by Saint Venant's torsion. As an example, consider an I beam with one end fully restrained as shown in Fig. A4.1. Evidently, part of the torsion is resisted as 'pure torsion' and the rest is resisted by the bending action of the flanges in their own plane. The torque resisted by the flanges is equal to the shear force in the flanges multiplied by the depth of the section. If the flange is in the x–y plane, with the element orientated along the x axis, the part of the torque resisted by bending action can be calculated as follows. The bending moment in the flange is $EI_f v''$. The shear force in the flange is $-EI_f v'''$ which is positive in the y direction. The torque resisted by the flanges equals the shear force in the flange multiplied by the distance h between the flanges. Thus the torque equals $-EI_f v''' h$. If the twist at a section is ψ, then the lateral displacement v of the flange is $0.5h\psi$. Therefore the flange bending torque can be expressed as

$$T = -EI_w \psi''' \tag{A4.2}$$

where $I_w = 0.5I_f h^2$ is the warping 'inertia,' I_f is the second moment of area of one flange about the y axis and h is the depth of the section.

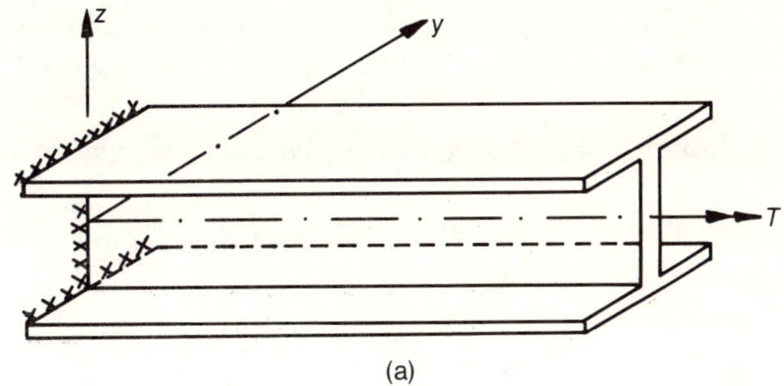

(a)

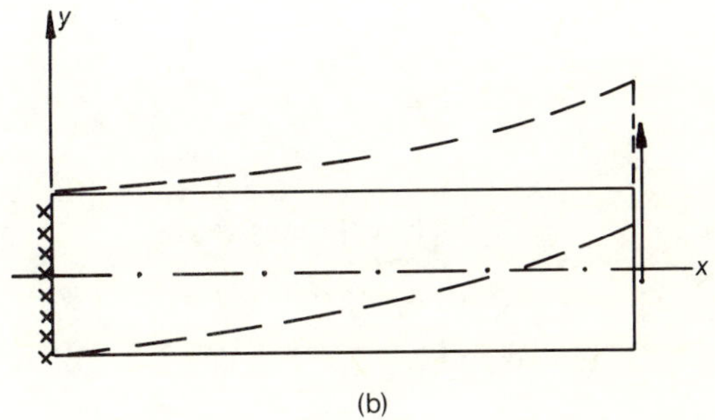

(b)

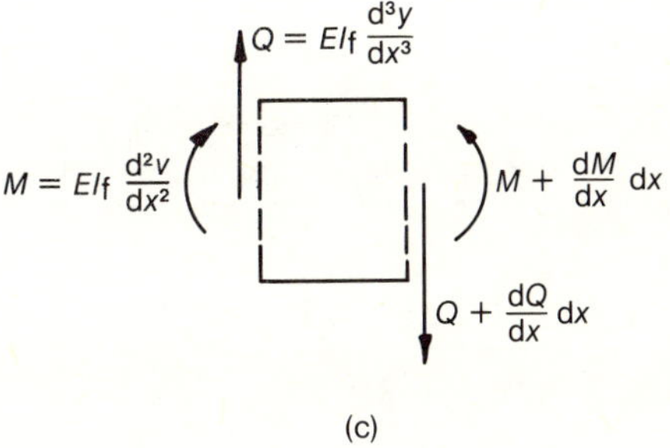

(c)

Fig. A4.1   Nonuniform torsion of an I beam: (a) cantilever I beam under torsional load; (b) deformation of the flange due to the flange shear force; (c) convention for the positive moment and the shear force in the flange.

Adding equations (A4.1) and (A4.2), we have

$$T = GJ\psi' - EI_w\psi''' \qquad\qquad (A4.3)$$

Differentiating equation (A4.3) w.r.t x, we have

$$\frac{dT}{dx} = GJ\psi'' - EI_w\psi'''' = t \qquad\qquad (A4.4)$$

where t is the intensity of applied torque on the element.

Note that, in the above sections, the expression for warping inertia was derived for a symmetrical I section. For other type of sections, see page 530 of the book by Timoshenko and Gere [5].

A4.2 DERIVATION OF THE STIFFNESS MATRIX CONSIDERING NONUNIFORM TORSION

The solution to equation (A4.3) when $t = 0$ is given by

$$\psi = C_1 + C_2 kx + C_3 \cosh kx + C_4 \sinh kx, \quad k^2 = \frac{GJ}{EI_w} \qquad (4.5)$$

The boundary conditions are best expressed in terms of the boundary conditions of the flange as follows:

$$\text{at end 1}, \quad x = 0, \quad v = v_1, \quad \frac{dv}{dx} = \theta_1$$

$$\text{at end 2}, \quad x = L, \quad v = v_2, \quad \frac{dv}{dx} = \theta_2$$

However, $v = 0.5h\psi$, where ψ is the twist of the section. Therefore the boundary conditions which will be used are

$$\text{at end 1}, \quad x = 0, \quad \psi = \psi_1, \quad \psi' = \psi'_1$$

$$\text{at end 2}, \quad x = L, \quad \psi = \psi_2, \quad \psi' = \psi'_2$$

where ψ' is the first derivative of ψ w.r.t. x.

The constants of integration are given by

$$C_1 = \frac{(SkL - C + 1)\psi_1 + (1 - C)\psi_2 + k^{-1}(kL - C - S)\psi'_1 + k^{-1}(S - kL)\psi'_2}{D}$$

$$C_2 = \frac{-S\psi_1 + S\psi_2 + k^{-1}(1 - C)\psi'_1 + k^{-1}(1 - C)\psi'_2}{D}$$

$$C_3 = \psi_1 - C_1$$

$$C_4 = k^{-1}\psi'_1 - C_2$$

where $D = 2(1 - C) + SkL$, $S = \sinh kL$, and $C = \cosh kL$.

The twist ψ can be expressed in terms of shape functions as

$$\psi = N_1\psi_1 + N_2\psi'_1 + N_3\psi_2 + N_4\psi'_2$$

$$N_1 = \frac{(kL - kx + \sinh kx)S + (1 - C)(1 + \cosh kx)}{D}$$

$$N_2 = \frac{kx(1 - C) - (CkL - S)(\cosh kx - 1) + (SkL + 1 - C)\sinh kx}{kD}$$

$$N_3 = \frac{(C - 1)(\cosh kx - 1) - S(\sinh kx - kx)}{D}$$

$$N_4 = \frac{(C - 1)(\sinh kx - kx) - (S - kL)(\cosh kx - 1)}{kD}$$

In considering the displacement of the flange, since the flange acts as a beam in its own plane the boundary conditions were expressed in terms of v and dv/dx. Naturally the forces corresponding to these displacements are the shear force and the bending moment in the flange. However, the shear force multiplied by the distance between the flanges gives only that part of the torque resisted by the flange bending action. Therefore it is better to express the boundary conditions in terms of the torque $T = GJ\psi' - EI_w\psi'''$ and, instead of the bending moment in the flange, the force considered is the bimoment defined as the bending moment in the flanges multiplied by the normal distance between them. The bimoment B is given by $B = hEI_f v'' = EI_w\psi''$. Using equation (A4.5) and the values of the constants of integration we can express the forces as

$$T = \frac{GJ}{LD}\left[(\psi_2 - \psi_1)SkL - (\psi_1' + \psi_2')(C - 1)L\right]$$

$$B = \frac{GJ}{D}\{(\psi_2 - \psi_1)[(C - 1)\cosh kx - S\sinh kx]$$

$$+ k^{-1}[(S - CkL)\cosh kx + (SkL + 1 - C)\sinh kx]\psi_1'$$

$$+ k^{-1}[(kL - S)\cosh kx + (C - 1)\sinh kx]\psi_2'\}$$

The boundary conditions for the forces are

$$\text{at end 1,} \quad x = 0, \quad T = -T_1, \quad B = -B_1$$

$$\text{at end 2,} \quad x = L, \quad T = T_2, \quad B = B_2$$

Using these boundary conditions, the following element stiffness matrix for a torsion element considering warping restraint can be derived:

$$\begin{bmatrix} T_1 \\ B_1 \\ T_2 \\ B_2 \end{bmatrix} = \frac{GJ}{L} \begin{bmatrix} TW_{11} & TW_{12} & -TW_{11} & TW_{12} \\ & TW_{22} & -TW_{12} & TW_{24} \\ & & TW_{11} & -TW_{12} \\ & \text{symmetric} & & TW_{22} \end{bmatrix} \begin{bmatrix} \psi_1 \\ \psi_1' \\ \psi_2 \\ \psi_2' \end{bmatrix} \qquad \text{(A4.6)}$$

$$TW_{11} = SkLD, \ TW_{12} = L(C - 1)D$$

$$TW_{22} = L^2\left(C - \frac{S}{kL}\right)D, \ TW_{24}L^2\left(\frac{S}{kL} - 1\right)D$$

where $D = 1.0/[2(1 - C) + SkL]$, $k^2 = GJ/EI_w$, $S = \sinh kL$ and $C = \cosh kL$.

Figure A4.2 shows the forces and displacements considered. It should be observed that the bimoment and the rotation of the flange are vectors along the z axis. In a plane grid structure in the x-y plane, compatibility demands that, at a joint, all the members must have the same value of flange rotation and equilibrium demands that the sum of the bimoments at a joint must be equal to zero.

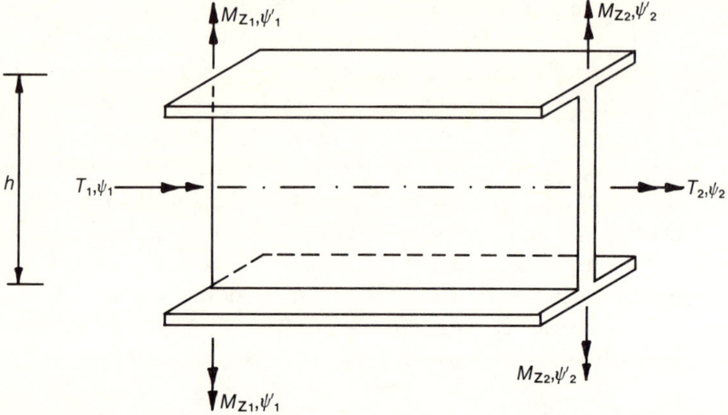

Fig. A4.2   End forces and displacements of a torsion element with warping restraint. Note that $B_1 = M_{z1}h$ and $B_2 = M_{z2}h$.

A4.3   THE TORSION ELEMENT STIFFNESS MATRIX WITH DISPLACEMENTS ALONG GENERAL COORDINATE DIRECTIONS

As discussed in Chapter 1, in a plane grid structure in the x–y plane, the displacements considered at a joint are rotations about the x and y axes and a translation along the z axis and, if the warping restraint is considered, then an additional vector equal to rate of twist has to be considered. The relationship between the twist ψ and the rotations about the x and y axes are given by

$$\psi_1 = \theta_{x1}l + \theta_{y1}m, \quad \psi_2 = \theta_{x2}l + \theta_{y2}m$$

Similarly the relationship between the torque T and the couples about x and y axes are given by

$$M_{x1} = T_1 l, \quad M_{y1} = T_1 m, \quad M_{x2} = T_2 l, \quad M_{y2} = T_2 m$$

Using the above relationships, the stiffness relationship in equation (A4.6) can be expressed as follows:

$$
\begin{bmatrix} M_{x1} \\ M_{y1} \\ B_1 \\ M_{x2} \\ M_{y2} \\ B_2 \end{bmatrix} = \frac{GJ}{L}
$$

$$
\times
\begin{bmatrix}
TW_{11}l^2 & TW_{11}lm & TW_{12}l & -TW_{11}l^2 & -TW_{11}lm & TW_{12}l \\
 & TW_{11}m^2 & TW_{12}m & -TW_{11}lm & -TW_{11}m^2 & TW_{12}m \\
 & & TW_{22} & -TW_{12}l & -TW_{12}m & TW_{24} \\
 & & & TW_{11}l^2 & TW_{11}lm & -TW_{12}l \\
 & & & & TW_{11}m^2 & -TW_{12}m \\
 \text{symmetric} & & & & & TW_{22}
\end{bmatrix}
\begin{bmatrix} \theta_{x1} \\ \theta_{y1} \\ \psi'_1 \\ \theta_{x2} \\ \theta_{y2} \\ \psi'_2 \end{bmatrix}
$$

A4.4 REFERENCES

[1]   S. P. Timoshenko, *Strength of Materials*, Vol. 2, Van Nostrand, 1956, Chapter 6, pp. 282–294.

[2]   Zbiroski-Koscia, *Thin Walled Beams*, Crosby–Lockwood, 1967.

[3]   R. J. Reilly, Stiffness analysis of grids including warping, *Journal of the Structural Division, Proceedings of the American Society of Civil Engineers*, July 1972, pp. 1511–1523.

[4]   J. N. Goodier and M. V. Barton, The effect of web deformations on the torsion of I beams, *Journal of Applied Mechanics*, March 1944, pp. A35–A40.

[5]   S. P. Timoshenko and J. M. Gere, *Theory of Elastic Stability*, McGraw-Hill, 1961.

[6]   J. S. Terrington, Combined bending and torsion of beams and girders, Publication No. 31, Part I (1968) and Part II (1970), British Constructional Steelwork Association.

APPENDIX 5   The Stability Stiffness Matrix Including Shear Deformation Effects

For a beam-column element, if shear deformation effects are ignored, then equation (1.27) is the element stability stiffness matrix. In the following, the matrix in equation (1.27) will be modified to take account of the effects of shearing deformations. For terms not defined in this appendix, the reader should refer to section 1.27.

A5.1   CORRECTION TO TRANSLATION STIFFNESS COEFFICIENTS

Consider a beam-column subjected at end 1 to a translation of V_{n1} without restraining the rotations at the ends, as shown in Fig. A5.1(a). Because of the presence of the axial load P, vertical forces equal to $P(V_{n1}/L)$ are needed at the ends to maintain equilibrium. The resultant of P and $P(V_{n1}/L)$ is an axial force in the member. The final displacements are at end 1 a translation of V_{n1} and a rotation of V_{n1}/L at the two ends, provided that the displacements are small. Considering the bending deformations only, the moment required to neutralize the rotations at the ends and the corresponding shear force considered positive 'anticlockwise' are from equation (1.27), as shown in Fig. A5.1(b),

$$M_1 = M_2 = -\frac{EI}{L^2} Q_{11} V_{n1}, \quad Q = \frac{M_1 + M_2}{L} = -2\frac{EI}{L^3} Q_{11} V_{n1}$$

It should be noted that Q is not the total vertical reaction but the shear force required to maintain equilibrium w.r.t moments M_1 and M_2 only. The total vertical reaction is equal to $Q - P(V_{n1}/L)$.

The shear deformation produces a translation at end 1 (as was explained in section 1.11, as shown in Fig. A5.1(c), equal to

$$v_s = -\frac{QL}{GA_s} = DV_{n1}$$

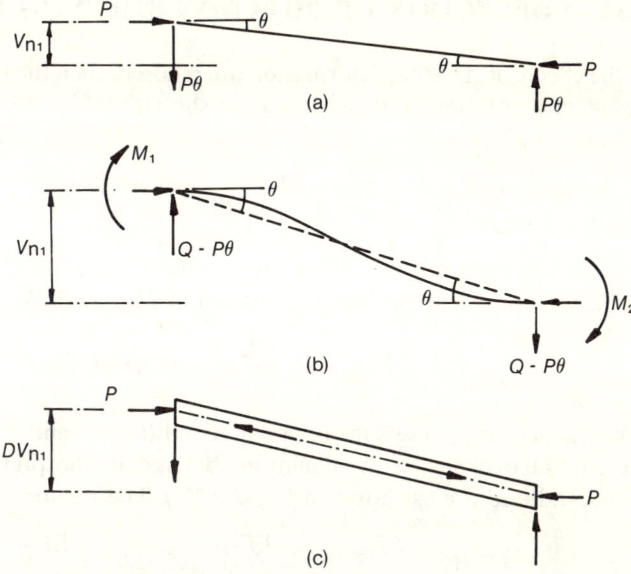

Fig. A5.1   Forces developed as a result of the translation at end 1 of a beam column: (a) 'Rigid-body' displacement (Note that $\sin \theta = \theta = V_{n1}/L$); (b) forces developed as a result of restraining the rotation at the ends; (c) additional translation due to shearing deformation.

where $D = (Q_{11}/6)\beta$ and $\beta = 12EI/GA_sL^2$. The shear deformation affects only the vertical translation but not the rotations because, during shear deformation, the vertical sections are not allowed to rotate as shown in Fig. 1.3. The net translation is equal to $V_{n1} + DV_{n1} = (1 + D)V_{n1}$. During the additional translation DV_{n1}, additional vertical forces equal to $P(DV_{n1}/L)$ are needed at the ends to maintain equilibrium as shown in Fig. A5.1(c). The net vertical forces are

$$F_{n1} = -F_{n2} = -P\frac{V_{n1}}{L} + \frac{2EI}{L^3}Q_{11}V_{n1} - P\frac{DV_{n1}}{L}$$

The three components of the vertical forces result from the 'rigid-body' translation of V_{n1} (Fig. A5.1(a)), the restraining of the rotations V_{n1}/L as in Fig. A5.1(b) and the 'rigid-body' translation of DV_{n1} as shown in Fig. A5.1(c). The moments induced are

$$M_1 = M_2 = -\frac{EI}{L^2}Q_{11}V_{n1}$$

The net translation is $(1 + D)V_{n1}$. Therefore, to have a net translation of V_{n1} only, it is necessary to divide the above forces by $1 + D$. The net forces are

$$M_1 = M_2 = -\frac{EI}{L^2}\overline{Q_{11}}V_{n1}, \quad F_{n1} = -F_{n2} = \frac{EI}{L^3}\overline{T_{11}}V_{n1}$$

where $\overline{Q_{11}} = Q_{11}/(1 + D)$, $\overline{T_{11}} = 2\overline{Q_{11}} + (kL)^2$, $k^2 = -P/EI$ and $D = (Q_{11}/6)\beta$.

A5.2 CORRECTION OF ROTATION STIFFNESS FACTORS

If the effects of shearing deformation are ignored, then the forces induced as a result of a rotation of θ_1 at end 1, as shown in Fig. A5.2(a), are given by equation (1.27) as

$$M_1 = \frac{EI}{L}S_{11}\theta_1, \quad M_2 = \frac{EI}{L}S_{12}\theta_1, \quad F_{n1} = -F_{n2} = -\frac{EI}{L^2}Q_{11}\theta_1$$

If shearing deformation effects are included, the shear force $Q = -F_{n1}$ induces a translation at end 1, as shown in Fig. A5.2(b), of

$$v_s = -\frac{QL}{GA_s} = -0.5DL\theta_1$$

The translation v_s upsets the boundary conditions at end 1. However, this can be corrected by using the equations derived in the previous section. The forces induced are as shown in Fig. A5.2(c). The net forces are given by

$$M_1 = \frac{EI}{L}S_{11}\theta_1 - \frac{EI}{L^2}\overline{Q_{11}}0.5DL\theta_1 = \frac{EI}{L}\overline{S_{11}}\theta_1$$

$$M_2 = \frac{EI}{L}S_{12}\theta_1 - \frac{EI}{L^2}\overline{Q_{11}}0.5DL\theta_1 = \frac{EI}{L}\overline{S_{12}}\theta_1$$

$$F_{n1} = -F_{n2} = -\frac{M_1 + M_2}{L} = -\frac{EI}{L^2}\overline{Q_{11}}\theta_1$$

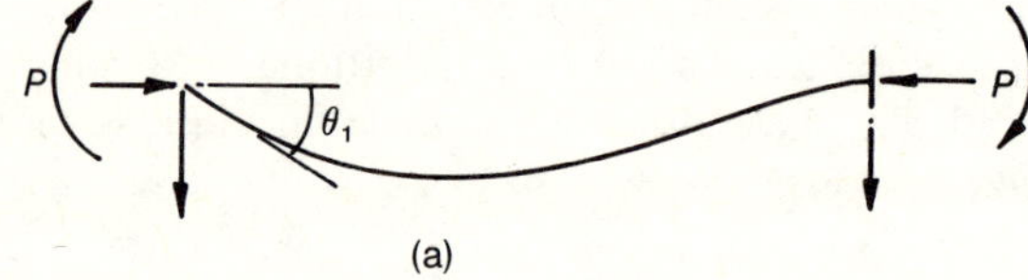

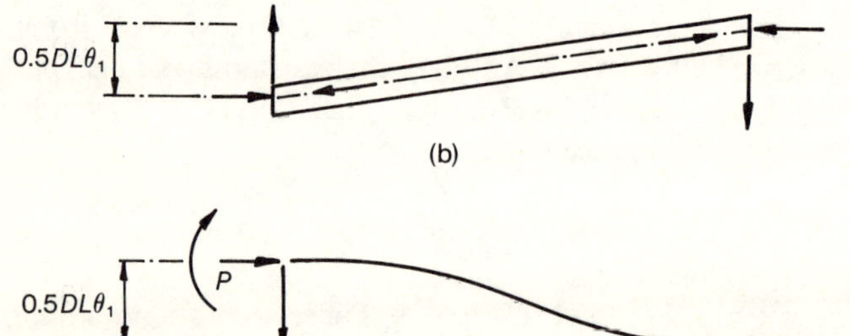

Fig. A5.2 Forces and deformations due to the end rotation of a beam-column: (a) forces developed as a result of bending action when end 1 is rotated by θ; (b) translation caused by shearing deformation; (c) forces developed as a result of compensation of the end translation.

where $\overline{S_{11}} = S_{11} - 0.5D\overline{Q_{11}}$, $\overline{S_{12}} = S_{12} - 0.5D\overline{Q_{11}}$, $\overline{Q_{11}} = Q_{11}/(1 + D)$ and $D = (Q_{11}/6)\beta$.

A5.3 THE ELEMENT STIFFNESS MATRIX

The element stiffness matrix with corrections for the effects of shearing deformation will be same as the matrix in equation (1.27), except that the unbarred quantities are replaced by their barred equivalents.

A5.4 REFERENCES

[1] A. K. Chugh, Stiiffness matrix for a beam element including transverse shear and axial load effects, *International Journal for Numerical Methods in Engineering*, Vol. 11, 1977, pp. 1681–1697.

Index